ak5

NMR and Chemistry

NMR and Chemistry

An introduction to modern
NMR spectroscopy

Fourth Edition

Dr J.W. Akitt

*Formerly Senior Research Officer at the
University of Leeds*

Professor B.E. Mann

*Professor of Chemistry
University of Sheffield*

STANLEY
THORNES

First edition published by Chapman & Hall 1973
Second edition published by Chapman & Hall 1983
Third edition published by Chapman & Hall 1992

Fourth edition published in 2000 by:
Stanley Thornes (Publishers) Ltd

Reprinted in 2002 by:
Nelson Thornes Ltd
Delta Place
27 Bath Road
CHELTENHAM
GL53 7TH
United Kingdom

03 04 05 / 10 9 8 7 6 5 4 3 2

A catalogue record for this book is available from the British Library

ISBN 0 7487 4344 8

Every effort has been made to contact copyright holders of any material reproduced in this book and we apologise if any have been overlooked.

Page make-up by Florence Productions Ltd

Printed and bound in Great Britain by Antony Rowe Ltd, Eastbourne

Contents

Abbreviations

APT	**A**ttached **P**roton **T**est
ASIS	**A**ssisted **S**olvent-**I**nduced **S**hifts
ATP	**A**denosine **T**ri**P**hosphate
BIRD	**BI**linear **R**otation **D**ecoupling
C222	$N(C_2H_4OC_2H_4OC_2H_4)_3N$
CAMELSPIN	a two dimensional NMR experiment now normally called ROESY
CHESS	**CHE**mical **S**hift **S**election imaging
COSY	**CO**rrelation **S**pectroscop**Y**
COSY-45	a COSY spectrum using a 45° pulse
COSY-90	a COSY spectrum using a 90° pulse
CP	**C**ross-**P**olarization
cpd	**c**omposite **p**ulse **d**ecoupling
CPMAS	**C**ross-**P**olarization **M**agic **A**ngle **S**pinning
CRAMPS	**C**ombined **R**otation **A**nd **M**ultiple-**P**ulse **S**pectroscopy
CSA	**C**hemical **S**hift **A**nisotropy
CW	**C**ontinuous-**W**ave
CYCLOPS	A method of cycling pulse and receiver phases to minimise artefacts
DANTE	**D**elays **A**lternating with **N**utation for **T**ailored **E**xcitation
dB	**d**eci**B**el scale
DEPT	**D**istortionless **E**nhancement by **P**olarization **T**ransfer
DSS	$Me_3SiCH_2CH_2CH_2SO_3Na$
EFG	quadrupolar **E**lectric **F**ield **G**radient
EXSY	**EX**change **S**pectroscop**Y**
FID	**F**ree **I**nduction **D**ecay
fod	6,6,7,7,8,8,8-heptafluoro-2,2-dimethyl-3,5-octanedionate
FT	**F**ourier **T**ransform
GARP	**G**lobally optimized **A**lternating-phase **R**ectangular **P**ulses, a broadband decoupling pulse sequence
HMBC	**H**eteronuclear **M**ultiple **B**ond **C**orrelation
HMQC	**H**eteronuclear **M**ultiple **Q**uantum **C**oherence

HOESY	**H**eteronuclear **O**verhauser **E**ffect **S**pectroscop**Y**
HOHAHA	**HO**monuclear **HA**rtman **HA**hn
HSQC	**H**eteronuclear **S**ingle **Q**uantum **C**oherence
Hz	Hertz, the unit of frequency in cycles per second
IF	**I**ntermediate **F**requency
INADEQUATE	**I**ncredible **N**atural **A**bundance **D**ouble **Q**uantum **T**ransfer **E**xperiment
INEPT	**I**nsensitive **N**uclei **E**nhanced by **P**olarization **T**ransfer
IR	**I**nfra**R**ed
IUPAC	**I**nternational **U**nion of **P**ure and **A**pplied **C**hemistry
JMOD	J **M**odulation
MAS	**M**agic **A**ngle **S**pinning
MHz	mega Hertz, 10^6 Hz
MLEV-17	A composite pulse sequence which can be used for spin locking or decoupling
MREV-8	A composite pulse sequence which can be used for decoupling
NMR	**N**uclear **M**agnetic **R**esonance
NOE	**N**uclear **O**verhauser **E**ffect
NOESY	**N**uclear **O**verhauser **E**ffect **S**pectroscop**Y**
PENDANT	**P**olarization **EN**hancement **D**uring **A**ttached **N**ucleus **T**esting
ppm	**p**arts **p**er **m**illion
PRESS	**P**oint **RES**olved **S**pectrosopy
PW	**P**ulse **W**idth or duration of the pulse
RF	**R**adio **F**requency
ROESY	**R**otating frame **O**verhauser **E**nhancement **S**pectroscop**Y**
STEAM	**ST**imulated **E**cho **A**cquisition **M**ode
TMS	**T**etra**M**ethyl**S**ilane
TOCSY	**TO**tal **C**orrelation **S**pectroscop**Y**
TSP	$Me_3SiCD_2CD_2CO_2Na$
UV	**U**ltra**V**iolet
WALTZ	a broad band decoupling pulse sequence
WATERGATE	**WATER** suppression by **G**r**A**dient **T**ailored **E**xitation
ZSM-5	a highly siliceous zeolite

Symbols and Abbreviations

A	Arrhenius pre-exponential factor
a_N	nuclear electron hyperfine interaction constant
B_0	applied magnetic field
B_1, B_2	radiofrequency magnetic fields
DW	dwell time
E	electric field

E_0	an energy barrier
E_a	Arrhenius activation energy
E_z	electric field along a bond
F	Nyquist frequency
G	field gradient
G_x, G_y, G_z	field gradients in the x, y, and z directions
$\hbar$	reduced Planck's constant, $h/2\pi$
I	moment of inertia
	nuclear spin quantum number
^nJ	coupling constant through n bonds (in Hz)
^nK	reduced coupling constant $4\pi^2 J/\hbar\gamma_A\gamma_B$, through n bonds
k	Boltzmann's constant, 1.3807×10^{-23} J/K
	rate constant
$K(\nu)$	field intensity at a frequency ν
m	angular momentum quantum number
$\boldsymbol{M}_x$	net magnetization along the x axis
$\boldsymbol{M}_{xy}$	net magnetization in the xy plane
$\boldsymbol{M}_y$	net magnetization along the y axis
$\boldsymbol{M}_z$	net magnetization along the z axis
P	power
	pressure
p_A, p_B	relative populations of sites A and B
Q	electric quadrupole moment
q_i	electric charge
R	relaxation rate, s^{-1}
	the gas constant, 8.3145 J K^{-1} mol^{-1}
r	a distance
R_1	spin-lattice relaxation rate, s^{-1}
R_{1DD}	dipole-dipole spin-lattice relaxation rate, s^{-1}
R_{1Q}	quadrupolar spin-lattice relaxation rate, s^{-1}
R_{1SR}	spin-rotation spin lattice relaxation rate, s^{-1}
R_2	spin-spin relaxation rate, s^{-1}
R_{2DD}	dipole-dipole spin-spin relaxation rate, s^{-1}
R_{2Q}	quadrupolar spin-spin relaxation rate, s^{-1}
S	electron spin quantum number
	nuclear spin quantum number of nucleus S,
T	relaxation time, s
τ	time, s
T_1	spin-lattice relaxation time, s
T_{1CSA}	chemical shift anisotropy spin-lattice relaxation time, s
T_{1DD}	dipole-dipole spin-lattice relaxation time, s
T_1^e	electron spin-lattice relaxation time, s
T_{1Q}	quadrupole spin-lattice relaxation time, s
T_{2Q}	quadrupole spin-spin relaxation time, s
T_{1SC}	scalar coupling spin-lattice relaxation time, s
T_{2SC}	scalar coupling spin-spin relaxation time, s
T_{1SR}	spin-rotation spin-lattice relaxation time, s
T_2	spin-spin relaxation time, s

T_2^*	apparent spin-spin relaxation time, s
T_{2DD}	dipole-dipole spin-spin relaxation time, s
T_p	time between pulses
v, u	real and imaginary components of the FID
V_1, V_2	voltage
V_{xx}	xx component of the quadrupolar electric field gradient tensor
V_{yy}	yy component of the quadrupolar electric field gradient tensor
V_{zz}	zz component of the quadrupolar electric field gradient tensor
W_0	zero quantum transition
W_1	single quantum transition
$W_{1/2}$	line width at half height
W_2	double quantum transition
Z	nuclear charge
α, β	nuclear spin states
α_0	pulse angle
δ	chemical shift
ΔE	energy separation
$\Delta G^{\ddagger}$	free energy of activation
$\Delta H^{\ddagger}$	enthalpy of activation
Δv	separation of two NMR signals
δN	population difference
$\Delta S^{\ddagger}$	entropy of activation
$\Delta V^{\ddagger}$	volume change of activation
ϕ	^{19}F chemical shift scale
γ	magnetogyric ratio
γ_e	magnetogyric ratio of the electron
η	asymmetry factor
	observed NOE
	the shear viscosity
	viscosity of the liquid
η_{max}	maximum NOE
v	frequency
v_0	spectrometer frequency
v_1, v_2	frequencies of individual signals
σ	screening constant
$\sigma_{11}, \sigma_{22}, \sigma_{33}$	components of the screening tensor
σ_d	diamagnetic term of the screening constant
σ_E	screening due to electric fields
σ_p	paramagnetic term of the screening constant
$\sigma_{\perp}$	$\perp$ component of the screening constant
$\sigma_{\parallel}$	$\parallel$ component of the screening constant
τ	a time in a pulse sequence
	an old ^{1}H chemical shift scale
τ_2	orientational correlation time

τ_c	rotational correlation time
τ_{ex}	exchange time
τ_{SR}	angular momentum correlation time
ω	frequency, rad s^{-1}
Ξ	absolute frequency reference relative to ^{1}H of TMS at 100.000 000 MHz.
μ	magnetic moment
μ_0	permeability of a vacuum
μ_z	the component of μ in the field direction

Preface to the Fourth Edition

It is some six years since J.W. Akitt wrote the preface to the third edition and now, in retirement, he welcomes the cooperation of an old colleague, Professor Brian Mann of the University of Sheffield, to update the text. The first edition appeared in 1972, over 26 years ago, and in this time NMR spectroscopy has seen immense changes and developments. The pace of technical change has perhaps slowed over the six years since the third edition appeared but the number of applications continues to increase and it has become time to make big changes in the presentation of the book while keeping the general layout of the subjects. High-field spectrometers have become commonplace and their operation is carried out via comprehensive computer control so that they are easy to use and are now operated even by undergraduate students as a teaching facility. This has led us to introduce rather more description of the workings and operation of spectrometers to help such debutant users to understand better what their commands actually cause to happen.

One topic that has been finally abandoned in this edition is the old method of continuous wave NMR spectroscopy. Also, to a large extent, the old CW NMR spectra, complete with wiggle beats have been replaced by Fourier transform ones. However, we should remember that the time-shared lock systems used on the FT equipment are best understood in CW terms.

Now that multinuclear NMR spectroscopy has become routine with a wide range of nuclei being easily observable, many examples of their use have been included, ranging from sensitive nuclei such as ^{31}P to very insensitive nuclei such as ^{187}Os. The discussion of multipulse and two-dimensional NMR spectroscopy has also been enlarged to reflect their increasing importance. Many such experiments have become routine, with fully automated instruments running them.

Examples and many new problems, mainly from the recent inorganic/organometallic chemistry literature, support the text throughout. Brief answers to all problems are provided in the text with fuller answers on the Thornes website at http://www.thorneseducation.com.

We are indebted to many people for their comments and assistance, especially to Professor K. Elsevier for commenting on early drafts of parts of the manuscript, Bruker Spectrospin Ltd for insights into present day electronics, and for permission to reproduce diagrams from their reports in Figs 5.4, 8.29, 10.3, 10.13, 11.2, 11.3, 11.4, and 11.10(b), Varian Associates for permission to reproduce the spectra used in Exercises 1 and 3 in Chapter 3, Dr A. Römer for the spectra used in Exercise 6 of Chapter 8, Professor B.L. Shaw for Fig. 7.29, Professor R.K. Harris for examples of solid state spectroscopy, and to him and Dr J. Klinowsky for comments about the solid state Chapter 11, to Drs M. Décorps, A. Ziegler and M. Raybaudi of INSERM, Grenoble, who showed J.W.A. around their laboratory and provided advice and several up-to-date examples for the imaging Chapter 10.

Finally, we are indebted to the numerous publishers and more numerous individuals who have given permission for illustrations from their publications to be used in this book and whose help is expressly acknowledged in the figure captions.

J.W.A.
Séez
B.E.M.
Sheffield
January 2000

The theory of nuclear magnetization

1

1.1 THE PROPERTIES OF THE NUCLEUS OF AN ATOM

The chemist normally thinks of the atomic nucleus as possessing only mass and charge and is concerned more with the interactions of the electrons that surround the nucleus, neutralize its charge and give rise to the chemical properties of the atom. Nuclei, however, possess several other properties that are of importance to chemistry, and to understand how we use them it is necessary to know something more about them.

Nuclei of certain natural isotopes of the majority of the elements possess intrinsic angular momentum or spin, of total magnitude $\hbar \sqrt{I(I+1)}$. The largest measurable component of this angular moment is $I\hbar$, where I is the nuclear spin quantum number and $\hbar$ is the reduced Planck's constant, $h/2\pi$. The spin quantum number I may have integral or half-integral values $(0, 1/2, 1, 3/2, \ldots)$, the actual value depending upon the isotope. Since I is quantized, several discrete values of angular momentum may be observable and their magnitudes are given by $\hbar m$ where the quantum number m can take the values I, $I-1$, $I-2$, $\ldots$, $-I$. There are thus $2I+1$ equally spaced spin states of a nucleus with angular momentum quantum number I.

A nucleus with spin also has an associated magnetic moment μ. We define the components of μ associated with the different spin states as $m\mu/I$, so that μ also has $2I+1$ components. In the absence of an external magnetic field the spin states all possess the same potential energy, but they take different energy values if a magnetic field is applied. The origin of the nuclear magnetic resonance (NMR) technique lies in these energy differences, though we must defer further discussion of this until we have defined some other basic nuclear properties.

The magnetic moment and angular momentum behave as if they were parallel or antiparallel vectors, i.e. pointing in the same or opposite directions. It is convenient to define a ratio between them which is called the magnetogyric ratio, γ:

$$\gamma = \frac{2\pi}{h}\frac{\mu}{I} = \frac{\mu}{\hbar I} \tag{1.1}$$

γ has a characteristic value for each magnetically active nucleus and is positive for parallel and negative for antiparallel vectors. We will see that the sign of γ influences both spin–spin coupling and the way energy is exchanged between spins.

If $I > 1/2$ the nucleus possesses in addition an electric quadrupole moment, Q. This means that the distribution of charge in the nucleus is non-spherical and that it can interact with electric field gradients arising from the electric charge distribution in the molecule. This interaction provides a means by which the nucleus can exchange energy with the molecule in which it is situated and affects certain NMR spectra profoundly.

Some nuclei have $I = 0$. Important examples are the major isotopes ^{12}C and ^{16}O, which are both magnetically inactive – a fact that leads to considerable simplification of the NMR spectra of organic molecules. Such nuclei are, of course, free to rotate in the classical sense, but this must not be confused with the concept of quantum-mechanical 'spin'. The nucleons, i.e. the particles such as neutrons and protons which make up the nucleus, possess intrinsic spin in the same way as do electrons in atoms. Nucleons of opposite spin can pair, just as do electrons, though they can only pair with nucleons of the same kind. Thus in a nucleus with even numbers of both protons and neutrons all the spins are paired and $I = 0$. If there are odd numbers of either or of both, then the spin is non-zero, though its actual value depends upon orbital-type internucleon interactions. Thus we build up a picture of the nucleus in which the different resolved angular momenta in a magnetic field imply different nucleon arrangements within the nucleus, the number of spin states depending upon the number of possible arrangements. If we add to this picture the concept that s bonding electrons have finite charge density within the nucleus and become partly nucleon in character, then we can see that these spin states might be perturbed by the hybridization of the bonding electrons and that information derived from the nuclear states might lead indirectly to information about the electronic system and its chemistry.

The most important properties of the elements relevant to NMR spectroscopy are listed in Table 1.1. This gives the atomic weight of the nuclear isotope of the element listed, and where there is more than one magnetically active isotope this is indicated by an atomic weight in parentheses. The isotope listed is the one most usually used, though the unlisted ones are in some cases equally usable. The next column gives the spin quantum number, I, followed by the natural abundance of the isotope. The receptivity or natural signal strength of the nucleus is given relative to that of ^{13}C, which itself gives a fairly weak signal, and this figure is made up of the intrinsic sensitivity of the nucleus (high if the magnetic moment is high) weighted by its natural abundance. Some elements are used in enriched forms (particularly ^{2}H and ^{17}O), when the receptivity is, of course, substantially higher. The data are completed by the quadrupole moment (where $I > 1/2$) and the resonant frequency in a particular magnetic field. This can be determined to a much higher precision than shown, for a resonance in an individual compound, and it should be remembered that

Table 1.1 Nuclear properties of some of the elements

Element	Atomic mass	Spin I	Natural abundance (%)	Receptivity ($^{13}C = 1.00$)	Quadrupole moment ($10^{30}\ m^2$)	Resonant frequency (MHz) at 2.348 T
Hydrogen	1	1/2	99.985	5670	None	100.00
Deuterium	2	1	0.015	0.0082	0.287	15.35
Tritium	3	1/2	Radioactive	–	None	106.66
Helium	3	1/2	0.00014	0.0035	None	76.18
Lithium	6	1	7.42	3.58	–0.064	14.72
Lithium	7	3/2	92.58	1540	–3.7	38.87
Beryllium	9	3/2	100	78.8	5.3	15.06
Boron	10	3	19.58	22.1	7.4	10.75
Boron	11	3/2	80.42	754	4.1	32.08
Carbon	13	1/2	1.108	1.00	None	25.15
Nitrogen	14	1	99.63	5.70	1.67	7.23
Nitrogen	15	1/2	0.37	0.022	None	10.14
Oxygen	17	5/2	0.037	0.061	–2.6	13.56
Fluorine	19	1/2	100	4730	None	94.09
Neon	21	3/2	0.257	0.0036	9	7.90
Sodium	23	3/2	100	524	10	26.43
Magnesium	25	5/2	10.13	1.54	22	6.13
Aluminium	27	5/2	100	1170	14	26.08
Silicon	29	1/2	4.7	2.1	None	19.87
Phosphorus	31	1/2	100	377	None	40.48
Sulfur	33	3/2	0.76	0.098	–6.4	7.67
Chlorine	35(37)	3/2	75.53	20.2	–8.2	9.81
Potassium	39	3/2	93.1	2.69	5.5	4.67
Calcium	43	7/2	0.145	0.053	–5	6.74
Scandium	45	7/2	100	1720	–22	24.33
Titanium	49(47)	7/2	5.51	1.18	24	5.64
Vanadium	51(50)	7/2	99.76	2170	–5.2	26.35
Chromium	53	3/2	9.55	0.49	–15	5.64
Manganese	55	5/2	100	1014	40	24.84
Iron	57	1/2	2.19	0.00425	None	3.24
Cobalt	59	7/2	100	1560	42	23.73
Nickel	61	3/2	1.19	0.24	16	8.93
Copper	63(65)	3/2	69.09	368	–22	26.51
Zinc	67	5/2	4.11	0.67	15	6.25
Gallium	71(69)	3/2	39.6	322	11	30.58
Germanium	73	9/2	7.76	0.62	–17	3.48
Arsenic	75	3/2	100	144	29	17.18
Selenium	77	1/2	7.58	3.02	None	19.07
Bromine	81(79)	3/2	49.46	279	27	27.10
Krypton	83	9/2	11.55	1.24	27	3.86
Rubidium	87(85)	3/2	27.85	280	13	32.84
Strontium	87	9/2	7.02	1.08	16	4.35
Yttrium	89	1/2	100	0.676	None	4.92
Zirconium	91	5/2	11.23	6.05	–21	9.34
Niobium	93	9/2	100	2770	–32	24.55
Molybdenum	95(97)	5/2	15.72	2.92	–1.5	6.55
Technetium	99	9/2	Radioactive	–	–0.13	22.51
Ruthenium	99(101)	5/2	12.72	0.815	7.6	4.61
Rhodium	103	1/2	100	0.18	None	3.16
Palladium	105	5/2	22.23	1.43	65	4.58
Silver	109(107)	1/2	48.18	0.28	None	4.65
Cadmium	113(111)	1/2	12.26	7.69	None	22.18

Table 1.1 continued

Element	Atomic mass	Spin I	Natural abundance (%)	Receptivity ($^{13}C = 1.00$)	Quadrupole moment ($10^{30}\ m^2$)	Resonant frequency (MHz) at 2.348 T
Indium	115(113)	9/2	95.72	1920	86	22.04
Tin	119(115,117)	1/2	8.58	25.7	None	37.29
Antimony	121 (123)	5/2	57.25	530	−33	24.09
Tellurium	125(123)	1/2	6.99	12.8	None	31.55
Iodine	127	5/2	100	541	−79	20.15
Xenon	129(131)	1/2	26.44	32.4	None	27.86
Caesium	133	7/2	100	275	−0.3	13.21
Barium	137(135)	3/2	11.32	45	28	11.19
Lanthanum	139(138)	7/2	99.91	343	22	14.24
Praseodymium	141	5/2	100	1620	−4.1	29.03
Neodymium	145(143)	7/2	8.3	0.393	−25	3.41
Samarium	149(147)	7/2	13.83	0.665	5.6	3.43
Europium	151(153)	5/2	47.82	464	114	24.48
Gadolinium	155(157)	3/2	14.73	0.124	160	3.09
Terbium	159	3/2	100	394	134	24.04
Dysprosium	163(161)	5/2	24.97	1.79	251	4.77
Holmium	165	7/2	100	1160	349	21.34
Erbium	167	7/2	22.94	0.665	283	2.90
Thulium	169	1/2	100	2.89	None	7.99
Ytterbium	171(173)	1/2	14.31	4.5	None	17.70
Lutetium	175(176)	7/2	97.41	173	346	11.43
Hafnium	177(179)	7/2	18.50	1.47	330	4.06
Tantalum	181	7/2	99.988	213	330	12.13
Tungsten	183	1/2	14.28	0.0608	None	4.22
Rhenium	187(185)	5/2	62.93	511	220	23.05
Osmium	187(189)	1/2	1.64	0.0015	None	2.28
Iridium	193(191)	3/2	62.7	0.122	78	1.90
Platinum	195	1/2	33.8	19.9	None	21.50
Gold	197	3/2	100	0.153	55	1.75
Mercury	199(201)	1/2	16.84	5.68	None	17.87
Thallium	205(203)	1/2	70.5	807	None	57.63
Lead	207	1/2	22.6	11.9	None	20.92
Bismuth	209	9/2	100	819	−37	16.36
Uranium	235	7/2	0.72	0.0054	455	1.84

Source: Mason (1987) *Multinuclear NMR*, Plenum, New York.

each nucleus will have a range of frequencies due to the chemical shift effect. The resonance frequency is proportional to the magnetogyric ratio.

The first point to note about this list is that almost all the elements are represented, the only missing stable ones being argon and cerium, and that, in principle, virtually the whole of the Periodic Table can be studied by NMR. Indeed, with modern instrumentation, this is now realizable, though there are some cases, notably nuclei with very high quadrupole moments, where study in the liquid state is not rewarding. The usefulness of a nucleus to the NMR spectroscopist depends in the first place upon the chemical importance of the atom it characterizes and then upon its receptivity. Thus the extreme importance of carbon spectroscopy for understanding the structures of organic molecules has led to technical developments that have overcome the disadvantages of its poor receptivity, so that ^{13}C NMR is now commonplace. Hydrogen, with its very high receptivity, has, of course, been studied right from the emergence of NMR spectroscopy as a technique useful to chemists, and proton spectroscopy, as it is often called (proton = ^1H, the term is commonly used by NMR spectroscopists when discussing the nucleus of neutral hydrogen), has been used for the identification of the majority of organic compounds and many inorganic ones, and for the physical study of diverse systems. Other much-studied nuclei are ^{11}B in the boron hydrides or carboranes, ^{19}F in the vast array of fluoro organics and inorganics, ^{27}Al and ^{29}Si in a wide range of inorganic and organic compounds, and ^{31}P in its many inorganic and biochemical guises. However, even quite low receptivity is no longer an insurmountable obstacle and, if a chemical problem is presented that is capable of solution by NMR spectroscopy, then whatever nucleus may require to be observed, the appropriate effort may well bring the desired rewards.

1.2 THE NUCLEUS IN A MAGNETIC FIELD

If we place a nucleus in a magnetic field B_0 it can take up $2I + 1$ orientations in the field, each one at a particular angle θ to the field direction and associated with a different potential energy. The energy of a nucleus of magnetic moment μ in field B_0 is $-\mu_z B_0$ where μ_z is the component of μ in the field direction. The energy of the various spin states is then

$$-\frac{m\mu}{I} B_0 \quad \text{or individually} \quad -\mu B_0, \quad -\frac{I-1}{I}\mu B_0, \quad \frac{I-2}{I}\mu B_0, \text{ etc.}$$

The energy separation between the levels is constant and equals $\mu B_0/I$. This is shown diagrammatically in Fig. 1.1 for a nucleus with $I = 1$ and positive magnetogyric ratio. The value of m changes sign as it is altered from I to $-I$ and accordingly the contribution of the magnetic moment to total nuclear energy can be either positive or negative, the energy being increased when m is positive. The energy is decreased if the nuclear magnetic vectors have a component aligned with the applied field in the classical sense. An increase in energy corresponds to aligning the vectors in opposition to the field. Quantum mechanics thus predicts a non-classical situation, which can only arise because of the existence of discrete energy states with the high-energy states indefinitely stable.

In common with other spectral phenomena, the presence of a series of states of differing energy in an atomic system provides a situation where interaction can take place with electromagnetic radiation of the correct frequency and cause transitions between the energy states. The frequency is obtained from the Bohr relation, namely

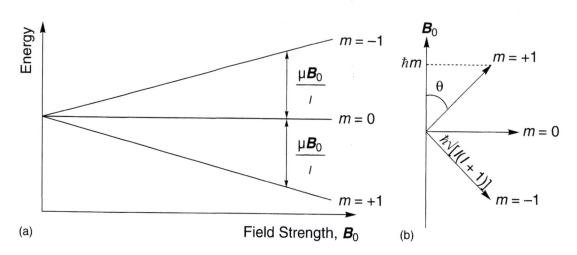

Figure 1.1 (a) The nuclear spin energy for a single nucleus with $I = 1$ (e.g. ^{14}N) plotted as a function of magnetic field B_0. The two degenerate transitions are shown for a particular value of B_0. (b) The alignment of the nuclear vectors relative to B_0 that correspond to each value of m. The vector length is $\hbar\sqrt{I(I + 1)}$ and its z component is $\hbar m$, whence $\cos\theta = m/\sqrt{I(I + 1)}$.

$$h\nu = \Delta E$$

where ΔE is the energy separation. For NMR

$$h\nu = \mu B_0/I$$

In this case, the transition for any nuclear isotope occurs at a single frequency since all the energy separations are equal and transitions are only allowed between adjacent levels (i.e. the selection rule $\Delta m = \pm 1$ operates). The frequency relation is normally written in terms of the magnetogyric ratio (1.1), giving

$$\nu = \gamma B_0/2\pi \qquad (1.2)$$

Thus the nucleus can interact with radiation whose frequency depends only on the applied magnetic field and the nature of the nucleus. Magnetic resonance spectroscopy is unique in that we can choose our spectrometer frequency at will, though within the limitation of available magnetic fields. The values of γ are such that for practical magnets the frequency for nuclei lies in the frequency range between a present maximum of 800 megahertz (MHz) and a minimum of a few kilohertz (kHz) using the earth's magnetic field.

1.3 THE SOURCE OF THE NMR SIGNAL

The low frequency of nuclear magnetic resonance absorption indicates that the energy separation of the spin states is quite small. Since the nuclei in each of the states are in equilibrium, this suggests that the numbers in the different spin states will be similar, though if there is a Boltzmann distribution among the spin states then we can expect more nuclei to reside in the lowest energy states. For a system of spin-1/2 nuclei, a Boltzmann distribution would give

$$N_u/N_l = \exp(-\Delta E/kT)$$

where N_u and N_l are the numbers of nuclei in the upper and lower energy states respectively, ΔE is the energy separation, k is the Boltzmann constant, and T is the absolute temperature. ΔE is given above as $\mu \boldsymbol{B}_0/I$, which, for $I = 1/2$, equals $2\mu \boldsymbol{B}_0$. Thus

$$N_u/N_l = \exp(-2\mu \boldsymbol{B}_0/kT)$$

which since $N_u \approx N_l$ can be simplified to

$$N_u/N_l = 1 - 2\mu \boldsymbol{B}_0/kT$$

For hydrogen nuclei in a magnetic field of 9.39 T (tesla), where the resonance frequency is 400 MHz, and at a temperature of 300 K, the quantity $2\mu \boldsymbol{B}_0/kT$ has a value of about 6×10^{-5}, which means that the excess population in the lower energy state is one nucleus per 300 000. This is extremely small. The z components of the magnetic moments of the nuclei in the upper energy state are all cancelled by those of the lower energy state, only the small number excess in the lower energy state being able to give rise to an observable magnetic effect. This is a weak nuclear paramagnetism, which is only observable at low temperatures where the population difference is a maximum. For this reason, we have to resort to a resonance technique in order to observe a signal.

If $I > 1/2$ one obtains a similar though more complex picture since the excess low-energy nuclei do not all have the same value of $m\mu/I$, and, for integral I, one energy level has no magnetic component in the z direction.

It should be noted that the size of the excess low-energy population is proportional to $\boldsymbol{B}_0$. For this reason the magnetic effect of the nuclei and therefore their signal intensity increases as the strength of the magnetic field is increased. Temperature also has an important effect, and the value of $2\mu \boldsymbol{B}_0/kT$ is increased to 7.2×10^{-5} at 250 K or decreased to 4.9×10^{-5} at 370 K at 9.39 T. This can produce detectable changes in signal strength relative to background noise in variable-temperature experiments.

We have so far built up a picture of the nuclear moments in a sample polarized with or against the magnetic field and lying at an angle θ to it. The total angular momentum (i.e. the length of the vectors of Fig. 1.1) is $\sqrt{I(I + 1)}$ and the angle θ is then given by

$$\cos \theta = \sqrt{\frac{m}{I(I + 1)}}$$

This angle can also be calculated classically by considering the motion of a magnet of moment μ in an applied magnetic field. It is found that the magnet axis becomes inclined to the field axis and rotates or precesses around it. The magnet thus describes a conical surface around the field axis. The half-apex angle of the cone is equal to θ and the angular velocity around the cone is γB_0, so that the frequency of complete rotations is $\gamma B_0/2\pi$, the nuclear resonant frequency (Fig. 1.2(a)). This precession is known as the Larmor precession.

For an assembly of nuclei with $I = 1/2$ there are two such precession cones, one for nuclei with $m = +1/2$ and one for $m = -1/2$ and pointing in opposite directions. It is usual, however, to consider only the precession cone of the excess low-energy nuclei, and this is shown in Fig. 1.2(b), which represents them as spread evenly over a conical surface and all rotating with the same angular velocity around the magnetic field axis, which is made the z axis. Since the excess low-energy nuclear spins all have components along the z axis pointing in the same direction, they add to give net magnetization M_z along the z axis. Individual nuclei also have a component μ_{xy} transverse to the field axis in the xy plane. However, because they are arranged evenly around the z axis, these components all average to zero, i.e. $M_x = M_y = 0$. The magnetism of the system is static and gives rise to no external effects other than a very small, usually undetectable, nuclear paramagnetism due to M_z.

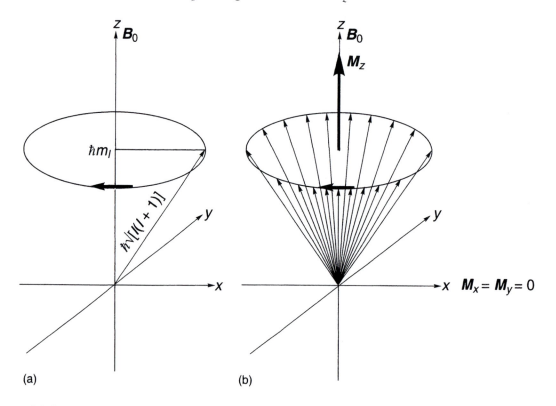

(a)

(b)

Figure 1.2 Freely precessing nuclei in a magnetic field B_0. (a) Larmor precession of a single nucleus. (b) The excess low-energy nuclei in a sample. The nuclear vectors can be regarded as being spread evenly over a conical surface. They arise from different atoms but are drawn with the same origin.

In order to detect a nuclear resonance we have to perturb the system. This is done by applying a sinusoidally oscillating magnetic field, B_1, in the xy plane with frequency $\gamma B_0/2\pi$. This can be thought of as stimulating both absorption and emission of energy by the spin system (i.e. as stimulating upward and downward spin transitions), but resulting in net absorption of energy, since more spins are in the low-energy state and are available to be promoted to the high-energy state. The perturbing field is generated by passing a radiofrequency (RF) alternating current through Helmholtz double coils placed on either side of the sample.

Classically we can analyse the oscillating magnetic field into a superposition of two magnetic vectors rotating in opposite directions. These add at different instants of time to give a zero, positive or negative resultant (Fig. 1.3). The vector B_1, which is rotating in the same sense as the nuclei (Fig. 1.4), is stationary relative to them, since we have arranged that it should have the same angular velocity. This induces transitions from one energy level to the other. The result is that the net nuclear magnetization, M, rotates about B_1. There is resultant

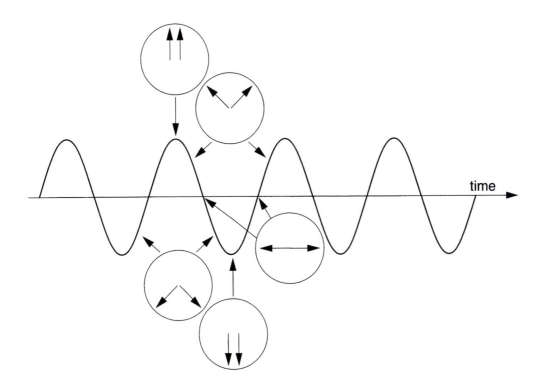

Figure 1.3 The RF current in the coil in Fig. 1.4 produces an oscillating magnetic field along the y axis. This can be equally well regarded as being composed of two rotating magnetic vectors rotating in opposite directions, whose resultant is the oscillating magnetic field. One of these vectors will rotate in the same sense as the nuclear precession and is conventionally called B_1.

magnetization M_{xy} transverse to the main field and rotating at the nuclear precession frequency. The magnetism of the system is no longer static and the rotating vector M_{xy} will induce a radiofrequency current in the coils placed around the sample, which we can detect, provided pick-up of B_1 can be avoided.

In order to follow the changes in the magnetization in the xy plane with time, it is convenient to use the rotating frame. In Fig. 1.4, the magnetic vector, B_1, is rotating at the Larmor frequency. In the rotating frame, the x and y axes rotate around the z axis at the Larmor frequency and B_1 is now apparently stationary. In order to differentiate the rotating frame from the laboratory frame used in Fig. 1.4, the x and y axes are described as x' and y'.

If B_1 is of large amplitude, then the magnetization, M, swings very rapidly around it. The magnetization in the xy plane increases and reaches a maximum when the nuclear magnets have precessed 90° around B_1 (Fig. 1.5). When B_1 is cut off, of course, this precession stops. In practice, it is possible to cause a 90° precession in very short times of between 2 and 50 μs (1 μs = 10^{-6} s), i.e. with a very short pulse of B_1 (Fig. 1.5).

If the B_1 frequency is substantially different from the Larmor frequency, B_1 does not rotate at the same frequency as the nuclei precess and the nuclear precession around B_1 is always changing direction, so M_{xy} can never become significant. It is this feature, whereby the signal is obtained from all the excess low-energy nuclei acting in concert, and only at a single frequency, that gives to the technique its name of 'resonance spectroscopy'. However, this applies strictly only to a B_1 pulse which is monochromatic. The fact that we have arranged for the B_1 pulse to be very short means that it is no longer monochromatic, but covers a band of frequencies and resonance occurs with

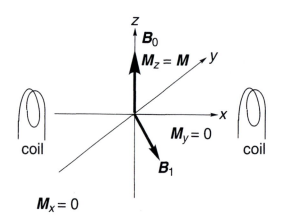

Figure 1.4 If a rotating magnetic vector B_1 with the same angular velocity as the nuclei is now added to the system, the nuclei will also tend to precess around B_1, and this causes the cone of vectors to tip and wobble at the nuclear precession frequency. The resulting rotating vector M_{xy} in the xy plane can induce a current in the coils that are wound beside the sample.

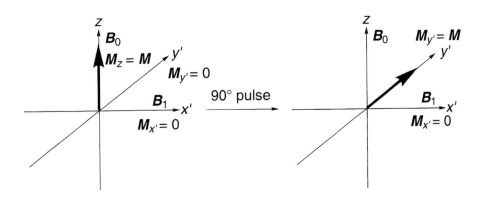

Figure 1.5 A sufficiently long or powerful B_1 along the x' axis will turn the magnetization into the xy $(x'y')$ plane and the whole of the nuclear magnetization lies along the y' axis. This is in the rotating frame and $M_{y'}$ rotates within the static laboratory frame, contributing to the signal picked up by the coil in Fig. 1.4 and the output is at a maximum. B_1 is thus applied in the form of a pulse, and a pulse that has the effect illustrated is known as a 90° pulse. In the rotating frame, B_1 remains pointing in a fixed direction and the magnetization rotates about this direction. In the laboratory frame, B_1 rotates in the xy plane and the magnetization follows a spiral path away from the B_0 axis during the pulse.

all nuclei with precession frequencies situated within this band. The strength of B_1 at frequencies around the B_1 frequency varies as $\sin(x)/x$ (Fig. 1.6), though the intensity distribution is reasonably constant near B_1. In fact, all nuclei within $\pm 1/4PW$ Hz of the spectrometer frequency (and therefore of the B_1 frequency) are almost equally affected. PW is the pulse width or duration. There are, in contrast, null points at $\pm n/PW$ Hz where the nuclei are not perturbed. The negative intensities at higher frequency separation indicate a 180° phase change where the nuclei would swing in the opposite direction.

Evidently, we are interested only in those nuclei in the central region of the frequency distribution. For these, the magnetization in the laboratory xy plane, M_{xy} precesses about B_0 with the Larmor frequency and continues to do so after B_1 has been switched off. This rotating

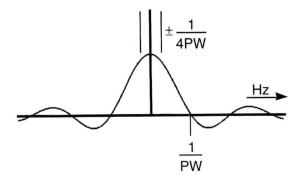

Figure 1.6 The effectiveness of a pulse of length, PW, against separation of the pulse frequency from that of the signal to be excited.

magnetization induces a signal in the coil which the spectrometer detects with no interference from B_1. The signal intensity diminishes to zero as the system returns to equilibrium with $M_{xy} = 0$. An output will be observable in general for between 1 ms and 10 s. If there is a single type of nucleus present then the result is a decaying cosine wave (Fig. 1.7). If there are several types of nuclei giving signals, then the result is the sum of a number of decaying cosine waves. The detected signals are subtracted from the spectrometer frequency giving a resulting low frequency cosine wave which only contains the difference frequencies between the spectrometer frequency and the nuclear frequencies (Fig. 1.7). This is called a free induction decay, FID. Fourier transformation of this function of time gives a function of frequency which is the conventional NMR spectrum with maxima at each nuclear difference frequency. This is discussed further in Chapter 5.

1.4 A BASIC NMR SPECTROMETER

We are now in a position to understand the principles underlying the construction of an NMR spectrometer. The object is to measure the frequency of a nuclear resonance with sufficient accuracy. The instrument (Fig. 1.8) comprises a strong, highly stable magnet in which the sample is placed and surrounded by transmitter/receiver coils. The magnet is normally a superconducting solenoid.

The magnetic field at the sample also inevitably varies throughout the bulk of the sample (i.e. the field is non-homogeneous), so that the signal frequency is not well defined. Two further sets of coils, known as shim coils, are placed around the sample in order to counteract these variations or field gradients and render the field as perfectly homogeneous as possible. The shim coils are not shown. Remaining inhomogeneities are minimized by spinning the sample tube about its long axis so that the sample molecules experience average fields. Very

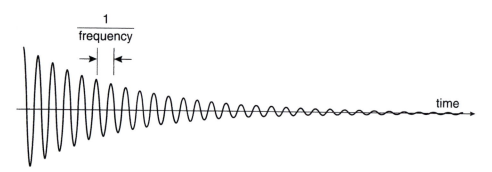

Figure 1.7 A free induction decay, FID, from a single type of nucleus. This signal contains the same information as a conventional NMR spectrum. The frequency of the signal relative to the spectrometer frequency is given by the frequency of the cosine wave. The intensity is given by the intensity of the cosine wave at its beginning, and the linewidth is given by the rate of decay of the cosine wave.

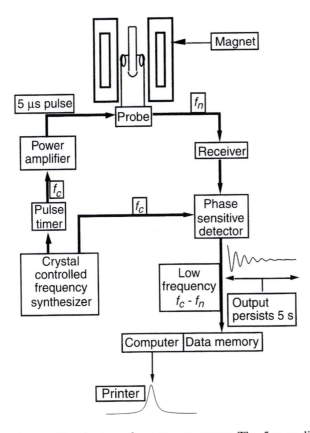

Figure 1.8 A basic Fourier transform spectrometer. The 5 μs radiofrequency (RF) pulse tips the nuclei in the sample by 90° provided their frequency lies within the bandwidth of the pulse. (The pulse switching in effect converts the monochromatic crystal frequency f_c into a band of frequencies of width $\pm 1/4PW$, where PW is the pulse length. In this case, the bandwidth is $\pm 50\,000$ Hz.) The nuclear output signal will be at a frequency f_n close to f_c and the difference frequency $f_c - f_n$ is obtained at the output of the phase-sensitive detector. The computer collects the output and then calculates the frequency of resonance relative to f_c. The resonance frequency has limited definition (i.e. it has width), which is related to the time for which the output signal persists. Note that a time-dependent output is converted to a frequency-dependent output for analysis. Note also the very different timescales for the RF pulse and the output, here given the arbitrary lengths of 5 μs and 5 s. The computer and pulser have to be linked in some way to synchronize pulse timing and data collection.

well defined frequencies, and so excellent resolution of close, narrow resonances, are obtained in this way. The $\boldsymbol{B}_1$ field is produced by a gated (switched input) power amplifier driven by a stable, crystal-controlled continuous oscillator. The nuclear signals following the $\boldsymbol{B}_1$ pulse are then amplified and detected in a device which compares them with the crystal oscillator output ($\boldsymbol{B}_1$ carrier f_c) and gives a low-frequency, time-dependent output containing frequency, phase and amplitude information. This output is digitized and collected in a computer memory for frequency analysis using a Fourier transform program, and the spectrum that results (a spectrum is a function of frequency) can be output and the resonance frequencies listed.

Not shown is a parallel spectrometer which can detect deuterium in the sample and use this signal to stabilize the magnetic field. This is a field-frequency lock, which will be described in detail later.

This type of spectrometer is known as a Fourier transform (FT) spectrometer. It is also possible to obtain spectra from the more receptive nuclei using a much simpler system, which dispenses with the pulser and computer and produces the recording of the spectrum directly. This is the continuous-wave (CW) spectrometer. All early high-resolution spectrometers were of this type, and indeed the modern FT instruments were developed from these.

1.5 QUESTIONS

1.1. A spectrometer operating at a fixed magnetic field is set up to observe the nucleus ^{13}C at a frequency of 25 000 000 Hz (25 MHz). The sample examined contains two resonances at 25 000 250 Hz and 25 001 000 Hz respectively. What frequencies will be present at the output to the computer digitizer? What would these be if the spectrometer frequency were changed to 25 001 250 Hz?

1.2. A particular NMR spectrometer used to obtain ^{13}C spectra at 25 MHz has a 90° pulse length of 100 μs. Calculate the bandwidth (or frequency coverage) of this pulse where uniform excitation of the nuclei is achieved. It is desired to obtain spectra over a range of 5000 Hz. Is the bandwidth sufficient for this? If we were able to increase the magnetic field of the spectrometer so that the ^{13}C operating frequency became 125 MHz (the frequency range needed would be 25 000 Hz), would the 100 μs pulse still be of sufficiently wide coverage? What is the maximum pulse angle that could be tolerated if the whole spectral range were to be stimulated uniformly?

1.3. A spectrometer observes ^{17}O at 54.256 MHz. A 90° pulse of length 40 μs was used. Explain why a signal 25 000 Hz from the spectrometer frequency was not observed. Suggest two different ways to change the spectrometer's operating conditions so that this signal can be observed.

1.4. In which of the following cases are M_z and/or M_{xy} zero: (a) following a 45° pulse; (b) following a 90° pulse; and (c) following a 180° pulse?

1.5. Which of the following nuclei are NMR inactive: ^{32}P, ^{32}S, ^{36}Cl, ^{40}Ar, ^{40}K, ^{64}Zn, ^{91}Zr, ^{76}As, and ^{71}Ge.

1.6. During a 90° pulse lasting 10 μs, how many times does the magnetization vector M rotate as it spirals down to the xy plane if the Larmor frequency is 400 MHz?

1.7. Calculate the angle at which a single proton will precess about a magnetic field.

1.8. What is the receptivity of the lithium nucleus, ^{6}Li, after enrichment to 100%?

1.9. The nucleus, ^{1}H, in water resonates at 400 MHz in a magnetic field of 9.39 T. The earth's magnetic field is 0.000 05 T. What is the ^{1}H precession frequency in the earth's magnetic field and what is the excess population of nuclei in the lower energy state in this field at 300 K?

1.10. A resonance in a given sample and spectrometer has a frequency of 100 000 000 Hz. A continuous monochromatic B_1 of frequency 100 000 002 Hz is applied in the xy plane to the sample. Because B_1 and nuclear frequencies are not the same, M_z will tend to rotate around the z axis and never leave it to reach the xy plane. What will be the frequency of rotation of M_z around the z axis in the rotating frame?

The magnetic field at the nucleus: nuclear screening and the chemical shift

2

2.1 EFFECTS DUE TO THE MOLECULE

So far we have shown that a single isotope gives rise to a single nuclear magnetic resonance in an applied magnetic field. This really would be of limited interest to the chemist except for the fact that the magnetic field at the nucleus is never equal to the applied field, but depends in many ways upon the structure of the molecule in which the atom carrying the nucleus resides.

The most obvious source of perturbation of the field is that which occurs directly through space due to nuclear magnets in other atoms in the molecule. If such nuclei have high magnetic moments, which means generally ^{1}H or ^{19}F, then in the solid state this interaction results in considerable broadening of the resonance, which obscures much information. However, in the liquid state, where the molecules rotate rapidly and randomly, the direct nuclear fields fluctuate wildly in both intensity and direction and have an average value that is exactly zero. The resonances are thus narrow and may show much structure. Thus spectroscopy of the liquid state has been a major preoccupation of chemists.

Since the magnetic nuclei do not directly perturb the field at the nucleus, we have therefore to consider the effect that the electrons in the molecule may have. We will concern ourselves only with diamagnetic molecules at this stage and will defer till later discussion of paramagnetic molecules possessing an unpaired electron. When an atom or molecule is placed in a magnetic field, the field induces motion of the electron cloud such that a secondary magnetic field is set up. We can think of the electrons as forming a current loop as in Fig. 2.1 centred on a positively charged atomic nucleus. The secondary field produced by this current loop opposes the main field at the nucleus and so reduces the nuclear frequency. The magnitude of the electronic current is proportional to B_0 and we say that the nucleus is screened

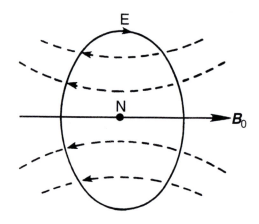

Figure 2.1 The motion of the electronic cloud E around the nucleus N gives rise to a magnetic field, shown by dashed lines, which opposes $\boldsymbol{B}_0$ at the nucleus.

(or shielded) from the applied field by its electrons. This concept is introduced into equation (1.2) relating field and nuclear frequency by the inclusion of a screening constant σ

$$\frac{\gamma \boldsymbol{B}_0}{2\pi}(1-\sigma) \tag{2.1}$$

σ is a small dimensionless quantity and is usually recorded in parts per million (ppm). The screening effect is related to the mechanism that gives rise to the diamagnetism of materials and is called diamagnetic screening.

The magnitude of the effect also depends upon the density of electrons in the current loop. This is a maximum for a free atom where the electrons can circulate freely, but in a molecule the free circulation around an individual nucleus is hindered by the bonding and by the presence of other positive centres, so that the screening is reduced and the nuclear frequency increased. Since this mechanism reduces the diamagnetic screening it is known as a paramagnetic effect. This is unfortunately a misleading term and it must be emphasized that it does not imply the presence of unpaired electrons. As used here, it merely indicates that there are two contributions to σ, the diamagnetic term σ_d and the paramagnetic term σ_p and that these are opposite in sign. Thus

$$\sigma = \sigma_d + \sigma_p \tag{2.2}$$

Since the magnitude of σ_d depends upon the density of circulating electrons, it is common to find in the literature discussion of the effect of inductive electron drifts on the screening of nuclei. The screening of protons in organic molecules, for instance, depends markedly on the substituents, and good linear correlations have been found between screening constants and substituent electronegativity, thus supporting the presence of an inductive effect. Currently, however, it is believed that most of these variations originate in the long-range effects to be described below and that the contribution of inductive effects is small, at least for σ-bonded systems.

The magnitude of the paramagnetic contribution σ_p is zero for ions with spherically symmetric S states but is substantial for atoms, particularly the heavier ones, with many electrons in the outer orbitals involved in chemical bonding. It is determined by several factors.

1. The inverse of the energy separation ΔE between ground and excited electronic states of the molecule. This means that correlations are found between screening constants and the frequency of absorption lines in the visible and ultraviolet.
2. The relative electron densities in the various p- and d-orbitals involved in bonding, i.e. upon the degree of asymmetry in electron distribution near the nucleus.
3. The value of $\langle 1/r^3 \rangle$, the average inverse cube distance from the nucleus to the orbitals concerned.

In the case of hydrogen, for which there are few electrons to contribute to the screening, and for which ΔE is large, σ_d and σ_p are both small and we observe only a small change in σ among its compounds, most of which fall within a range of 20×10^{-6} or 20 ppm. In the case of elements of higher atomic number, ΔE tends to be smaller and more electrons are present, so that, while both σ_d and σ_p increase, σ_p increases disproportionately and dominates the screening. Thus changes in σ_p probably account for a major part of the screening changes observed for boron in its compounds (a range of 200 ppm), and σ_p almost certainly predominates for fluorine (where the range is 1300 ppm) or for thallium (where it is 5500 ppm). We see that the changes observed for the proton are thus unusually small.

In the case of screening of the fluorine nucleus, the values for fluorine compounds are known relative to the bare, unscreened fluorine nucleus. In certain of its compounds, fluorine is less screened even than the bare nucleus, some examples being F_2, UF_6 or FOOF. Presumably this arises because of electronic circulation near to, but not centred upon, the nucleus.

The observable changes in screening of each nucleus do not increase continuously with atomic number but exhibit a periodicity, increasing steadily along each period but then falling markedly at the start of the next. The behaviour down each group is similar, as is shown in Fig. 2.2. This periodicity follows closely the values of $\langle 1/r^3 \rangle$ for each element.

Because we are dealing with the effects of electronic circulation in a three-dimensional molecule relative to a unidirectional magnetic field, it is easy to see that the nature of the circulation will change as the orientation of the molecule changes in the magnetic field. Thus the screening at any instant is a function of the attitude of the molecule relative to B_0. These orientational effects are fully described by

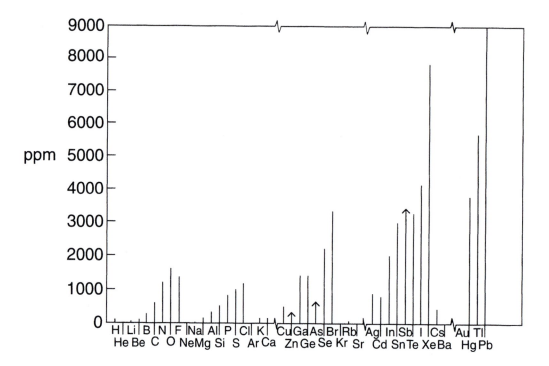

Figure 2.2 Ranges of screening constants for nuclei of main-group and post-transition elements. It should be remembered that, while the ranges shown reflect the periodicity of the $\langle 1/r^3 \rangle$ term, they also are influenced by how many compounds of a given element have been measured and, indeed, by the extent of its chemistry. (After Jameson and Mason (1987) *Multinuclear NMR*, Plenum, New York, with permission.)

the screening tensor, which has nine components, though only three influence the observed screening. These, the diagonal components of the tensor, are called σ_{11}, σ_{22} and σ_{33} and all three are required if the nucleus is in a site with no symmetry. If the site is axially symmetric, then only two values are needed to describe the screening since $\sigma_{11} = \sigma_{22}$. These components are then called $\sigma_{\perp}$ and σ_{33} is denoted $\sigma_{\parallel}$. This is called the screening anisotropy, and the differences between the values of the components can be substantial.

Screening anisotropy of a nucleus can be observed by taking its spectrum from a powdered solid where the solid particles all have different orientations. Two types of spectrum are shown diagrammatically in Fig. 2.3, and the values of the tensor components are given by the discontinuities on the curves. The effect is another source of line broadening in solids but also allows a full description of the screening mechanism to be obtained. In liquid samples, the isotropic rotation of the molecules produces an average σ where $\sigma = (\sigma_{11} + \sigma_{22} + \sigma_{33})/3$ and the lines become narrow. The high-resolution spectrum that is thus obtained is generally more useful to the chemist, but it should not be forgotten that some fundamental information is lost in the process.

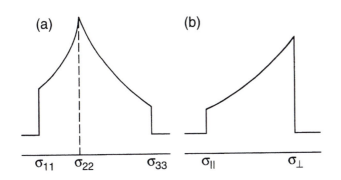

Figure 2.3 Representations of the powder spectra of two different solid samples (a) of a nucleus in an asymmetric environment, and (b) of a nucleus in an axially symmetric environment. Broadening due to through-space magnetic interaction with other nuclei in the samples is avoided by choosing samples where protons, for example, are relatively distant.

The variations in screening among the compounds of a given element depend upon all the factors summarized above, but in a way that is very difficult to separate into dominance by any particular influence. For instance, it seems certain that increased charge density upon an atom results in increased screening of its nucleus but that this occurs principally because the $\langle 1/r^3 \rangle$ term is increased; in other words, the change in orbital radius determines the change in screening. Two examples serve to emphasize the complexity of the situation. As expected from simple considerations of charge density, the ^{14}N nucleus in NH_3 is 25 ppm more shielded than in $[NH_4]^+$, the influence of the apparent change in coordination number being slight since NH_3 has an electron lone pair as effective fourth ligand. In contrast, the ^{14}N nucleus in the pyridinium ion, (**2.1**), is 100 ppm more shielded than that in pyridine, (**2.2**). In this case, the nitrogen is part of a π system and protonation removes an accessible low-energy transition and so modifies the ΔE term.

Oxidation state also affects screening via electron density changes. The influence of electron imbalance in the bonds is manifested by the common effect in which increased screening follows increases in coordination number; for example ^{31}P screening in PCl_3 $< [PCl_4]^+ < PCl_5 < [PCl_6]^-$, and several similar trends will be observed in the scales of Fig. 2.14. It must, however, be emphasized that this effect is not universally true and other considerations operate in some cases. In all the normal cases, the increase in coordination number can be regarded as leading to an increase in local symmetry and so a more balanced structure. Electron imbalance is also related to bond ionicity, π-bond order or s character and so to bond angles. Thus the ^{31}P nucleus is 125 ppm more screened in trimethylphosphine, PMe_3, than in the sterically crowded tri-t-butylphosphine, PBu^t_3, because of the increased C–P–C bond angles in the latter.

(**2.1**)

(**2.2**)

Substituents also exhibit a marked effect upon nuclear screening, which often correlates with changes in substituent electronegativity. However, the direction of the correlation is not uniform, and screening may increase or decrease with increase in electronegativity depending upon the sequence of ligands chosen. Again, several effects can operate simultaneously. For instance, the ^{27}Al nucleus is increasingly screened in the series of tetrahalo anions $[AlCl_4]^- < [AlBr_4]^- < [AlI_4]^-$. This is known as the normal halogen dependence, since it is in the opposite sense for certain transition metals. The change in screening between chloride and bromide (22 ppm) is less than that between bromide and iodide (47 ppm). The changes are ascribed first to the nephelauxetic effect of the halogen, the larger halogens expanding the electron orbitals on the aluminium and so decreasing the $\langle 1/r^3 \rangle$ term and so σ_p. In addition, the heavier halogens produce the heavy-atom screening effect, which is proportional to Z^4, where Z is the nuclear charge of the halogen, and this enhances the apparent nephelauxetic effect and explains the bigger increase in screening between the bromide and iodide. The heavy-atom effect operates via a complex relativistic spin–orbit coupling mechanism.

Usually the contributions to σ_d and σ_p for a nucleus are considered only for the electrons immediately neighbouring, or local to, that nucleus. More distant electrons give rise to long-range effects on both σ_d and σ_p, which are large but cancel to make only a small net contribution to σ. It is therefore more convenient to separate the long-range effects into net contributions from different, quite localized, parts of the rest of the molecule. Two types of contribution to screening can be recognized and, though they are small, they are of particular importance for the proton resonance.

2.1.1 Neighbour anisotropy effects

We have already mentioned that in liquid samples, owing to the rapid and random motion of the molecules, the magnetic fields at each nucleus due to all other magnetic dipoles average to zero. This is only true if the magnet (e.g. a nucleus) has the same dipole strength whatever the orientation of the molecule relative to the field direction. If the source of magnetism is anisotropic and the dipole strength varies with orientation in the applied field, then a finite magnetic field appears at the nucleus.

Such anisotropic magnets are formed in the chemical bonds in the molecule, since the bonding electrons support different current circulation at different orientations of the bond axis to the field. The result is that nuclei in some parts of the space near a bond are descreened while in other parts the screening is increased. Figures 2.4 and 2.5 show the way screening varies around some bonds.

A special case of anisotropic screening where the source of the anisotropy is clearly evident occurs in aromatic compounds, which exhibit what is called ring current anisotropy. The benzene structure,

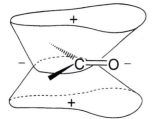

Figure 2.4 Screened and descreened volumes of space around a carbonyl bond. The sign + indicates that a nucleus in the space indicated would be more highly screened. The magnitude of the screening falls off with increasing distance from the group and is zero in the surface of the solid figure.

for instance, can support a large electronic ring current around the conjugated π-bond system when the plane of the ring is transverse to the field axis but very little when the ring lies parallel to the field axis. This results in large average descreening of benzene protons since the average secondary magnetic field, which must oppose the applied field within the current loop, acts to increase the field outside the loop in the region of the benzene protons (Fig. 2.6).

2.1.2 Through-space electric field effects

Molecules that contain electric dipoles or point charges possess an electric field whose direction is fixed relative to the rest of the molecule. Such electric fields can perturb the molecular orbitals by causing electron drifts at the nuclei in the bond directions and by altering the electronic symmetry. It has been shown that the screening σ_E due to such electric fields is given by

$$\sigma_E = - AE_z - BE^2 \tag{2.3}$$

where A and B are constants, $A \gg B$, E_z is the electric field along a bond to the atom whose nuclear screening we require and E is the maximum electric field at the atom. The first term produces an increase in screening if the field causes an electron drift from the bond onto the atom and a decrease if the drift is away from the atom. The second term leads always to descreening. It is only important for proton screening in the solvation complexes of highly charged ions where

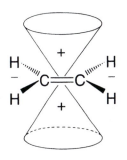

Figure 2.5 Screened and descreened volumes of space around a carbon–carbon double bond. The significance of the signs is the same as in Fig. 2.4. The cone axis is perpendicular to the plane containing the carbon and hydrogen atoms.

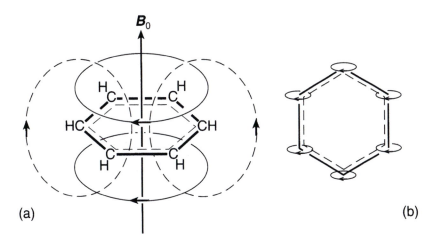

Figure 2.6 Ring current descreening in benzene. (a) The plane of the benzene ring is at right angles to the applied magnetic field B_0. A current is induced in the delocalized π-orbitals generating an opposing magnetic field in the centre of the ring and a reinforcing magnetic field outside the ring. The result is that protons above the ring move to lower frequency, while those outside the ring, in the plane of the ring, move to higher frequency. (b) When the applied magnetic field lies in the plane of the benzene ring, the area of the ring current loops due to the electrons is much smaller.

E can be very large, though it is of greater importance for the nuclei of the heavier elements.

The electric field effect is, of course, attenuated with increasing distance. It is an intramolecular effect since the effect of external fields, for which the BE^2 term can be neglected, averages to zero as the molecule tumbles and E continually reverses direction along the bond.

The descreening of protons that occurs in many organic compounds containing electronegative substituents X probably occurs because of the electric field set up by the polar C–X bond. This will also increase with X electronegativity and produce a similar result to an inductive electron drift. The effect typically produces proton descreening in molecules containing keto, ester, or ether groups and in halides, some values being given below.

Molecule	CH_3Cl	$CH_3-C(O)O-CH_3$		$(CH_3)_2CO$	$(CH_3)_2O$
ppm less screened than methane	2.83	1.78	3.44	1.86	3.01

The closer the protons are to the bond generating the electric field, then the more they are descreened. This also shows up as a fall-off in descreening along an alkyl chain for the protons further away from the substituent, and, for instance, in 1-chloropropane (n-propyl chloride) the comparable descreening figures are α-CH_2 3.24 ppm, β-CH_2 1.58 ppm and CH_3 0.83 ppm.

2.2 ISOTOPE EFFECTS

Because the bonds in molecules are not rigid fixed entities but have dimensions determined by vibrational phenomena, the substitution of an atom by one of its isotopes of different mass alters the vibrational energies in the bonds to that atom and so alters the electron distribution about it. This necessarily implies changes in nuclear screening following the substitution, both near the substitution site and at some distance from it. Such changes are small but measurable in many cases, and provide useful spectroscopic data. It is necessary to distinguish between primary isotope effects and secondary isotope effects. The former are the effects experienced by the isotopically substituted nucleus itself. For instance, for $^{14}N/^{15}N$ substitutions, the change in ^{14}N screening between $^{14}NH_3$ and $CH_3{}^{14}NO_2$ is not exactly the same as that for ^{15}N between $^{15}NH_3$ and $CH_3{}^{15}NO_2$, given identical conditions. Clearly, such changes are not easy to measure and for many elements are within experimental error, so that primary effects will not be considered further. The secondary effects are those observed on the nuclei in the rest of the molecule; in the above example the proton screening changes, say, between $^{14}NH_3$ and $^{15}NH_3$. The trends observed are as follows.

1. The isotope shift is greatest when the substitution causes the greatest fractional change in mass; thus $^2D/^1H$ substitution is the most effective.
2. Screening is greatest for nuclei near the heavier isotope. This is not always true and in some cases the opposite holds. The nearer the nucleus observed is to the substituted site, the greater is the effect.
3. The magnitude of the effect reflects the overall range of screening experienced by the observed nucleus (cf. Fig. 2.2).
4. The effect of multiple isotopic substitutions is additive, or approximately so.

Thus the ^{15}N signal obtained from the nitrite ion, $[^{15}NO_2]^-$, in water and which has had a partial substitution of ^{16}O by ^{18}O, fully randomized over the two positions, consists of three signals as shown in Fig. 2.7. This spectrum shows the additive effect and the greater screening due to the ^{18}O. It also proves unequivocally that the nitrite anion contains two oxygen atoms. Similarly, the ^{31}P spectrum of partially ^{18}O substituted $[PO_4]^{3-}$ is an equally spaced quintet with spacing of 0.0206 ppm and proves that the orthophosphate contains four oxygen atoms.

This property of partial isotopic substitution provides an almost digital technique for measuring numbers of exchangeable atoms in a

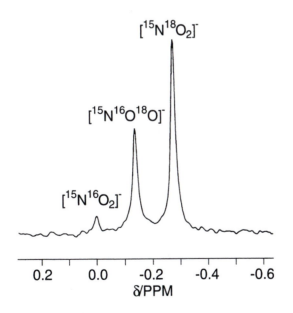

Figure 2.7 The ^{15}N NMR spectrum of the nitrite ion, $[NO_2]^-$, in water. The ion was enriched to 95% in ^{15}N and 77% in ^{18}O. The three resonances arise from ions of isotopic composition $^{15}N^{16}O_2$, $^{15}N^{16}O^{18}O$ and $^{15}N^{18}O_2$ in the concentration ratios $6:33:61$. Replacement of ^{16}O by ^{18}O causes a low frequency shift of 0.138 ppm. (From Van Etten and Risley (1981) *J. Am. Chem. Soc.*, **103**, 5634; copyright (1981) American Chemical Society, reprinted with permission.)

structure. It is, for instance, now possible to count the number of water molecules in aqua complexes such as $[Al(H_2O)]^{3+}$. Analytical methods never give whole numbers, and while the hexaaqua cation is known to exist in solids from X-ray structural analysis, it is possible to argue that the hydration number may not be so definite in solution. If $Al(H_2O)_6(ClO_4)_3$ is dissolved in acetone, it is possible to observe the water proton signal and note that this is highly descreened relative to free water. This is a consequence of the strong electric field of the cation. If a proportion of the water is replaced by D_2O, then the complex will contain H_2O, HOD and D_2O. The signal of the HOD protons shows a strong isotope effect, though this is abnormal as they are less screened even than the H_2O. This probably arises because the deuterium substitution shortens the Al–O bond in that molecule and so increases the electric field effect.

More importantly, the two proton resonances (HOD and HOH) show fine structure as in Fig. 2.8. This arises because of long-distance

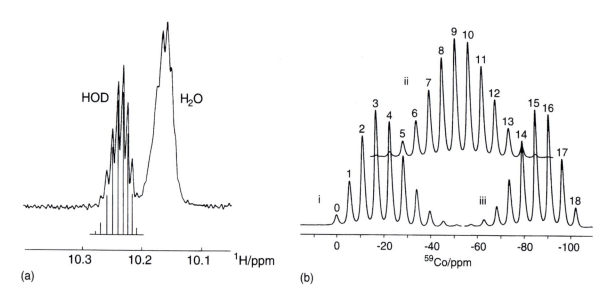

(a)   (b)

Figure 2.8 (a) The 400 MHz ^{1}H NMR spectrum of the water complexed to the cation $[Al(H_2O)]^{3+}$ in ($^2H_6$)acetone (deuterioacetone or acetone-d_6) taken at –30°C. The complex had been partially deuterated so as to contain 35% ^{2}H. The resonance at 10.23 ppm is due to all the HOD molecules in the complex and that at 10.17 ppm is due to all the HOH molecules. The fine structure arises because different molecules have different total numbers of ^{2}H, each giving a smaller isotope effect due to the more distant substitution. The stick diagram gives the calculated intensities obtained from the deuterium content and assumes completely random distribution throughout the sample. (After Akitt *et al.* (1986) *J. Chem. Soc. Chem. Commun.*, 1047, with permission.) (b) The 95.7 MHz ^{59}Co NMR spectrum of $[Co(NH_3)_6]^{3+}$ in various mixtures of D_2O/H_2O to generate the complete range of nineteen deuterated compounds from $[Co(NH_3)_6]^{3+}$ to $[Co(ND_3)_6]^{3+}$. The introduction of each deuterium moves the signal approximately 6 ppm to low frequency. (i) 15% D_2O/85% H_2O, (ii) 50% D_2O/50% H_2O, (iii) 85% D_2O/15% H_2O. The number besides each signal refers to the number of hydrogen atoms that have been replaced by deuterium. (Remeasured, but based on Russell and Bryant (1983) *Anal. Chim. Acta*, **151**, 227, copyright (1994), with permission from Elsevier Science.)

isotope effects between a given HOD and the other ten replaceable sites in the complex. A 13-line pattern is theoretically possible, though some lines are too weak to detect, and one has to calculate an intensity distribution from the known level of deuteration.

The changes can be even more dramatic when a nucleus with a larger chemical shift range is observed. This is shown in Fig. 2.8(b) for ^{59}Co in $[Co(NH_3)_6]^{3+}$, where replacement of hydrogen by deuterium produces a shift of about 6 ppm for each replacement.

2.3 EFFECTS DUE TO UNPAIRED ELECTRONS

The electron (spin = 1/2) has a very large magnetic moment and if, for instance, paramagnetic transition-metal ions are present in the molecule, large effects are observed. The NMR signal of the nuclei present may be undetectable, but under certain circumstances, when the lifetime of the individual electron in each spin state is short, so that its through-space effect averages to near zero, NMR spectra can be observed. The screening constants measured in such systems, however, cover a very much larger range than is normal for the nucleus, and this arises because the electronic spins can be apparently delocalized throughout a molecule and appear at, or contact, nuclei. The large resonance displacements that result are known as contact shifts and the ligands in certain transition metal-ion complexes exhibit proton contact shifts indicating several hundred ppm changes in σ. In addition, if the magnetic moment of the ion is anisotropic, one gets a through-space contribution to the contact interaction similar to the neighbour anisotropy effect, and this is called a pseudo-contact shift.

The NMR signals are also broadened by the presence of the unpaired electron(s). The broadening is proportional to γ^2/r^6, where γ is the gyromagnetic ratio of the nucleus being observed and r is the distance between the unpaired electron(s) and the observed nucleus.

2.3.1 Paramagnetic transition metal compounds

For the vast majority of paramagnetic transition metal compounds, the presence of unpaired electrons produce substantial line broadening. In favourable cases, the line broadening is small compared with the shifts and well resolved signals are observed. For instance, in Fig 2.9 the ^1H NMR spectrum of $[Co(4,6-Me_2-phenanthroline)_3]^{2+}$, (2.3), is shown. The ion consists of two isomers depending on whether the nitrogen atoms in the 1 position on the phenanthroline ring are arranged *mer* or *fac*. In the *fac*-isomer all the three phenanthroline ligands are equivalent giving rise to one set of signals, but in the *mer*-isomer they are all inequivalent. Hence four sets of ligand signals are observed in a mixture of both isomers. The signals are substantially shifted due to the presence of the unpaired electrons. This is particularly marked

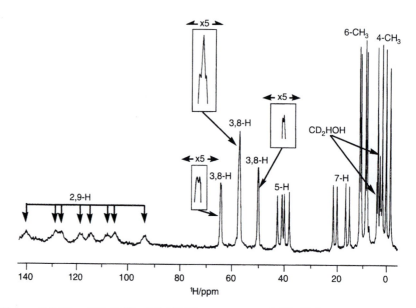

Figure 2.9 The 100 MHz ^{1}H NMR spectrum of [Co(4,6-Me$_2$-phenanthroline)$_3$]$^{2+}$, (**2.3**), in CD$_3$OD at –20°C. Note that the compound consists of two isomers, *mer* and *fac*. (Reproduced by permission of the American Chemical Society from La Mar and Van Hecke (1970) *Inorg. Chem.*, **9**, 1546.)

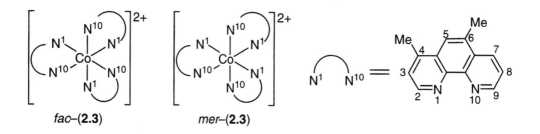

fac–(**2.3**)   *mer*–(**2.3**)

for the H^2 and H^9 signals which are moved by some 110 ppm. Note that despite the presence of the unpaired electron, the signals are relatively sharp.

However, in other cases, the ^{1}H NMR signals may be so broad that separate signals cannot be resolved. Even then all is not lost. As the broadening depends on γ^2, the solution to the problem is to observe a low γ nucleus. $\gamma(^2$H$)^2$ is only 0.024 that of ^{1}H resulting in much sharper signals, and the result is that deuterated samples of Cr^{3+} and Cu^{2+} compounds which fail to give usable ^{1}H NMR spectra can give useful information in their ^{2}H NMR spectra. For example, in [Cr$_3$(μ-OH)$_2$(μ-O$_2$CCR$_3$)$_4$(O$_2$CCR$_3$)$_2$(bipy)$_2$][ClO$_4$], R = H or D, the resolution of the acetate signals is greatly improved when R = D and the ^{2}H NMR spectrum is recorded compared with the ^{1}H NMR spectrum, when R = H (Fig. 2.10), where all the acetate signals are observable giving five peaks in the ratio 2 : 1 : 1 : 1 : 1.

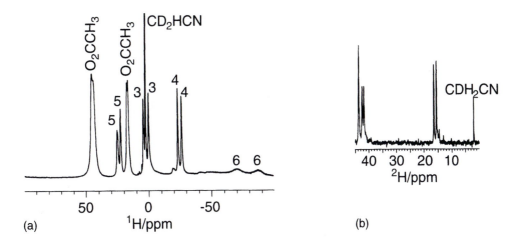

Figure 2.10 The 360 MHz ^{1}H, (a), and 76.75 MHz ^{2}H, (b), NMR spectra of $[Cr_3(\mu\text{-}OH)_2(\mu\text{-}O_2CCR_3)_4$ $(O_2CCR_3)_2(bipy)_2][ClO_4]$, R = H or D, in CD$_3$CN, (a), or CH$_3$CN, (b). In spectrum (b), only the signals due to the deuteriated acetate ligands and the residual deuterium in the solvent are observed. Note the better resolution observed for the acetate signals at 16 and 42 ppm in the ^{2}H NMR spectrum when compared with the ^{1}H NMR spectrum. The numbered resonances in (a) arise from the bipyridyl ligand. (Reproduced by permission of the American Chemical Society from Harton *et al.* (1997) *Inorg Chem*, **36**, 4875.)

As the broadening by the unpaired electrons falls off rapidly with distance, varying as r^{-6}, only the nuclei close to the metal are badly broadened. The result is that for large molecules the ^{1}H NMR spectrum can be run for the metal free molecule, and then a metal such as Mn^{2+} or Cu^{2+} added, so that the ^{1}H NMR signals for protons close to the metal binding site become badly broadened and in effect vanish from the spectrum. This approach has been used to identify protons close to metal binding sites of enzymes.

2.3.2 Paramagnetic lanthanide and actinide compounds

Many of the paramagnetic lanthanides and actinides give relatively sharp ^{1}H NMR spectra. For example, many U^{IV} compounds give sharp signals (Fig. 2.11).

Such effects are particularly marked with the lanthanide cations, whose complexes are commonly used to simplify the spectra of organic compounds. These substances are called shift reagents and consist of a lanthanide element (Pr, Eu, Dy, or Yb) complexed with an organic ligand chosen, among other things, to make the complex soluble in organic solvents so that it can be codissolved with the compound to be investigated. They are octahedral complexes, but the lanthanides are capable of assuming higher coordination numbers than six, so that if the organic molecule possesses a suitable donor site, such as oxygen or nitrogen, it can interact with the shift reagent. This produces a pseudo-contact shift of all the proton and carbon nuclei in the

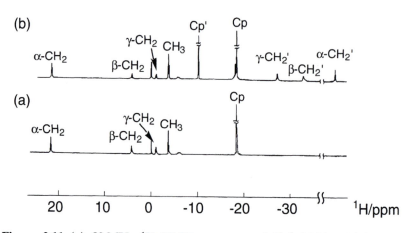

Figure 2.11 (a) 80 MHz ^{1}H NMR spectrum of $[(\eta^5\text{-}C_5H_5)_3UC(O)CH_2CH_2$ $CH_2CH_3]$ in C_6D_6. (b) ^{1}H NMR spectrum of a mixture of $[(\eta^5\text{-}C_5H_5)_3UC$ $(O)CH_2CH_2CH_2CH_3]$ and $[(\eta^5\text{-}C_5H_5)_3UCH_2CH_2CH_2CH_3]$ in C_6D_6, the latter giving the primed signals. (Reproduced from Paolucci *et al.* (1984) *J. Organomet. Chem.*, **272**, 363, copyright (1984), with permission from Elsevier Science.)

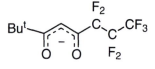

(2.4)

molecule, which can represent very large changes in screening. For instance, the normal proton spectrum of pentanol consists of five signals with the signals due to the γ-CH$_2$ and δ-CH$_2$ protons being coincident (Fig. 2.12). If [Eu(fod)$_3$], fod = 6,6,7,7,8,8,8-heptafluoro-2,2-dimethyl-3,5-octanedionate, (**2.4**), is added, all the signals move to high frequency. Alternatively, if [Pr(fod)$_3$] is added, all the signals move to low frequency. The extent of the movement depends on the distance from the lanthanide to the proton, so the α-CH$_2$ moves most and the CH$_3$ group moves the least. The OH proton has moved out of the spectral region shown in Fig. 2.12. In both cases, the overlapped signals due to the γ- and δ-CH$_2$ protons are resolved. The fine structure is due to spin–spin coupling, which we will meet in the next chapter.

If the organic molecule is rigid, then the magnitude of the shifts can be used to calculate its geometry using an interactive method that optimizes both geometry and position of the lanthanide. This is normally supported by also using the corresponding gadolinium complex to broaden the signals, and the signal broadening is proportional to r^{-6}, where r is the distance between the broadened proton and the metal.

A second use of these reagents is in determining chirality or optical purity of organic substrates. This is achieved by preparing chiral shift reagents, i.e. ones in which the ligands are themselves chiral. Such reagents produce slightly different screening effects in substrates of different handedness. It is, for instance, possible to obtain separate signals for the two optical isomers of C$_5$H$_{11}$CHDOH, which can be prepared using optically active reducing agents. Figure 2.13 shows the very small separation between the signals of the two forms of some 0.07 ppm.

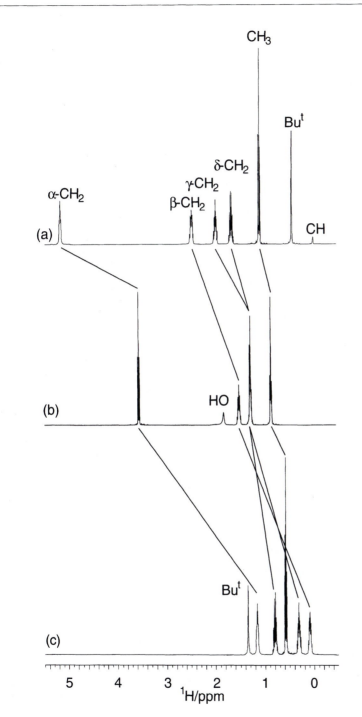

Figure 2.12 Three 400 MHz ^{1}H NMR spectra of 1-pentanol in CDCl$_3$. (a) With added [Eu(fod)$_3$]. (b) No added shift reagent. (c) With added [Pr(fod)$_3$]. Note that fod is 6,6,7,7,8,8,8-heptfluoro-2,2-dimethyl-3,5-octanedionate (**2.4**). Signals due to the fod ligand are also observed.

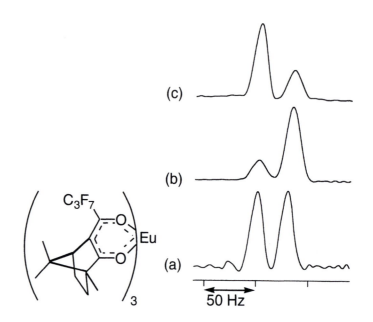

Figure 2.13 The ^2H NMR signals of the CHD deuterons in various samples of the optically active hexanol, $C_5H_{11}CHDOH$, made using an optically active reducing agent and obtained in the presence of a chiral shift reagent, in which the ligands are hexafluoropropyl camphorate (see formula). The lower trace shows the racemic mixture and the two upper traces are from materials made from reducing agents of opposite chirality.

2.4 THE CHEMICAL SHIFT

So far we have talked only in terms of the changes in the screening of nuclei in different environments. It is, however, more usual to describe these changes as 'chemical shifts' although the word 'screening' will still be encountered from time to time. It is usual to calibrate chemical shifts by using a suitable compound as a marker resonance. Thus if we have unknown and marker resonances in different environments with screening constants σ_1, and σ_2, then the two nuclear frequencies in a given magnetic field $\boldsymbol{B}_0$ are

$$\nu_1 = \frac{\gamma \boldsymbol{B}_0}{2\pi} (1 - \sigma_1) \tag{2.4a}$$

$$\nu_2 = \frac{\gamma \boldsymbol{B}_0}{2\pi} (1 - \sigma_2) \tag{2.4b}$$

whence

$$\nu_1 - \nu_2 = \frac{\gamma \boldsymbol{B}_0}{2\pi} (\sigma_2 - \sigma_1) \tag{2.5}$$

It is possible to measure a frequency with very high precision but not a magnetic field strength so this expression is not of great use. We thus eliminate field from the equation by dividing through by v_1 (equation (2.4a)). This gives us the frequency change as a fraction of v_1

$$\frac{v_1 - v_2}{v_1} = \frac{\sigma_2 - \sigma_1}{1 - \sigma_1} \qquad (2.6)$$

which since $\sigma \ll 1$ in many cases reduces to

$$\frac{v_1 - v_2}{v_1} = \sigma_2 - \sigma_1 \qquad (2.7)$$

when the chemical shift range is small, although the approximation cannot be used for large chemical shift differences as is found for many of the heavier nuclei.

Thus the fractional frequency change is the same as the difference in screening in the two nuclear environments. This is called the chemical shift and is given the symbol δ. Its value is expressed in parts per million (ppm). It can be determined with high accuracy since it is possible to resolve shifts of 0.001 ppm for spin-1/2 nuclei, or even less in favourable cases.

In order to establish a chemical shift scale for a given nucleus, it is necessary to choose some substance as a standard and define its chemical shift as zero. The usual, almost universal, standard for the four nuclei, the proton (^{1}H), deuterium (^{2}H), carbon (^{13}C) and silicon (^{29}Si), is tetramethylsilane, $(CH_3)_4Si$, usually called TMS. It gives a narrow singlet resonance for protons, and for the other three nuclei if decoupling is used (see Chapter 7 for an explanation of this), which in all cases is outside the normal range of chemical shifts of the compounds studied. It is miscible with most organic solvents, it is inert and, being highly volatile, can easily be removed after measurements have been made.

In the case of ^{1}H and ^{13}C spectroscopy, the resonances of interest come predominantly from nuclei that are less screened than is TMS, the main exceptions occurring where metal atoms are present in the compounds studied.

It is usual in recording spectra to depict the descreened region to the left-hand side of the spectrum with TMS to the right. The frequency then increases to the left of TMS since the magnetic fields at descreened nuclei are apparently higher and the nuclear precession frequency is higher. For historical reasons, however, this region is very commonly referred to as 'low field', the aptness of this name being apparent from equation (2.1). Both the names 'low field' and 'high frequency' are met in practice, the latter being the most logical since it is indeed the frequencies that are measured. The older name, though, does explain the apparently eccentric way in which the shifts are displayed increasing to the left. This is summarized in Fig. 2.14. Because we have chosen our standard arbitrarily, we also find that we have introduced sign into the scale. Thus in ^{1}H spectroscopy all those protons low frequency of TMS (high field) have negative shift values. This scale is called the δ scale, and it is important to note that the

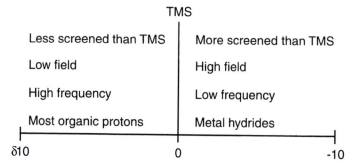

Figure 2.14 Summary of proton chemical shift scales.

symbol δ may either refer to this scale or simply be used as short-hand for 'chemical shift'.

In early work, the sign convention that low frequency (high field) is positive was often used and this led to the 'tau' scale for ^1H NMR spectroscopy, where $\tau = 10 - \delta$. During the 1970s the sign convention that high frequency (low field) is positive became generally accepted, and became the rule as a result of decisions by IUPAC in 1972 and 1976. Unfortunately, the literature prior to this ruling can be very confusing with chemical shifts quoted using either sign convention. The result is that great caution is necessary concerning the sign of chemical shifts published prior to about 1975. The problem is particularly severe for ^{19}F and ^{31}P chemical shifts.

It is, of course, possible to use subsidiary standards, which are referred in turn to TMS. ^1H spectroscopy in aqueous solution thus often has recourse to either $(CH_3)_3Si(CH_2)_3COOH$ or $[(CH_3)_4N]^+$.

Calculation of a chemical shift is a simple matter. Modern spectrometers give the frequency separation of signals and it is only necessary to divide by the spectrometer operating frequency. If in a given solution the signal of benzene is 2901 Hz higher in frequency than TMS and the frequency of TMS is 400.134 394 MHz (i.e. $B_0 = 9.39$ T), then the chemical shift is

$$\delta = \frac{2901}{400.134\ 394} \times 10^6 = 7.25 \text{ ppm} \tag{2.8}$$

Or, more easily remembered, it is the frequency difference in hertz divided by the frequency of the reference in megahertz. We see also that the frequency separations will be different in spectrometers operating at different frequencies. Thus in a 200 MHz spectrometer, the frequency separation above will be 1450 Hz.

The conventions adopted for other nuclei are less firm. The shifts are usually large, so that it is not quite so important to be able to compare different workers' results with high accuracy, and the standard substance is often chosen according to the dictates of

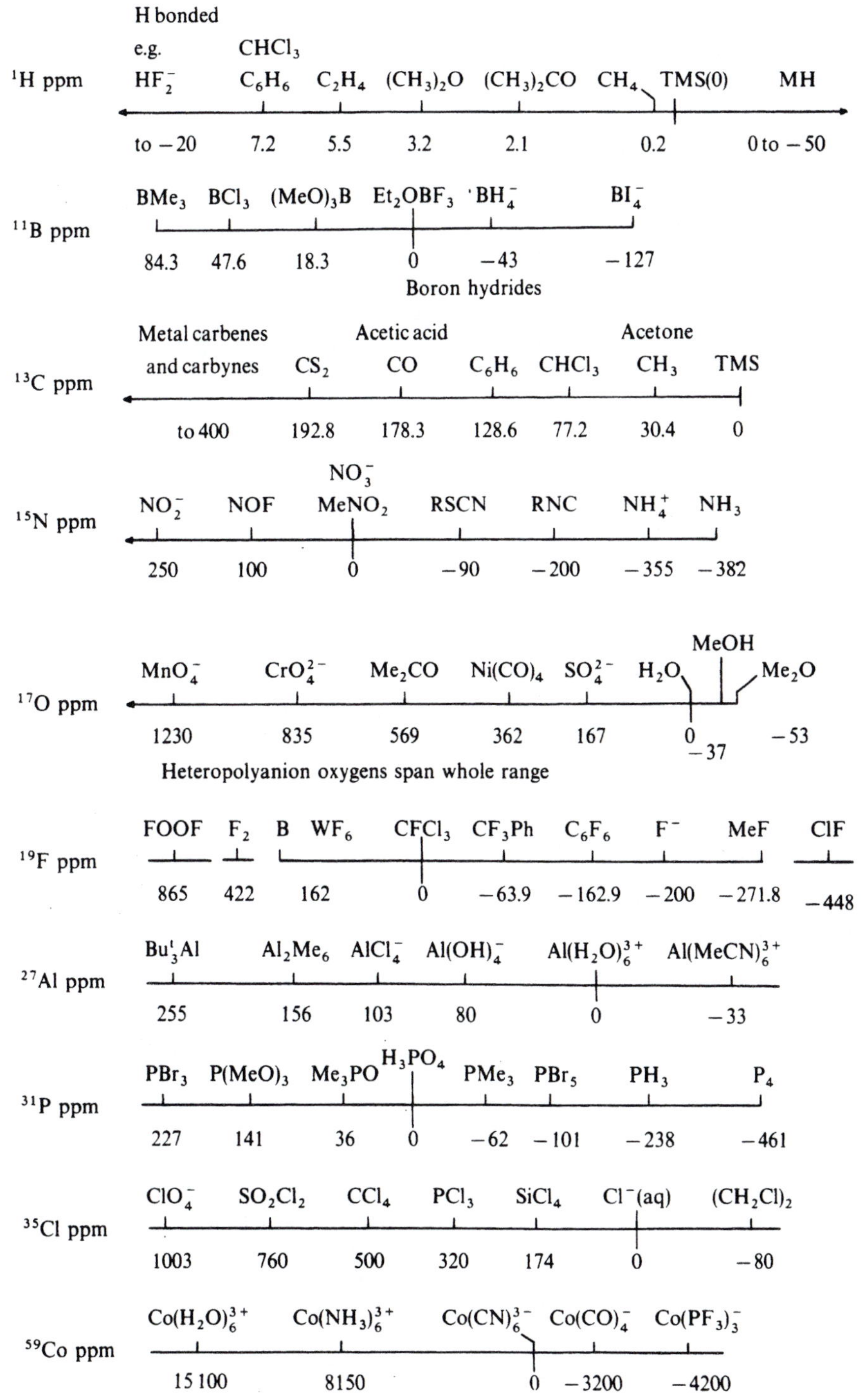

^{1}H ppm

H bonded
e.g. CHCl$_3$
HF$_2^-$ C$_6$H$_6$ C$_2$H$_4$ (CH$_3$)$_2$O (CH$_3$)$_2$CO CH$_4$ TMS(0) MH

to -20 7.2 5.5 3.2 2.1 0.2 0 to -50

^{11}B ppm

BMe$_3$ BCl$_3$ (MeO)$_3$B Et$_2$OBF$_3$ ·BH$_4^-$ BI$_4^-$

84.3 47.6 18.3 0 -43 -127

Boron hydrides

^{13}C ppm

Metal carbenes
and carbynes CS$_2$ CO C$_6$H$_6$ CHCl$_3$ CH$_3$ TMS
Acetic acid   Acetone

to 400 192.8 178.3 128.6 77.2 30.4 0

^{15}N ppm

NO$_2^-$ NOF NO$_3^-$ MeNO$_2$ RSCN RNC NH$_4^+$ NH$_3$

250 100 0 -90 -200 -355 -382

^{17}O ppm

MnO$_4^-$ CrO$_4^{2-}$ Me$_2$CO Ni(CO)$_4$ SO$_4^{2-}$ H$_2$O MeOH Me$_2$O

1230 835 569 362 167 0 -37 -53

Heteropolyanion oxygens span whole range

^{19}F ppm

FOOF F$_2$ B WF$_6$ CFCl$_3$ CF$_3$Ph C$_6$F$_6$ F$^-$ MeF ClF

865 422 162 0 -63.9 -162.9 -200 -271.8 -448

^{27}Al ppm

Bu$_3^t$Al Al$_2$Me$_6$ AlCl$_4^-$ Al(OH)$_4^-$ Al(H$_2$O)$_6^{3+}$ Al(MeCN)$_6^{3+}$

255 156 103 80 0 -33

^{31}P ppm

PBr$_3$ P(MeO)$_3$ Me$_3$PO H$_3$PO$_4$ PMe$_3$ PBr$_5$ PH$_3$ P$_4$

227 141 36 0 -62 -101 -238 -461

^{35}Cl ppm

ClO$_4^-$ SO$_2$Cl$_2$ CCl$_4$ PCl$_3$ SiCl$_4$ Cl$^-$(aq) (CH$_2$Cl)$_2$

1003 760 500 320 174 0 -80

^{59}Co ppm

Co(H$_2$O)$_6^{3+}$ Co(NH$_3$)$_6^{3+}$ Co(CN)$_6^{3-}$ Co(CO)$_4^-$ Co(PF$_3$)$_3^-$

15 100 8150 0 -3200 -4200

Figure 2.15 Various chemical shift scales and chemical shifts of some compounds.

convenience. The standard is assigned 0 ppm and the δ scales are then as in Fig. 2.15. An aqueous salt solution is often used as standard for groups 1, 2, 3, and 17: $^7Li^+$ (aq) for lithium, $[^{27}Al(H_2O)_6]^{3+}$ for aluminium and $^{35}Cl^-$ (aq) for chlorine, for instance. Some much used references are: $(CH_3O)_3B$ and $(CH_3CH_2)_2O \rightarrow BF_3$ for ^{11}B spectroscopy; nitromethane, CH_3NO_2, or nitrate ion, $[NO_3]^-$, for ^{14}N and ^{15}N spectroscopy; H_2O for ^{17}O spectroscopy; 85% orthophosphoric acid, H_3PO_4, for ^{31}P spectroscopy; and the refrigerant $CFCl_3$ is commonly used as standard for ^{19}F work. This ^{19}F chemical shift scale is often called the ϕ scale. Other standards used for ^{19}F spectroscopy are hexafluorobenzene, C_6F_6, or trifluoroacetic acid.

These and some other chemical shift scales are illustrated in Fig. 2.15, which shows the standards used (0 ppm) and the chemical shifts of some compounds, chosen to cover the full range of shifts for a given element rather than to pick out any particular trends with composition. For many of the heavier elements, the shift is very medium sensitive and the spot value given is purely illustrative. For the ^{19}F scale, the shift marked B is that of the bare nucleus (189 ppm). Note also the very large chemical shift range for the transition-metal nucleus ^{59}Co.

2.5 NOTES ON SAMPLE PREPARATION, STANDARDIZATION AND SOLVENT AND TEMPERATURE EFFECTS

An NMR experiment involves a highly sophisticated instrument capable of resolving resonances and making measurements to one part in 10^{11} in favourable circumstances. (This is equivalent to comparing the lengths of two steel rods each 1 km long to within 10 Å or three atoms of iron). One must, therefore, accept the responsibility of preparing a sample that will not degrade the spectrometer performance. However, any sample placed in the magnet will distort the magnetic field. Fortunately, the distortion occurs externally to a cylindrical sample and the field remains homogeneous within it except at the ends, though its magnitude is changed by an amount that depends both on the shape of the sample and on the bulk magnetic susceptibility of the tube glass and of the sample itself. Imperfections in the glass, variations in wall thickness, variations in diameter, or curvature of the cylinder along its length all lead to degradation of the field homogeneity within the sample, with consequent line broadening. For this reason high-precision bore sample tubes are always used for NMR. Since solid particles distort the field around them, suspended solids must also be filtered from the liquid sample prior to measurement. Paramagnetic metal ions and complexes also cause signal broadening and if present must be removed, often by chromatography.

As the NMR tube distorts the magnetic field, adjustment of the magnetic field using the shim coils can be kept to a minimum by always setting the NMR tube in the spinner at the same depth, using a depth gauge. Provided enough solvent is used so that the top of the solution is well above the coil, there will be only minor changes to the field homogeneity, and this is the normal procedure for samples with strong signals. Where sensitivity is low and the quantity of compound is limiting, it is advisable to have as much compound as possible inside the detector coils. This is done by adjusting the position of the bottom of the tube and the top of the solution to match the bottom and top of the detector coil, but considerable time will be required to reshim the magnet to obtain best homogeneity.

2.5.1 Standardization

We have shown that chemical shifts are invariably measured relative to a standard of some sort. There are five ways of standardizing a resonance, which are now given.

2.5.1.1 Internal standardization

The standard substance is dissolved in the sample solution and its resonance appears in the spectrum. This method has the advantage that the magnetic field is exactly the same at sample and standard molecules. The standard must be chosen so as not to obscure sample resonances and also must be inert to the sample. Internal standardization is the method normally used for ^{1}H, ^{2}H, ^{13}C, ^{14}N, ^{15}N, ^{19}F, and ^{29}Si. The main disadvantage of the method is that weak interactions with the solvent produce small chemical shifts, which are difficult to predict and which reduce the accuracy of the measurements by an unknown amount. These shifts are said to arise from solvent effects. TMS is reasonably free from solvent effects and is the reference of choice. For aqueous solutions, TMS is not soluble and either TSP, $Me_3SiCD_2CD_2CO_2Na$, or DSS, $Me_3SiCH_2CH_2CH_2SO_3Na$, are used as water soluble references. Generally, the former is now preferred, as on account of the deuteration only the Me_3Si group gives a ^{1}H NMR signal.

For most other nuclei, the normal reference material cannot be added to the solution without severe risk of reaction or strong interactions.

2.5.1.2 Reference to an absolute frequency

On modern spectrometers, it is easy to reference one nucleus to the frequency of another. There are a number of reports of the absolute frequencies of reference substances. For example, the usual reference for ^{199}Hg is [HgMe$_2$]. This compound is highly toxic and in practice is rarely used – recently, a fatality in a laboratory has been attributed to its use as an NMR reference. However it has been reported that when TMS is exactly at 100 000 000.00 Hz, the signal of ^{199}Hg in

[HgMe$_2$] is at 17 910 841 Hz. This absolute frequency referencing is given the symbol Ξ, and it and others are quoted as precise numbers, scaling the frequency to TMS being at exactly 100 MHz. Referencing to a standard compound presents a major problem for many nuclei with chemical shifts which are very temperature, solvent and concentration dependent. For instance, a temperature dependence of the shift of 1 ppm/°C is common. Hence referencing to ^{1}H of TMS is far more accurate than referencing to some compound of the heavier nuclei.

The use of frequency referencing is simple. If a sample containing TMS and *mer*-[RhCl$_3$(SMe$_2$)$_3$] is examined, giving the frequency of ^{1}H in TMS as 400.134 394 MHz and ^{103}Rh in *mer*-[RhCl$_3$(SMe$_2$)$_3$] as 12.693 503 MHz, then the absolute frequency of ^{103}Rh in *mer*-[RhCl$_3$(SMe$_2$)$_3$] can then be calculated as Ξ = 3.172 310 MHz as

$$\Xi \ (mer\text{-}[RhCl_3(SMe_2)_3]) = \frac{12.693\ 503 \times 100.000\ 000}{400.134\ 394}$$

$$= 3.172\ 310 \text{ MHz} \tag{2.9}$$

In practice, the use of the large numbers of the Ξ scale is inconvenient. Frequently, these values are used to calculate the spectrometer reference frequency and then the usual δ scale is used. For example, the ^{1}H NMR spectrum of a mercury sample containing TMS is measured. The frequency of the TMS is determined exactly, say for example 400.134 394 MHz. The frequency of the [^{199}HgMe$_2$] reference is then calculated as

$$^{199}\text{Hg reference frequency} = \frac{400.134\ 394}{100.000\ 000} \times 17.910\ 841 \text{ MHz}$$

$$= 71.667\ 435 \text{ MHz} \tag{2.10}$$

and this frequency is put into the spectrometer computer memory as the ^{199}Hg reference frequency. The chemical shifts are quoted in ppm relative to this absolute frequency.

In a few cases it has been proposed that an arbitrary absolute frequency is used in preference to the frequency for a reference compound. This is done for ^{103}Rh at Ξ = 3.16 MHz and ^{195}Pt at Ξ = 21.4 MHz. This avoids the problems associated with Ξ for a reference compound being solvent, concentration and temperature dependent, and in many cases the necessity to correct for magnetic susceptibility differences when the reference is external, see section 2.5.1.4.

2.5.1.3 Standardization using the ^{2}H NMR signal of the lock

Rather than using the ^{1}H NMR signal of TMS as the frequency reference, the frequency of the ^{2}H lock is often used. In most solutions the relationship between the lock frequency and the frequency of the ^{1}H NMR signal of TMS is constant to much better than 0.1 ppm. This is far better than is achieved in referencing many nuclei. Hence this method is commonly used to reference signals. The computer systems of modern NMR spectrometers are set up to automatically reference to the ^{2}H lock and it is then up to the operator to select a more appropriate method of referencing. At best, this method is only as good as the frequencies which have been input into the computer and it is important that the calibration is checked by running some reference compounds.

2.5.1.4 External standardization

The standard is sealed into a capillary tube that is placed coaxially within the sample tube. The main disadvantage of the method is that, since the volume magnetic susceptibilities of the sample and standard will differ by several tenths of a ppm, the magnetic fields in each will be different and a correction will have to be made for this. Since the volume susceptibilities of solutions are often not known, these must be measured, so that a single shift determination becomes quite difficult if accurate work is required. The magnitude of the correction depends upon the sample shape and is zero for spherical samples, so that this disadvantage can be minimized by constructing special concentric spherical sample holders. The method is used with very reactive samples or with samples where lack of contamination is important.

The change from the old permanent and electromagnets to the current superconducting solenoids has resulted in a change of field direction from being at right angles to the NMR tube to being coaxial with it. This has resulted in a change in sign of the magnetic susceptibility correction. For a perfect pair of coaxial tubes, the susceptibility correction in a permanent or electromagnet with the magnetic field direction at right angles to the sample is

$$\delta_{int} = \delta_{obs}^{\perp} - \frac{2\pi}{3}(\chi_{ref} - \chi_{sample}) \tag{2.11}$$

while in a superconducting solenoid with the magnetic field parallel to the sample is

$$\delta_{int} = \delta_{obs}^{\parallel} + \frac{4\pi}{3}(\chi_{ref} - \chi_{sample}) \tag{2.12}$$

where δ_{int} is the corrected chemical shift, $\delta_{obs}^{\perp}$ and $\delta_{obs}^{\parallel}$ are the measured chemical shifts, χ_{ref} and χ_{sample} are the magnetic susceptibilities of the reference capillary and the sample, respectively. Most of the results quoted in the literature using an external reference have not been corrected for susceptibility effects, e.g. for ^{31}P, an external reference of 85% H_3PO_4 is normally used, and the result is a discrepancy of up to 1 ppm between the older work recorded using a permanent or electromagnet and the more recent work using a superconducting solenoid.

Because the capillary holding the standard distorts the magnetic field around it, the field homogeneity in the annular outer part of the sample is destroyed. This can be restored by spinning, which is essential with this type of standardization. Distorted capillaries can even then degrade the resolution, and if there is any asymmetry in the annular region this will be averaged by spinning to give a field different in value from the true average, i.e. a small shift error will result. The method is nevertheless the only one suitable for measuring solvent shifts.

2.5.2 Solvent effects

The solvent shift effects mentioned under internal standardization (section 2.5.1.1), while a nuisance to those interested simply in structure determination, are of interest in their own right, since they tell us something about the weak interactions that occur between solvent and solutes. The effect is particularly large for aromatic solvents or where specific interactions occur. The chemical shifts of a number of substances relative to TMS in a series of solvents are given in Table 2.1.

The variation in δ between solvents, of course, contains contributions from the solvent effect on both solute and standard. Table 2.1 nevertheless is useful in indicating the existence of certain interactions involving the solutes. Thus the low frequency shifts obtained in the aromatic solvent benzene and for all solutes but chloroform in the aromatic solvent pyridine are obvious. Even the relatively inert cyclohexane is shifted to low frequency by 0.05 ppm. This arises because solutes tend to spend a larger amount of time face on to the disc-shaped aromatic molecules and so on average are shifted low frequency by the ring current anisotropy. The magnitude of the effect depends upon molecular shape and is also increased if there is any tendency for polar groups in the molecule to interact with the aromatic π electrons.

In the case of complex solutes, each type of proton in the molecule suffers a solvent shift, but because the proximity of each to solvent depends upon the shape of the molecule, each suffers a different

Table 2.1 Internal chemical shifts δ of solutes in different solvents

	Solvent				
Solute	$CDCl_3$	$(CD_3)_2SO$	Pyridine	Benzene	CF_3COOH
Acetone, $(CH_3)_2CO$	2.17	2.12	2.00	1.62	2.41
Chloroform, $CHCl_3$	7.27	8.35	8.41	6.41	7.25
Dimethyl-sulfoxide, $(CH_3)_2SO$	2.62	2.52	2.49	1.91	2.98
Cyclohexane, C_6H_{12}	1.43	1.42	1.38	1.40	1.47

solvent shift. For this reason, a complex solute may have quite different spectra in, for instance, chloroform and benzene, and a change from one solvent to the other may remove some degeneracy or avoid the overlapping of signals that would otherwise be difficult to disentangle. Figure 2.16 shows the quite large changes that can occur. This technique is often known by the acronym ASIS, or **A**ssisted **S**olvent-**I**nduced **S**hifts.

Chloroform as a solute suffers considerable solvent shifts. The pure liquid is self-associated by hydrogen bonding, but upon progressive dilution in an inert solvent the proportion of hydrogen-bonded molecules is reduced and its resonance is shifted 0.29 ppm to low frequency. The shifts noted in Table 2.1 are in excess of this and we must consider the existence of several other types of interaction. Thus specific interactions with the Lewis bases, dimethylsulfoxide and pyridine, result in high-frequency shifts. In the case of pyridine, this implies a preference for an edge-on approach to the aromatic ring and

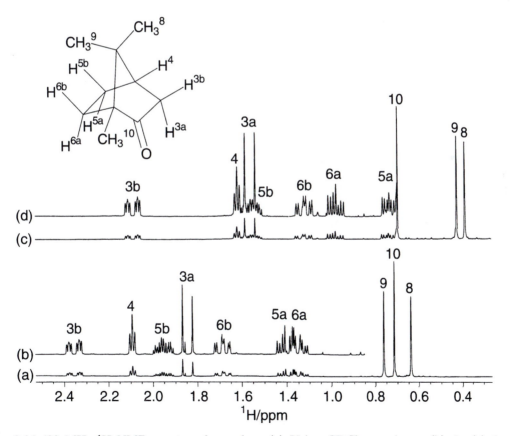

Figure 2.16 400 MHz ^{1}H NMR spectra of camphor. (a) Using CDCl$_3$ as solvent. (b) As (a), but with increased gain. (c) Using C$_6$D$_6$ as solvent. (d) As (c), but with increased gain. Note that in CDCl$_3$, the signals due to H^{5a} and H^{6a} overlap, while they are separated in C$_6$D$_6$. However, in C$_6$D$_6$, the signals due to H^{3a}, H^4, and H^{5b} now overlap. The positions of the signals are markedly solvent dependent. The fine structure is due to spin–spin coupling, to be discussed in the next chapter.

therefore some ring current deshielding. In benzene, on the other hand, hydrogen bonding is reduced, and there is probably face-wise interaction of the chloroform with the benzene π electrons. Both processes tend to increase the screening, so that the chloroform is shifted strongly to low frequency. In addition, the chloroform molecule is polar and its dipole electric field will polarize the surrounding solvent by an amount related to the solvent dielectric constant ε. This induced charge gives rise to an electric field, which is called the reaction field and which will also produce chemical shifts of the chloroform solute. Thus some of the variation observed in Table 2.1 will originate from differences in solvent dielectric constant.

Two further contributions to solvent shifts are also usually considered. One arises from the van der Waals interactions and is responsible for vapour to liquid shifts of 0.1 to 0.5 ppm. The other arises in the case of external standardization and is due to bulk diamagnetic susceptibility differences between solvents. These susceptibility shifts can be comparable in magnitude with those due to the other effects and so must be considered in any interpretation of solvent shifts, but they do not, of course, arise from any chemical interaction.

The various contributions to the solvent shift δ_S can be summarized by a five-term equation

$$\delta_S = \delta_B + \delta_A + \delta_E + \delta_H + \delta_W \tag{2.13}$$

where δ_B is the bulk susceptibility contribution, δ_A is the anisotropy contribution, δ_E is the reaction field contribution, δ_H is the contribution of hydrogen bonding and specific interactions, and δ_W is the van der Waals contribution.

Although for ^1H chemical shifts, the solvent shift is usually less than 1 ppm, the effect for other nuclei can be very substantial. For example, the ^{195}Pt reference compound, $[PtCl_6]^{2-}$, changes its chemical shift by 11 ppm simply by changing the solvent from H_2O to D_2O, while a wide range of solvents produces a variation in chemical shifts over a range of 400 ppm.

2.5.3 Temperature effects

Chemical shifts are temperature dependent. As noted above, this makes referencing of nuclei difficult and ^1H in TMS is often preferred as the nucleus to reference all nuclei. Many heavy nuclei change their chemical shift by more than 1 ppm per degree C. For example, the temperature dependence of the ^{195}Pt chemical shift of aqueous $[PtCl_6]^{2-}$ is 1.1 ppm/°C. At 9.4 T, this corresponds to 94 Hz/°C! Where this effect is most detrimental is in the way it affects the resolution of the NMR signal. In an NMR experiment, a number of NMR spectra are added together to produce the final result. If during the experiment the temperature changes or if there is a temperature gradient in the NMR tube, the signal is broadened. This is a major factor in producing line broadening for nuclei other than ^1H. For example, if a sample gives a resolution of 0.2 Hz for ^{1}H, it should give a resolution of better than 0.1 Hz for ^{13}C and ^{31}P. However, a resolution of better than 1 Hz is difficult to achieve due in part to temperature changes during the course of the experiment. The problem can be alleviated by using the spectrometer's variable temperature accessory to thermostat the sample and by incorporating a substantial number of dummy scans where the spectrometer goes through the process of spectrum acquisition without actually storing the FID. This enables the sample to come to thermal equilibrium. In addition to the probe often being at a different temperature to the room, ^{1}H decoupling, see Chapter 5, is frequently used which also warms the sample.

Normally temperature effects do not have a major effect on the linewidths of ^{1}H spectra. There is one major exception. When D_2O is the solvent and lock, temperature variations can have a disastrous effect on resolution. Any ^{1}H or ^{2}H NMR signal from hydrogen or deuterium atoms involved in hydrogen bonding shows a substantial temperature dependence. Consequently if D_2O is used as the lock substance, any temperature changes cause the lock signal to move and with it all the other signals in the spectrum due to the consequential magnetic field changes. This problem is particularly severe away from room temperature. It is often then necessary to wait a long time for temperature stability to be achieved, monitoring the stability by measuring the position and lineshape of a reference compound, e.g. TSP or DSS.

2.6 QUESTIONS

2.1. Which of the two chemically different types of protons in $CH_2ClCHCl_2$ resonate at higher frequency?

2.2. A proton spectrometer operating at 100 MHz was used to measure the frequency separation of the resonances of chloroform, $CHCl_3$, and TMS, which was found to be 730 Hz, the $CHCl_3$ being to high frequency. What is the chemical shift of chloroform on the δ scale? What would the frequency separation and chemical shift be if the sample were measured in a spectrometer operating at 300 MHz?

2.3. The 100 MHz spectrometer above is used to produce the proton spectrum of a complex molecule. Would the resolution be better or worse if the ^{2}H NMR spectrum of the fully deuterated form of the same molecule were obtained? Assume that the linewidths of ^{1}H and ^{2}H resonances are the same. The frequency used for ^{2}H spectra is 15.35 MHz. Assume also that there is no primary isotope effect and ignore spin–spin coupling.

2.4. The value of ΔE for the separation of the highest occupied and the lowest unoccupied orbitals of the cobalt in $[Co(OH_2)_6]^{3+}$, $[Co(NH_3)_6]^{3+}$ and $[Co(CN)_6]^{3-}$ are 16 600, 21 000 and 32 400 cm^{-1} respectively. Show that this is consistent with the chemical shifts given in Fig. 2.15. Use the data to estimate σ_d for ^{59}Co. Assume that ΔE^{-1} is the only significant term contributing to the σ_p variations.

2.5. In annulene,

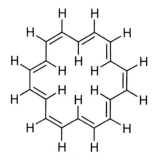

the inner protons have a chemical shift of δ −3.0 while the outer ones come at δ 9.3. Explain why this happens when the ^{1}H chemical shift of ethene is δ 5.5.

2.6. Figure. 2.17 shows the ^{195}Pt NMR spectrum of $[PtCl_6]^{2-}$. Explain why the signal is a multiplet and account for the number of signals and their intensities.

2.7. Many old reports involving the use of NMR spectrometers with electromagnets report the ^{31}P chemical shift of PPh_3 as being between δ −6 and −7. More recent reports using NMR spectrometers with superconducting magnets report the ^{31}P chemical shift of PPh_3 as being between δ −5 and −6. Explain why the apparent chemical shift has changed.

2.8. Ξ for ^{125}Te in the reference compound Me_2Te has been reported as being 31.549 802 MHz. If TMS in a sample in $CDCl_3$ has been measured as being at 400.134 394 MHz, what frequency should be used to reference a ^{125}Te signal from this solution with Me_2Te at zero? The ^{125}Te NMR signal of a sample of Ph_2Te was determined as being at 126.365 452 MHz. What is the chemical shift in ppm relative to Me_2Te?

2.9. There are two common references for ^{195}Pt, namely Ξ = 21.4 MHz and $[PtCl_6]^{2-}$ which has Ξ = 21.496 700 MHz. The compound $[Pt(CN)_4]^{2-}$ has Ξ = 21.395 634 MHz. Calculate the chemical shifts of $[PtCl_6]^{2-}$ and $[Pt(CN)_4]^{2-}$ using first Ξ = 21.4 MHz and then $[PtCl_6]^{2-}$ as the reference. Explain why when using equation (2.6), the chemical shift separation of $[PtCl_6]^{2-}$ and $[Pt(CN)_4]^{2-}$ in ppm depends on the chosen reference.

2.10. Using equations (2.11) and (2.12), work out an expression which allows the magnetic susceptibility differences between two liquids to be obtained from measurements of the chemical shifts of resonances in the two liquids in both a superconducting magnet and an electromagnet.

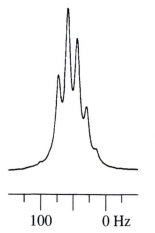

100   0 Hz

Figure 2.17 The 85.6 MHz ^{195}Pt NMR spectrum of 2M $Na_2[PtCl_6]$ in D_2O. (Reproduced by permission of The Royal Society of Chemistry from Sadler *et al.* (1980) *J. Chem. Soc., Chem. Commun.*, 1175.)

Internuclear spin–spin coupling $\boxed{\textbf{3}}$

3.1 THE MUTUAL EFFECTS OF NUCLEAR MAGNETS ON RESONANCE POSITIONS

The Brownian motion in liquid samples averages the through-space effect of nuclear magnets to zero. However, in solutions of $POCl_2F$, for example, the phosphorus nucleus gives two resonances whose separation does not depend upon the magnetic field strength. (The chlorine nuclei ($I = 3/2$) have no effect. This is explained in Chapter 4.) The two resonances correspond to the two spin orientations of the fluorine nucleus so that the nuclei are nevertheless able to sense one another's magnetic fields. Theoretical considerations show that the interaction occurs via the bonding electrons. The contact between one nucleus and its s electrons perturbs the electronic orbitals around the atom and so carries information about the nuclear energy to other nearby nuclei in the molecule and perturbs their nuclear frequency. The effect is mutual and in $POCl_2F$ both the fluorine ($I = 1/2$) and the phosphorus ($I = 1/2$) resonances are split into doublets of equal hertz separation. The magnitude of the effect for a particular pair of nuclei depends on the following factors:

1. The nature of the bonding system, i.e. upon the number and bond order of the bonds intervening between the nuclei and upon the angles between the bonds. The interaction is not usually observed over more than five or six bonds and tends to be attenuated as the number of bonds increases, though many cases are known where coupling over two bonds is less than coupling over three bonds.
2. The magnetic moments of the two nuclei. The interaction is directly proportional to the product $\gamma_A \gamma_B$ where γ_A and γ_B are the magnetogyric ratios of the interacting nuclei.
3. The valence s electron density at the nucleus and therefore upon the s character of the bonding orbitals. This factor also means that the interaction increases periodically as the atomic number of either or both nuclei is increased in the same way as does the chemical shift range.

The magnitude of the coupling interaction is measured in hertz (Hz) since it is independent of the magnetic field strength. It is called the coupling constant and is given the symbol J; its magnitude is very variable and values have been reported from 0.05 Hz up to thousands of hertz. The value of J gives information about the bonding system but this is obscured by the contribution of γ_A and γ_B to J. For this reason correlations between the bonding system and spin–spin coupling often use the reduced coupling constant, K, which is equal to $4\pi^2 J/\hbar\gamma_A\gamma_B$.

It is important to understand that coupling constants can be either positive or negative and that the frequency of one nucleus may be either increased or decreased by a particular orientation of a coupled nucleus, the sign depending upon the bonding system and upon the sign of the product $\gamma_A\gamma_B$.

Considerable data are available upon the magnitudes of interproton spin coupling constants from the mass of data accumulated for organic compounds. Interproton coupling is usually (though not always) largest between geminal protons (H–C–H), and depends upon the angle between the two carbon–hydrogen bonds. J_{gem} is typically –12 Hz in saturated systems. J falls rapidly as the number of intervening bonds is increased, being 7 to 8 Hz for vicinal protons (H–C–C–H) and near zero across four or more single bonds. The same rules apply if oxygen or nitrogen forms part of the coupling path, and methoxy protons (H_3COCHR_2) do not usually show resolvable coupling to the rest of the molecule, though alcoholic or amino protons may do so to vicinal protons in, for example, $HOCH_3$. On the other hand, coupling may be enhanced if there is an unsaturated bond in the coupling path, due to a $\sigma-\pi$ configuration interaction, and may be resolved over up to as many as nine bonds, e.g. $^9J(\text{H–H}) = 0.4$ Hz between the hydrocarbon protons in $H_3C(C{\equiv}C)_3CH_2OH$. In saturated molecules, a planar zig-zag configuration of the bonds may also lead to resolvable coupling over four or five single bonds. Note the use of the super-script 9 in the acetylene example to indicate the number of bonds over which the interaction occurs.

Karplus has calculated the values of the vicinal interproton coupling constants and shown that these depend upon the dihedral angle ϕ between the carbon–hydrogen bonds (Fig. 3.1). Unfortunately, the magnitude of coupling is also influenced by such factors as the nature of the other substituents on the two carbon atoms, their electronegativity, their orientation, the hybridization at the carbon, the bond angles other than the dihedral angle and the bond lengths. For this reason, two curves are shown, which give an idea of the range of coupling constants that may be encountered. If the geometry of the molecule is fixed, then the value of the coupling constant enables an estimate to be made of the value of ϕ. If the geometry is not fixed, then the coupling constant is an average over the possible values of ϕ. If free rotation is possible, then the average is over the whole of the curves illustrated. Thus in fluoroethane (ethyl fluoride) and ethyllithium the vicinal coupling is 6.9 and 8.4 Hz respectively, the value increasing with decreasing substituent electronegativity. If the geometry is static, then the values of the coupling constants may

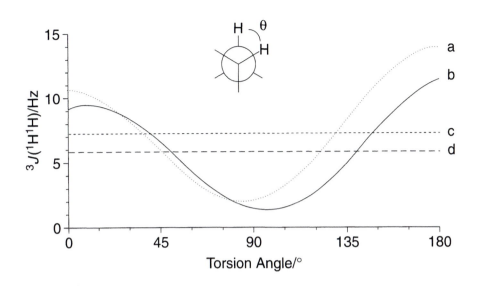

Figure 3.1 Karplus curves, using the equation modified by Altona, relating the dihedral angle ϕ in a HC–CH fragment and the vicinal proton–proton coupling constant. The inset shows a view along the carbon–carbon bond. Two curves are shown relating to differently substituted molecules, CH_3CH_3, dashed line, (a) and CH_3CH_2F, solid line, (b). The horizontal lines, (c) and (d), show the average value obtained when a group can rotate freely, giving rise to an averaged J_{vic}.

show quite clearly what the conformation of the molecule studied has to be. Thus, in Fig. 3.2, part of the proton spectrum of menthol is given in which various doublet splittings due to spin–spin coupling are evident. The resonance of H^1 is split into a doublet of triplets with a small J of 4.2 Hz and two large ones of 10.5 Hz. H^1 is easily identifiable by its chemical shift, being next to the electronegative OH group, and it is the large couplings $J(H^1–H^{2ax})$ and $J(H^1–H^6)$ that are due to these axial–axial pairs with a dihedral angle of near 180° and large predicted J. The other lesser interaction is $J(H^1–H^{2eq})$, H^1 and H^2 being an axial–equatorial pair with dihedral angle of near 60° and so small predicted J.

This vicinal ^3J dependence on dihedral or torsion angle seems to be quite general and Karplus-type curves have been established for the coupling paths $^{13}C–C–C–^1H$, $^{31}P–C–C–^{31}P$ and $^{13}C–C–C–^{31}P$ and are also likely for coupling between ^{77}Se or ^{125}Te and ^1H or ^{13}C or for the more exotic systems such as $^3J(^{199}Hg–^{13}C)$ in alkylmercury compounds.

3.2 THE APPEARANCE OF MULTIPLETS ARISING FROM SPIN–SPIN COUPLING

The appearance of these multiplets is very characteristic and contains much information additional to that gained from chemical shift data.

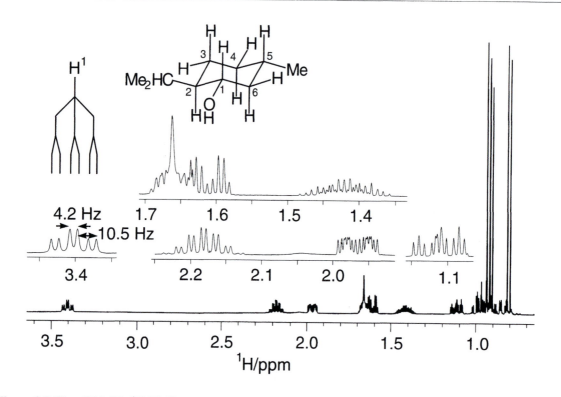

Figure 3.2 The 400 MHz ^{1}H NMR spectrum of menthol whose structure is shown in the inset. The numbers represent hydrogen atoms attached to the ring. The resonances are split by spin coupling with adjacent hydrogen atoms. In particular, H^1, at δ 3.4, is split into a doublet, 4.2 Hz, of triplets, 10.5 Hz, by coupling with the equatorial H^6 and the two axial H^2 and H^6 nuclei.

The coupling patterns and constants enable connectivity to be established between atoms in the molecules and hence enable the structure of the molecule to be deduced. We have already seen one such application in Fig. 3.2 and can see the patterns produced by coupling can be very complex. The simplest case to consider is the effect that a single chemically unique nucleus of $I = 1/2$ has on other nuclei in the molecule that are sufficiently closely bonded. In half the molecules in the sample, the spin of our nucleus A will be oriented in the same direction as the field and all the other nuclei in these molecules will have corresponding resonance positions. In the remainder of the molecules, the spin of A will be opposed to the field and all the other nuclei in this half of the sample will resonate at slightly different frequencies to their fellows in the first half. Thus when observing the sample as a whole, each of the nuclei coupled to A gives rise to two lines. The line intensities appear equal since the populations of A in its two states only differ by < 0.01%, which is not detectable. We say that A splits the other resonances into 1:1 doublets. Because the z component of magnetization of A has the same magnitude in both spin states, the lines are equally displaced from the chemical shift positions of each nucleus, which are therefore at the centres of the doublets.

Let us in illustration consider the molecule $CHCl_2CH_2Cl$ and its proton resonance. This contains two sorts of hydrogen, with the $CHCl_2$ proton resonating to high frequency of the CH_2Cl protons due to the greater electric field effect of two geminal chlorine–carbon bonds. The two CH_2Cl protons have the same frequency since rotation around the carbon–carbon bond averages their environments and makes them chemically equivalent. It also averages the vicinal coupling to the single proton to be the average of the Karplus curve. These two protons thus have the same chemical shift and the same coupling constant to all other magnetically active nuclei in the sample; a trivial condition in this case since there is only one other proton and the spectra are indifferent to the chlorine nuclei for reasons that we shall see later. We say that the CH_2 nuclei are isochronous and magnetically equivalent. They are, of course, coupled quite strongly, but because they are isochronous they resonate as if they were a unit and give a singlet resonance unless coupled to other nuclei. This is a consequence of the second-order effects to be considered later.

In the present example, however, the CH_2Cl resonance is split into a 1:1 doublet because of coupling to the non-isochronous $CHCl_2$ proton. Equally, since the coupling interaction is mutual, the $CHCl_2$ proton is split by the two CH_2Cl protons, though the splitting pattern is more complex. We can discover the shape of the $CHCl_2$ multiplet in several ways.

1. By an arrow diagram (Fig. 3.3). In some molecules, both the CH_2Cl spins will oppose the field, in others both spins will lie with the field, while in the remainder they will be oriented in opposite directions. The $CHCl_2$ protons in the sample can each experience one of three different perturbations and their resonance will be split into a triplet. Since the CH_2Cl spins can be paired in opposition in two different ways, there will be twice as many molecules with them in this state as there are with them in each of the other two. The $CHCl_2$ resonance will therefore appear as a 1:2:1 triplet with the spacing between the lines the same as that of the CH_2Cl 1:1 doublet. An actual spectrum is shown in Fig. 3.4, where it should be noted that the intensities are not exactly as predicted by the simple first-order theory but are perturbed by second-order effects. Note also that the total multiplet intensity is proportional to the number of protons giving rise to each multiplet. The coupling pattern thus counts precisely the number of protons in interaction and the intensity enables us to relate different, non-coupled groups of lines in the spectrum.

2. We can also work out the multiplicity by considering the possible values of the total magnetic quantum number Σm of the two CH_2Cl protons. We have $I = 1/2$, and m can be $\pm 1/2$.

Therefore for two protons (Fig. 3.3)

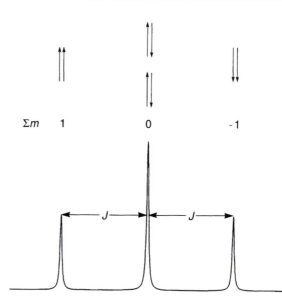

Figure 3.3 A simulated signal demonstrating the splitting due to two $I = 1/2$ nuclei. When $\Sigma m = 0$ there is no perturbation of the coupled resonance so that the centre line corresponds to the chemical shift position. This holds for all multiplets with an odd number of lines. The spacing J between the middle of the lines corresponds to a change in Σm of unity.

$$\Sigma m = + \frac{1}{2} + \frac{1}{2} = +1 \tag{3.1}$$

or

$$\Sigma m = \begin{cases} +\dfrac{1}{2} - \dfrac{1}{2} = 0 \\[2mm] -\dfrac{1}{2} + \dfrac{1}{2} = 0 \end{cases} \tag{3.2}$$

or

$$\Sigma m = - \frac{1}{2} - \frac{1}{2} = -1 \tag{3.3}$$

Therefore we have three lines.

Methods 1 and 2 are equivalent, but method 2 is particularly useful when considering multiplets due to nuclei with $I > 1/2$ where arrow diagrams become rather difficult to write down clearly. Note that when $\Sigma m = 0$ there is no perturbation of the chemical shift of the coupled group so that the centre of the spin multiplet corresponds to the chemical shift of the group.

Next let us consider the very commonly encountered pattern given by the ethyl group CH_3CH_2-. The isochronous pair of CH_2 protons are usually found to high frequency of the CH_3 protons and are spin-coupled to them. The CH_3 protons therefore resonate as a 1 : 2 : 1

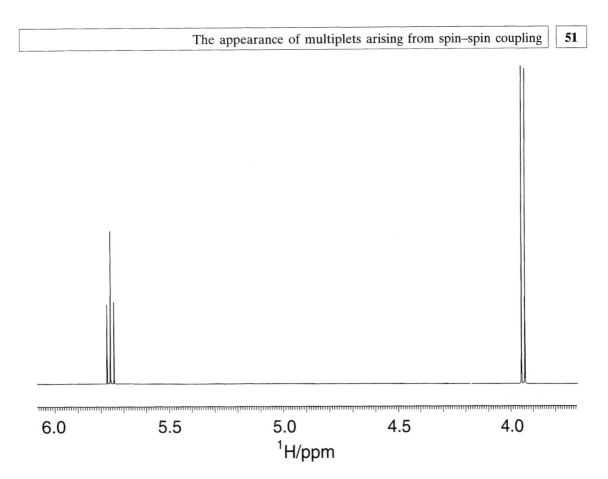

Figure 3.4 The 400 MHz ^{1}H NMR spectrum of $CH_2ClCHCl_2$ in $CDCl_3$.

triplet. The splitting of the CH_2X resonance caused by the CH_3 group can be found from Fig. 3.5, and is a 1 : 3 : 3 : 1 quartet. A typical ethyl group spectrum is shown in Fig. 3.6.

We have done enough now to formulate a simple rule for splitting due to groups of $I = 1/2$ nuclei. Thus the number of lines due to coupling to n equivalent $I = 1/2$ nuclei is $n + 1$. The intensities of the lines are given by the binomial coefficients of $(a + 1)^n$ or by Pascal's triangle, which can be built up as required. This is shown in Fig. 3.7. A new line of the triangle is started by writing a 1 under and to the left of the 1 in the previous line and then continued by adding adjacent figures from the old line in pairs and writing down the sum as shown. The multiplicity enables us to count the number of $I = 1/2$ nuclei in a group and the intensity rule enables us to check our assignment in complex cases where doubt may exist, since the outer components of resonances coupled to large groups of nuclei, e.g. the CH of $(CH_3)_2CH$ may be too weak to observe in a given spectrum.

Coupling to nuclei with $I > 1/2$ leads to different relative intensities and multiplicities. In the case of a single nucleus the total number of spin states is equal to $2I + 1$ and this equals the multiplicity. If $I = 1/2$

we get two lines, $I = 1$ gives three lines, $I = 3/2$ gives four lines, and so on. The spin populations of each state are virtually equal and so the lines are all of equal intensity and of equal spacing (Fig. 3.8).

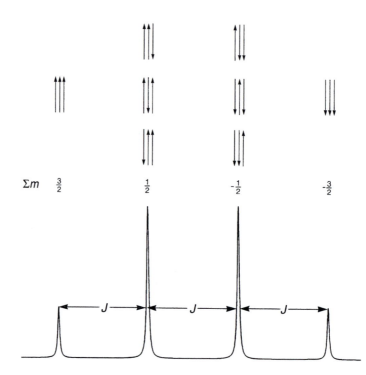

Figure 3.5 The splitting due to three $I = 1/2$ nuclei. There is no line that corresponds to $\Sigma m = 0$. The multiplet is, however, arranged symmetrically about the $\Sigma m = 0$ position, so that the centre of the multiplet corresponds to the chemical shift. This rule holds for all multiplets with an even number of lines.

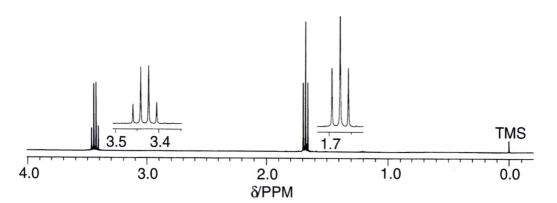

Figure 3.6 The 400 MHz ^{1}H NMR spectrum of ethyl bromide CH_3CH_2Br in $CDCl_3$. The CH_2 and CH_3 signals are expanded.

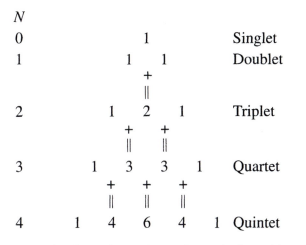

N						
0			1			Singlet
1			1 1			Doublet
			+			
			‖			
2		1	2	1		Triplet
		+	+			
		‖	‖			
3	1	3	3	1		Quartet
	+	+	+			
	‖	‖	‖			
4	1	4	6	4	1	Quintet

Figure 3.7 Pascal's triangle can be used to estimate the intensities of the lines resulting from coupling to different numbers, N, of equivalent $I = 1/2$ nuclei. The numbers in each line are obtained by adding adjacent pairs of numbers in the line above.

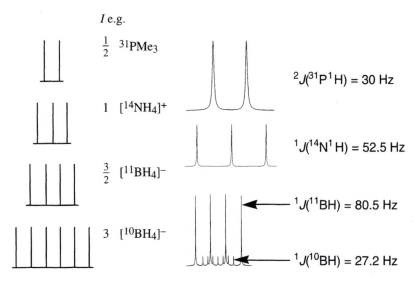

Figure 3.8 Multiplets observed in the proton spectra of a variety of species due to coupling to single nuclei with $I = 1/2$ ($^{31}PMe_3$), $I = 1$ ($[^{14}NH_4]^+$), $I = 3/2$ ($[^{11}BH_4]^-$) and $I = 3$ ($[^{10}BH_4]^-$). In the latter case, the two spectra are obtained simultaneously in the same spectrum since natural boron contains both isotopes, in the ratio 80.22 : 19.78. Note the smaller coupling constant to the isotope with the smaller magnetogyric ratio.

Splitting due to multiple combinations of $I > 1/2$ nuclei is much less common, but a few examples have been recorded. Figure 3.9 illustrates the pattern commonly encountered when obtaining ^1H spectra in the solvent ($^2H_6$) acetone {deuterioacetone, $(CD_3)_2CO$}. In practice,

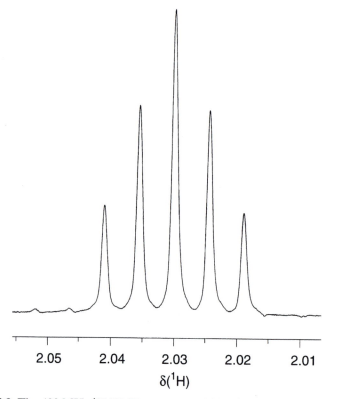

2.05   2.04   2.03   2.02   2.01

$\delta(^1H)$

Figure 3.9 The 400 MHz ^{1}H NMR spectrum of $(CD_3)C(O)(CHD_2)$. The multiplet pattern obtained for a proton coupled to two deuterium nuclei with $I = 1$. The value of $^2J(^2H^1H)$ is 2.2 Hz. The value of $^2J(^1H^1H)$ can be calculated from this spectrum. This coupling exists but cannot be measured in the all-hydrogen form.

there will always be a small amount of hydrogen present in these molecules and some of the methyl groups will be CD_2H-groups. The proton sees two equivalent deuterons with $I = 1$. The maximum total Σm is $1 + 1 = 2$. Thus $\Delta m = \pm 1$ and there are therefore five spin states and the proton resonance will be a quintet. In order to determine the line intensities, we have to determine the number of ways each value of Σm can be obtained. This is shown in Table 3.1 and indicates relative intensities of $1 : 2 : 3 : 2 : 1$, a distribution that differs from the simple binomial.

Table 3.1 Line intensities for coupling to two nuclei with $I = 1$

Σm	Possible spin combinations	Number of spin combinations
2	(+1, +1)	1
1	(+1, 0), (0, +1)	2
0	(+1, −1), (0, 0), (−1, +1)	3
−1	(−1, 0), (0, −1)	2
−2	(−1, −1)	1

Thus the rule given for $I = 1/2$ nuclei can be generalized to include groups of nuclei of any given I, the number of lines observed for coupling to n equivalent nuclei of spin I being $2nI + 1$. Obtaining the relative intensities is, however, tedious and it is better to use a 'Pascal's' triangle type of construction as shown in Fig. 3.10 for when $I = 1$ and Fig. 3.11 for when $I = 3/2$.

In the case of $I = 1$, the numbers in the triangle are placed immediately under the previous ones, since there is always a line in the centre of the multiplet. For $I = 3/2$ (and all half-integral spins), the numbers are staggered, since only if n is even is there a line at the centre of the multiplet. The triangles are constructed by moving a box, which can enclose $2I + 1$ numbers of a line, along the line, enclosing first the 1, then two numbers, then three, then four, and so on. The numbers enclosed by the box are added to give a number for the next line. Thus for $I = 3/2$ the box can enclose four numbers, and if we sweep through the line for $n = 1$, we enclose first a 1, then $1 + 1 = 2$, then $1 + 1 + 1 = 3$, then $1 + 1 + 1 + 1 = 4$ and reducing again to 3, 2, 1. The construction given above for $I = 1/2$ nuclei is the first example of this series.

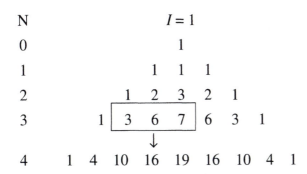

Figure 3.10 A version of 'Pascal's triangle' to determine the intensities of lines resulting from coupling to different numbers, N, of equivalent $I = 1$ nuclei.

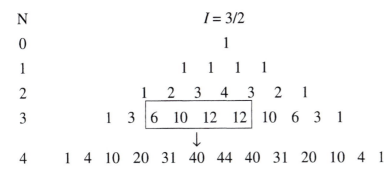

Figure 3.11 A version of 'Pascal's triangle' to determine the intensities of lines resulting from coupling to different numbers, N, of equivalent $I = 3/2$ nuclei.

More complex coupling situations also arise where a nucleus may be coupled simultaneously to chemically different groups of nuclei of the same or of different isotopes or species. The patterns are found by building up spectra, introducing the interactions with each group of nuclei one at a time. Thus Fig. 3.12 shows how a group M coupled to two chemically different $I = 1/2$ nuclei A and X is first split by $J(A–M)$ into a doublet, and shows that each doublet line is further split by $J(M–X)$. If $J(A–M) = J(M–X)$, a $1:2:1$ triplet is obtained; but if $J(A–M) \neq J(M–X)$, then a doublet of doublets with all lines of equal amplitude arises. This can be distinguished from a ^{11}B coupling because the line separations are irregular, and of course the preparative chemist is usually aware whether or not there should be boron in the compound.

The same technique can be used to predict the shapes of multiplets due to several equivalent nuclei with $I > 1/2$, i.e. introducing the splitting due to one nucleus several times as shown in Fig. 3.13 for two $I = 3/2$ nuclei. Comparison with the 'Pascal's triangle' diagrams will show that the two methods are exactly equivalent.

When analysing such multiplets, it always has to be borne in mind that overlap of lines may occur so that fewer than the theoretical number of lines are observed and the intensities are unusual. Such a case is illustrated in Fig. 3.14, which shows the ^{9}Be spectrum of the complex $[(\eta^5\text{-}C_5H_5)BeBH_4]$. The ^{9}Be resonance is split by coupling to

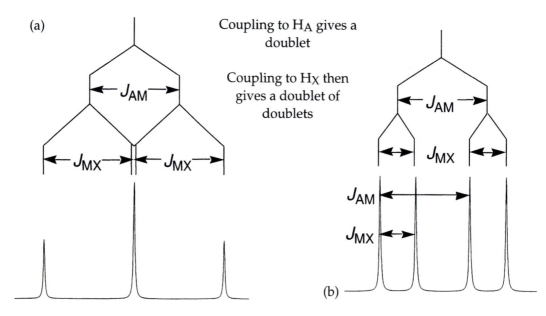

(a)

Coupling to H_A gives a doublet

Coupling to H_X then gives a doublet of doublets

Figure 3.12 Splitting of the H_M protons of $Z_2CH_A\text{-}C(H_M)_2\text{-}CH_XY_2$ due to coupling to H_A and H_X. (a) $J(A–M) = J(X–M)$, the centre lines overlap and the multiplet is a $1:2:1$ triplet just as if H_M were coupled to a CH_2 group. (b) $J(A–M) \neq J(M–X)$ and we get a doublet of doublets from which both $J(A–M)$ and $J(M–X)$ can be measured.

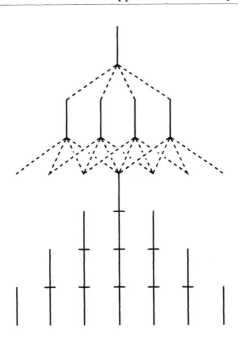

Figure 3.13 An illustration showing how by successively introducing the $1:1:1:1$ quartet splitting due to coupling to two equivalent nuclei with $I = 3/2$ the correct $1:2:3:4:3:2:1$ pattern is obtained.

the nuclei of the BH_4 ligand only, with $J(^9Be–H) = 10.2$ Hz and $J(^9Be-^{11}B) = 3.6$ Hz. The first coupling causes splitting to a $1:4:6:4:1$ quintet, each line of which is further split into a $1:1:1:1$ quartet by the ^{11}B. So 20 lines are expected, but because the ratio of the coupling constants is almost integral (it is 2.8) some lines overlap and only 16 are observed. The overlap is not exact but the close pairs of lines are not resolvable because of the appreciable natural linewidth of the ^9Be resonance.

3.2.1 Rules for the analysis of any first-order multiplet arising from $I = 1/2$ coupling

The analysis of any well resolved first-order multiplet due to spin 1/2 nuclei is actually a trivial exercise. There are some very simple rules which make the analysis straightforward.

1. Check that the multiplet is centro-symmetric. If it is not, then it consists of either overlapping multiplets or is second order, see section 3.7.
2. Assign an intensity to each line in the multiplet, starting with 1 for the outermost lines and using only integers. The distribution of the intensities must be centro-symmetric.
3. Add up the assigned intensities. The sum must be 2^n, where n is the number of $I = 1/2$ nuclei coupling. If the sum does not come

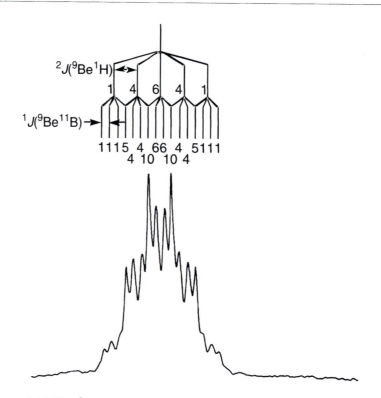

Figure 3.14 The ^{9}Be spectrum at 14.06 MHz of $[(\eta^5\text{-}C_5H_5)BeBH_4]$ in C_6F_6 solution. Coupling is observed to both ^{11}B ($I = 3/2$) ($J = 3.6$ Hz) and ^{1}H ($I = 1/2$) ($I = 10.2$ Hz) of the BH_4 group. Some lines overlap and the multiplicity is less than the maximum given by the basic rules. (From Gaines *et al.* (1981) *J. Magn. Reson.*, **44**, 84, with permission.)

to 2^n, then go through rule 2 again paying particular attention to lines where there could be doubt about the intensity.

4. The separation of the outermost pair of lines to one side of the multiplet must be a coupling constant. The intensities of the outermost pair of lines give the pattern. Thus $1:1$ shows that this first coupling constant arises from coupling to one nucleus. $1:2$ shows that this first coupling constant arises from coupling to two nuclei with the same coupling constant, and is part of a $1:2:1$ pattern. $1:3$ shows that this first coupling constant arises from coupling to three nuclei with the same coupling constant, and is part of a $1:3:3:1$ pattern. This argument continues through the 'Pascal triangle'.

5. Draw the first coupling pattern on the paper as an inverted stick diagram. Write the intensity of the first line of the multiplet on the central line of the stick pattern. In the case of coupling to two or more nuclei, remember that the separation of each adjacent pair of lines is equal, so that once the first pair of lines and their intensities have been identified, the rest of that pattern has been identified.

6. Subtract the intensities of the first coupling pattern from those generated in 3.

7. Go to the next line in the original coupling pattern which still has intensity after instruction 6. Treat this as the outermost line of the coupling pattern generated in 5. The outermost line may now have an intensity greater than 1. If so all the lines of the pattern must be multiplied by this intensity, which will be an integer. Draw the first coupling pattern on the paper as an inverted stick diagram. Write the intensity of the first line of the multiplet on the central line of the stick pattern.

8. Subtract the intensities of this coupling pattern from those generated in 6.

9. Keep repeating 7 and 8 until all the lines in the original multiplet have been accounted for. You will have now removed the smallest coupling constant, and generated a simpler multiplet. Go back to 3 and keep repeating the procedure until a singlet is generated. The stick diagram will have been completed.

(These rules to analyse multiplets have been reproduced from Mann (1995) *J. Chem. Ed.*, **72**, 614, with permission copyright © 1995, Division of Chemical Education, Inc.)

The application of these rules is illustrated in Fig. 3.15 for their application to the =CH– proton of allyl bromide, $CH_2=CHCH_2Br$.

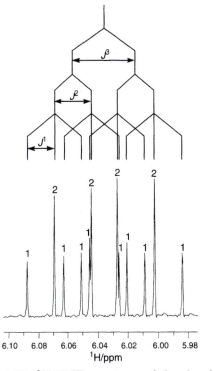

Figure 3.15 The 400 MHz ^{1}H NMR spectrum of the signal at δ 6.04 of allyl bromide, $CH_2=CHCH_2Br$.

The intensities are put on each line and they sum to $16 = 2^4$. Hence there are four $I = 1/2$ nuclei coupling. The separation of the first pair of lines is a coupling constant, and their intensities are $1:2$. Hence, from 'Pascal's triangle', this is part of a $1:2:1$ triplet. The separation between this pair of lines must equal that between the central one and the third line of intensity one, so the first, second, and fourth lines are selected as making up the triplet. The triplet is drawn above the multiplet. We now move to the next unassigned line, the third one and repeat the same triplet, using the same coupling constant. We keep repeating this until all the lines are accounted for. A $1:1:1:1$ quartet has been generated. This rapidly reduces to a doublet, and then a singlet and the pattern is analysed.

3.2.2 Rules for the analysis of any first order multiplet arising from both $I = 1/2$ and $I > 1/2$ coupling

The rules for determining the sum of intensities of the lines need modifying for nuclei with $I > 1/2$.

$$\text{Sum of intensities} = \prod_{j=0}^{i} (2I_j + 1)^{n_j}$$

where there are n_j nuclei with spin I_j. The appropriate intensity triangle for the nuclear spin has to be used. We return to the case of $[(\eta^5\text{-}C_5H_5)BeBH_4]$ to illustrate this (Fig. 3.14). The intensity ratio of the lines is $1:1:1:5:4:4:10:6:6:10:4:4:5:1:1:1$. (Note that the intensities are not obvious from the spectrum in this case due to the overlap of the lines.) The sum of the intensities is 64. The coupling is to four protons and one ^{11}B. The sum of intensities should therefore be $(2 \times \frac{3}{2} + 1)(2 \times \frac{1}{2} + 1)^4 = 4 \times 2^4 = 64$.

The separation of the outer pair of lines is a coupling constant. This could be to either ^{11}B or to ^1H, but coupling to the four protons gives a $1:4:6:4:1$ quintet, and the multiplet starts $1:1:1$. Hence the smallest coupling constant is to ^{11}B, giving a $1:1:1:1$ quartet. This $1:1:1:1$ pattern is drawn together by drawing a vertical line above the middle of it and the intensity of the outermost line, 1, is attached. The $1:1:1:1$ intensity is taken away from the first four lines to leave residual intensities $0:0:0:4:4:4:10:6:6:10:4:4:5:1:1:1$. The same pattern is subtracted, but the intensity starts 4, so the intensities to be subtracted are $4:4:4:4$, and 4 is written beside the central vertical line. These intensities are then subtracted to leave residual intensities $0:0:0:0:0:0:6:6:6:10:4:4:5:1:1:1$. The same pattern is subtracted again, but the intensity starts 6, so the intensities to be subtracted are $6:6:6:6$, and 6 is written beside the central vertical line. These intensities are then subtracted to leave residual intensities $0:0:0:0:0:0:0:0:0:4:4:4:5:1:1:1$. The procedure is continued reducing the multiplet to a 1:4:6:4:1 quintet, consistent with four protons coupling.

3.3 SPIN–SPIN COUPLING SATELLITES

A glance at the table of nuclear properties (Table 1.1) will show that certain elements have as principal isotope a magnetically non-active species, but that they have also a more or less small proportion of magnetically active species, some with $I = 1/2$. Examples are platinum with 33.8% of the active nucleus ^{195}Pt and carbon with just 1.1% of the active ^{13}C. The spectra of other nuclei in compounds of these elements will thus arise from differentiable molecular species: those with the non-active isotope, which will give rise to intense patterns depending upon the other nuclei present, and those with the active isotope, which will give a weaker sub-spectrum in which spin–spin coupling will be seen arising from interaction with this less abundant isotope. These weak doublets are centred approximately around the corresponding lines in the spectrum of the main species and are thus called satellites. They are usually particularly obvious in ^1H or ^{31}P spectra of platinum compounds, since their intensity is about a quarter of that of the central line of the major species. We will give two examples of compounds containing one and two platinum atoms.

First, we will consider the proton spectrum of the mononuclear square planar complex ion, $[PtMe(PMe_2Ph)_3]^+$. The methyl groups attached to platinum and to phosphorus are both coupled to ^{195}Pt, and their spectra, in the low frequency region, are shown in Fig. 3.16, with the effect of the ^{31}P atoms removed by a process of double irradiation, which we will discuss in detail in Chapter 5. There are three major methyl resonances which originate from the complexes with magnetically inactive platinum, and each is flanked by two lines of one quarter intensity, which are the doublets due to coupling to one

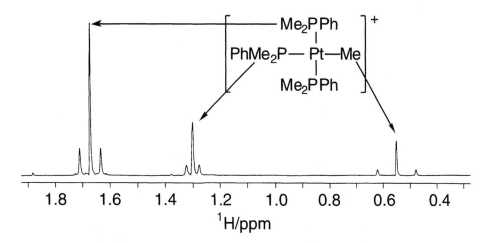

Figure 3.16 The 400 MHz ^1H spectrum of $[PtMe(PMe_2Ph)_3][PF_6]$ in $CDCl_3$, with the effect of the ^{31}P nuclei, $I = 1/2$, removed by a double irradiation technique. The P–Me and Pt–Me resonances are $1:4:1$ triplets, the weaker lines in each triplet being spin coupling satellites due to protons coupled to the minor platinum isotope ^{195}Pt.

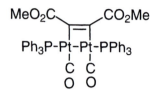

(3.1)

spin-1/2 nucleus, ^{195}Pt. The shorter-distance coupling path gives an appreciably stronger coupling than the longer-distance one via phosphorus. Evidently, the values of the coupling constants give structural information and the existence of the 1:4:1 pattern is good proof of the presence of ^{195}Pt in the complex.

If we now consider what sort of molecular populations we might expect if we have a complex that contains two platinum atoms, we will realize that we should have three sub-spectra, one each from those complexes which contain no ^{195}Pt, those which contain one and the smaller number which contain two. If the complex is symmetric, such as **(3.1)**, then we get three interlaced spectra as shown in Fig. 3.17, which gives the ^{31}P spectrum with the couplings to the protons now suppressed by double irradiation. This complex contains 43.8% of its molecules with no ^{195}Pt and in which the phosphorus atoms are entirely equivalent and give a central, strong singlet. Some 44.8% of the molecules contain one ^{195}Pt atom and this is coupled unequally to the two phosphorus nuclei with $^1J(^{195}\text{Pt}^{31}\text{P}) = 2409$ Hz and $^2J(^{195}\text{Pt}^{31}\text{P}) = 783$ Hz. This makes the two phosphorus atoms non-equivalent and they also can couple to one another with $^2J(^{31}\text{P}^{31}\text{P}) = 163$ Hz. The resulting sub-spectrum contains two doublets, one widely spaced due to the ^{31}P directly bonded to ^{195}Pt and one with a smaller spacing due to the more distant ^{31}P nucleus. The phosphorus nuclei then split mutually

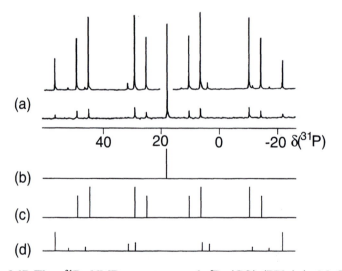

Figure 3.17 The ^{31}P NMR spectrum of [Pt$_2$(CO)$_2$(PPh$_3$)$_2$(μ-MeO$_2$CC≡CCO$_2$Me)], **(3.1)**, with the effect of the ^{1}H nuclei removed by a double irradiation technique. (a) The experimental spectrum with signals due to coupling to ^{195}Pt, $I = \frac{1}{2}$, shown above with increased gain. (b) The simulated spectrum for the molecules containing no ^{195}Pt. (c) The simulated spectrum for the molecules containing one ^{195}Pt atom. (d) The simulated spectrum for the molecules containing two ^{195}Pt atoms. (Reproduced with permission from Kose *et al.* (1981) *Inorg. Chem.*, **20**, 4408, copyright (1981) American Chemical Society.)

into doublets, to give eight lines in all. The remaining 11.4% of the molecules contain two ^{195}Pt atoms and give a weak spectrum. This sub-spectrum is a second-order spectrum type, $[AX]_2$, to be discussed in section 3.5.4. Analysis of this sub-spectrum yields $^1J(^{195}Pt^{195}Pt)$ = 786 Hz. The spectra become much more complex for bigger clusters and are of considerable use in studying the structures of this interesting class of compound.

Another nucleus that provides interesting examples of spin coupling satellites is ^{183}W, which has a natural abundance of 14.28%. The ^{19}F spectrum of the binuclear complex $[W_2O_2F_9]^-$ is shown in Fig. 3.18, and consists principally of a doublet of intensity 8 and a nonet of intensity 1. The nine fluorine atoms can thus be divided into an isochronous set of eight atoms and one unique atom. The main spectrum arises from those molecules in which both tungsten atoms are the magnetically inactive tungsten isotopes but satellite lines are also observed due to the molecules with one ^{183}W atom in their structure.

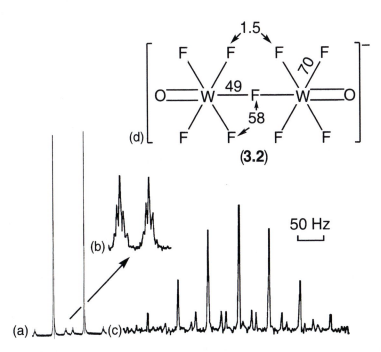

Figure 3.18 The 56.4 MHz ^{19}F spectrum of $[W_2O_2F_9]^-$, (**3.2**), in $(MeO)_2SO$, showing spin satellites due to 14.28% of ^{183}W. (a) The multiplet due to the terminal fluorides. (b) Expansions of the central signals which are the ^{183}W satellites. (c) The multiplet of the bridging fluoride recorded at higher gain. (d) The structure of the molecule showing the various coupling constants in hertz. The outer lines of the nonet are lost in the baseline noise and the student should confirm that the intensity ratios of the observed lines correspond to those expected for the inner seven of the nonet rather than to those expected for a septet. (After McFarlane *et al.* (1971) *J. Chem. Soc. A*, 948, with permission.)

In principle, lines should also exist due to those molecules with two ^{183}W atoms, but their proportion is low and their resonances are too weak to observe. Each of the lines of the intense doublet has two ^{183}W satellites, each of which is further split into a $1:4:6:4:1$ quintet. This pattern must arise from coupling to four fluorine atoms. We can therefore conclude that we have four of the eight isochronous fluorine atoms associated with the ^{183}W atom and therefore split into a satellite doublet and then further coupled to the remaining four, which are equally associated with the NMR inactive tungsten isotopes. This provides considerable confirmatory evidence that the structure is $[OWF_4 \bullet F \bullet WF_4O]^-$ with a fluorine atom bridging the tungsten atoms.

Finally we must consider the effect that the 1.1% of naturally occurring ^{13}C has on the proton spectra of organic compounds, which contain principally the inactive ^{12}C. Some of the molecules will however

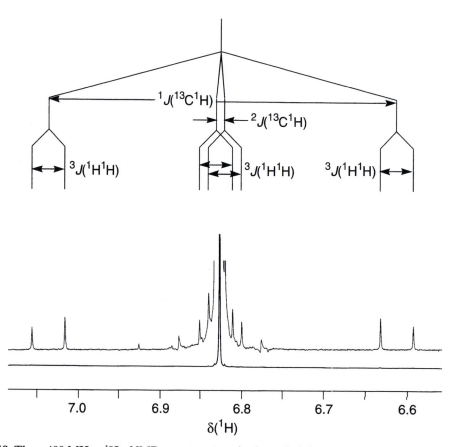

Figure 3.19 The 400 MHz ^{1}H NMR spectrum of the olefinic protons of dimethyl fumarate, $MeO_2CCH=CHCO_2Me$ in $CDCl_3$. In addition to a spectrum at normal gain, above there is a spectrum with increased gain to show the ^{13}C satellites. The molecules, $MeO_2C^{12}CH=^{13}CHCO_2Me$ show coupling. The ^{13}CH proton is split into a large doublet, $^1J(^{13}C^1H) = 167.9$ Hz, which is further split by the ^{12}CH proton with $^3J(^1H^1H) = 15.8$ Hz. The ^{12}CH proton is split into a small doublet, $^2J(^{13}C^1H) = 5.9$ Hz, which is further split by the ^{13}CH proton with $^3J(^1H^1H) = 15.8$ Hz.

contain one ^{13}C atom, distributed randomly. In fact, the spin satellites due to this minor isotopic component are not easy to observe among the intense ^1H–^{12}C resonances, but have proved to be of considerable use. In simple compounds such as acetone, $(CH_3)_2CO$, for instance, the proton resonance has two pairs of spin coupling satellites due to molecules with one ^{13}C atom in the methyl group $\{^1J(^1H$–$^{13}C) = 126\ Hz\}$ and to those with ^{13}C in the carbonyl group $\{^2J(^1H$–$^{13}C) = 5.9\ Hz\}$. Thus we can measure proton–carbon coupling constants, and with double-resonance techniques we will see later that we can discover correlations between the proton and carbon spectra of a molecule, see Chapter 8.

The reduction of symmetry in a molecule caused by the presence of ^{13}C can also prove valuable. A simple example is to determine whether dimethyl fumarate, $MeO_2CCH=CHCO_2Me$, has a *cis*- or *trans*-configuration at the double bond. The compound gives two singlets in the ratio $1:3$ in the ^1H NMR spectrum due to the olefinic CH and the methyl protons. In the ^{12}C compounds, the olefinic protons are equivalent and give a singlet. However, the presence of one olefinic ^{13}C carbon atom removes the equivalence and the coupling between the two olefinic protons can be measured as 15.8 Hz (Fig. 3.19). The 15.8 Hz coupling is in the range to be expected for a *trans*-stereochemistry about the double bond in view of the electronegativity of the substituents.

3.4 THE DESCRIPTION OF SPIN SYSTEMS

3.4.1 The alphabetic description of spin systems

A nomenclature has been adopted for these cases in which the magnetically non-equivalent sets of spins are labelled with letters from the alphabet, choosing letters that are well separated in the alphabetic sequence to signify large chemical shift separation. Thus $CH_2ClCHCl_2$ is an A_2X system, CH_3CH_2R is an A_3X_2 system, and CH_3CH_2F is an A_3M_2X system. Their spectra are called first order.

If the chemical shift between coupled groups is reduced, then the rules given above no longer apply, the coupling patterns become distorted and more complex and the spectra are said to become second order. To signify this and the fact that the chemical shifts between the coupled nuclei are relatively small, the spins are labelled with letters close together in the alphabet. Thus, for example, two coupled protons resonating close together are given the letters AB, and an ethyl group in $(CH_3CH_2)_3Ga$, where the methyl and methylene protons resonate close together, is described as an A_3B_2 grouping. Mixed systems are also possible and a commonly encountered one is the three-spin ABX grouping where two nuclei resonate close together and a third is well shifted or is of a different nuclear species.

(3.3)

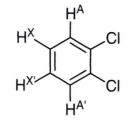

(3.4)

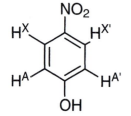

(3.5)

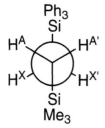

Figure 3.20 A Newman projection of Ph$_3$SiCH$_2$ CH$_2$SiMe$_3$, showing the *trans*-arrangement of the Ph$_3$Si and SiMe$_3$ groups. The result is that $J(H^A–H^X) \neq J(H^A–H^{X'})$ and the spin system is [AX]$_2$.

3.4.2 Magnetic equivalence

Coupling is only observed between nuclei which are magnetically inequivalent. What is meant by magnetic equivalence? Nuclei are magnetically equivalent if they satisfy two conditions:

1. They have the same chemical shift. In most cases this is achieved as a result of symmetry and/or dynamic behaviour which averages their properties.
2. They have the same coupling constant to any other magnetically inequivalent nucleus.

The first condition is easily understood, but the second is quite subtle. A classic example is 1,1-difluoroethene, (**3.3**). The two protons have the same chemical shift as do the two fluorines. In both cases they are related by both a C_2 axis and a mirror plane. However, they do not fulfil the second condition. $J(H^A–F^A) \neq J(H^B–F^A)$. Similarly, $J(H^A–F^A) \neq J(H^A–F^B)$. Hence the two protons and the two fluorines are magnetically inequivalent. The spectrum is shown in Fig. 3.33.

As the two protons are magnetically inequivalent, they can couple to each other. As they have the same chemical shift, and are as close as they can be in chemical shift terms, the spectrum is second order and complex. This gives rise to an [AX]$_2$ spin system, see section 3.5.4. The symbolism AA'XX' with primed letters is frequently used as an alternative description for this spin system. However, this notation is now being used to differentiate certain slightly different types of system. The [AX]$_2$ notation requires the two A nuclei and the two X nuclei each to have the same chemical shift as a result of the symmetry of the molecule. The AA'XX' notation is used when the two A nuclei and the two X nuclei each have the same chemical shift as a result of accidental degeneracy. An example of this will be given in section 3.5.3.

[AX]$_2$ spectra are frequently encountered. Some common examples are 1,2-dichlorobenzene, (**3.4**), 4-nitrophenol, (**3.5**), and Bu^{t}CH$_2$CH$_2$ SiMe$_3$. In the latter case, the bulk of the Bu^t and SiMe$_3$ groups results in the *trans*-configuration predominating (Fig. 3.20).

The resulting [AX]$_2$ spectra are second order and cannot be analysed using simple first-order rules given above (Fig. 3.21). A procedure to analyse [AX]$_2$ NMR spectra is given in section 3.5.4.

3.4.3 Diastereotopic atoms and groups

Apparently chemically equivalent atoms and groups can be inequivalent due to diastereotopism. The simplest example is when a CH$_2$R group is attached to a chiral centre, –CXYZ. There are three rotamers possible, (**3.6**), (**3.7**), and (**3.8**). If one rotamer predominates, say (**3.6**), then it is easy to see that H^A and H^B are in different environments, with H^A spending most of its time between groups X and Z, while H^B spends most of its time between groups X and Y. Even if there are

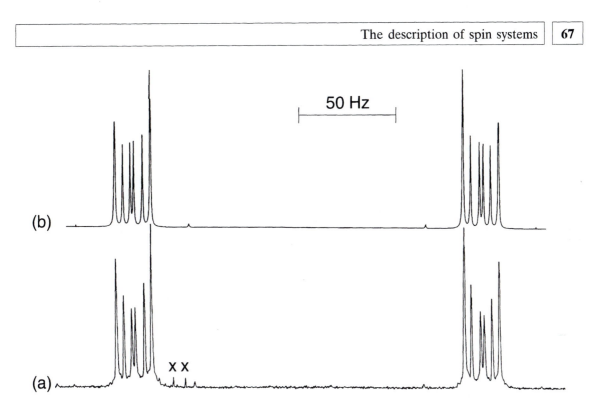

Figure 3.21 A partial 250 MHz ^{1}H NMR spectrum of Ph$_3$SiCH$_2$CH$_2$SiMe$_3$ in CDCl$_3$, showing the CH$_2$CH$_2$ protons. (a) Experimental spectrum, x = impurity. (b) Calculated spectrum. (Reproduced with permission, V.E. McGrath, PhD thesis, Sheffield, 1993.)

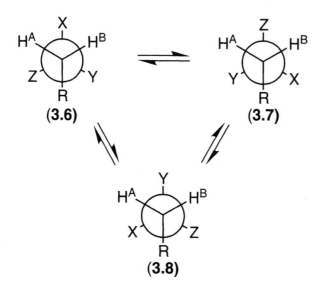

equal populations of all three rotamers, the chemical shifts of H^A and H^B can still be different as the X–C–C–H^B torsion angle in (**3.6**) does not have to equal the X–C–C–H^A torsion angle in (**3.8**).

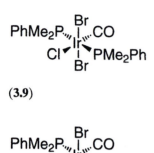

(3.9)

(3.10)

Diastereotopism is not restricted to –CH₂R groups, and is observed in any group of the type –MA₂B, where M is a tetravalent atom. The methyl resonances of –SiMe₂Buᵗ and –PMe₂Ph groups are commonly used to probe the symmetry of molecules using this effect. The diastereotopic centre can be several bonds removed in the molecule, and the only test that need be applied is whether or not there is a plane of symmetry through the bond joining the –MA₂B moiety to the rest of the molecule. Hence in **(3.9)** the PMe₂Ph methyl groups are equivalent, while in **(3.10)** they are inequivalent.

Molecules need not be chiral to produce diastereotopic atoms or groups. This is the case in **(3.10)**. Another common example is where two identical groups are attached to one carbon atom, e.g. MeCH(OCH₂CH₃)₂. From the viewpoint of one –OCH₂ group, there are three different groups attached to the tertiary carbon, and X = H, Y = Me, and Z = OEt in the rotamers **(3.6)** to **(3.8)**. Consequently, the CH₂ protons are diasterotopic, the spectrum being shown in Fig. 3.22.

The common presence of diastereotopism in molecules leads to coupling between apparently equivalent CH₂ protons being frequently observed.

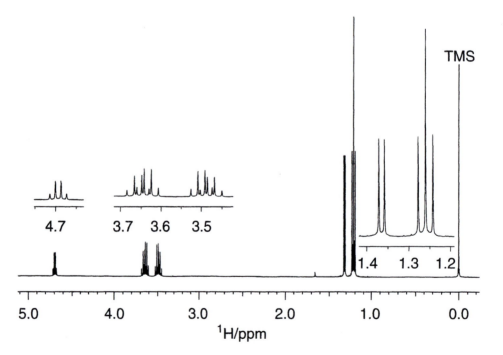

Figure 3.22 The 400 MHz ¹H NMR spectrum of MeCH(OCH₂CH₃)₂ in CDCl₃. The methyl protons are seen between 1.2 and 1.4 ppm, the CHMe being a doublet and the methyl groups of the ethoxys a triplet. The CH proton gives a quartet at 4.69 ppm. The geminal CH₂ protons give individual resonances at 3.65 and 3.49 ppm which are split into multiplets by coupling to each other and to the vicinal methyl protons.

3.5 SECOND-ORDER EFFECTS

The rules so far discussed apply to spectra of nuclei of the same species where the separation between multiplets in hertz (i.e. the chemical shift) is large compared to the value of the coupling constant between them, or to coupling between nuclei of different elements or isotopes where the differences in NMR frequency are invariably large.

Second-order spectra arise when the frequency separation between multiplets due to different magnetically equivalent sets of nuclei is similar in magnitude to the coupling constant between them; under these circumstances, the effects due to spin coupling and chemical shift have similar energy and become intermingled, leading to alterations in relative line intensities and in line positions. Because it is the ratio between J and the frequency separation that is important, chemical shifts are always expressed in hertz (Hz) and not in parts per million (ppm) when discussing second-order spectra. The hertz separation is obtained by multiplying the chemical shift δ by the spectrometer operating frequency and is written $\nu_0\delta$. The perturbation of the spectra from the first-order appearance is then a function of the ratio $J/\nu_0\delta$ and is different for spectrometers operating at different frequencies. If a high enough frequency is used, many second-order spectra approach their first-order limit in appearance and are then much more easily interpreted. This is one of the advantages of high-field instrumentation. We will describe first the simplest possible system consisting of two $I = 1/2$ nuclei.

3.5.1 The AB second-order system

We consider a system consisting of two isolated but mutually spin-coupled $I = 1/2$ nuclei with different chemical shifts. When the chemical shift between them is large, much larger than the coupling, then we see two doublets. A typical arrangement with $J = 10$ Hz and $\nu_0\delta = 200$ Hz is shown in Fig. 3.23(a). If we reduce the chemical shift progressively to zero, we can imagine the two doublets approaching one another until they coincide. However, we know that an isolated pair of equivalent nuclei give rise to a singlet, and the problem is how can two doublets collapse to give a singlet. If the coupling constant remains at 10 Hz, why do we not get a doublet A_2 spectrum?

The behaviour of the multiplets can be predicted using a quantum-mechanical argument. The system can have any one of four energy states which are characterized by the spin orientations of the two nuclei. The wave functions of the two orientations are normally written α for $I = +1/2$ and β for $I = -1/2$. There are four such spin states with the wave functions $\alpha\alpha$, $\alpha\beta$, $\beta\alpha$, and $\beta\beta$, where the first symbol in each pair refers to the state of the A nucleus and the second symbol to the state of the B nucleus. The energies of the states $\alpha\alpha$ and $\beta\beta$ can be calculated straightforwardly, but the two states $\alpha\beta$ and $\beta\alpha$ have the same total spin angular momentum and it is found that the

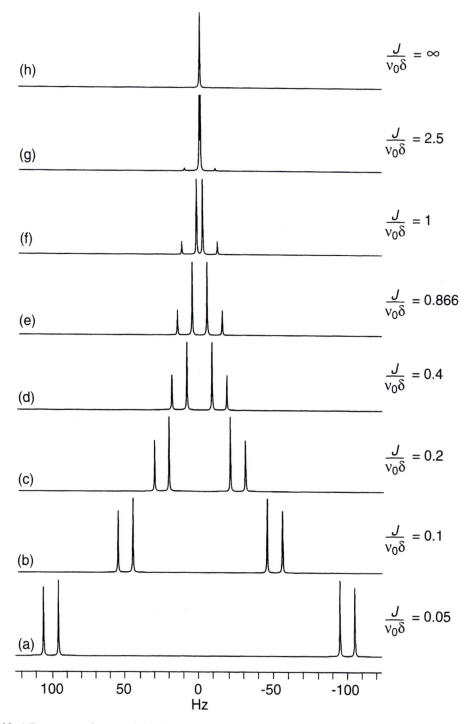

Figure 3.23 AB quartets for several values of the ratio $J/v_0\delta$. In each case, the coupling constant is kept at 10 Hz, and the chemical shift separation reduced in the sequence 200, 100, 50, 25, 17.32, 10, 4, and 0 Hz. The result is the inner lines become more dominant and the outer lines become weaker.

quantum-mechanical equations can only be solved for two linear combinations of $\alpha\beta$ with $\beta\alpha$. This is described as a mixing of the states and means that none of the observed transitions corresponds to a pure A or a pure B transition. The form of the wave functions and the energy levels derived are shown in Table 3.2. C_1 and C_2 are constants, ν_A and ν_B are the A and B nuclear frequencies in the absence of coupling and $\nu_0\delta = |\nu_A - \nu_B|$. The transition energies are the differences between four pairs of energy states, 3–4, 2–4, 1–3 and 1–2, each transition involving a mixed energy level. Because of the mixing, the transition probabilities are no longer equal as in the first-order case and intensity is transferred from lines in the outer parts of the total multiplet into the central region. The transition energies relative to the centre of the multiplet, i.e. to the mean frequency $\frac{1}{2}(\nu_A - \nu_B)$, and the intensities are shown in Table 3.3.

There are thus four lines as in the AX spectrum but with perturbed intensities and resonance frequencies. The resulting spectral pattern and the corresponding energy level diagram are shown in Fig. 3.24.

Table 3.2 Wave functions and energy levels for the AB second-order spin system

State	Wave function	Energy level (Hz)
1	$\alpha\alpha$	$\frac{1}{2}(\nu_A + \nu_B) + \frac{1}{4}J$
2	$C_1(\alpha\beta) + C_2(\beta\alpha)$	$\frac{1}{2}\sqrt{(\nu_0\delta)^2 + J^2} - \frac{1}{4}J$
3	$-C_2(\alpha\beta) + C_1(\beta\alpha)$	$-\frac{1}{2}\sqrt{(\nu_0\delta)^2 + J^2} - \frac{1}{4}J$
4	$\beta\beta$	$-\frac{1}{2}(\nu_A + \nu_B) + \frac{1}{4}J$

Table 3.3 Transition energies relative to the centre of the multiplet and relative intensities for the AB second-order spin system.

Transition	Energy (Hz)	Relative intensity
a $\;3 \rightarrow 1$	$+\frac{1}{2}J + \frac{1}{2}\sqrt{(\nu_0\delta)^2 + J^2}$	$1 - \dfrac{J}{\sqrt{(\nu_0\delta)^2 + J^2}}$
b $\;4 \rightarrow 2$	$-\frac{1}{2}J + \frac{1}{2}\sqrt{(\nu_0\delta)^2 + J^2}$	$1 + \dfrac{J}{\sqrt{(\nu_0\delta)^2 + J^2}}$
c $\;2 \rightarrow 1$	$+\frac{1}{2}J - \frac{1}{2}\sqrt{(\nu_0\delta)^2 + J^2}$	$1 + \dfrac{J}{\sqrt{(\nu_0\delta)^2 + J^2}}$
d $\;4 \rightarrow 3$	$-\frac{1}{2}J - \frac{1}{2}\sqrt{(\nu_0\delta)^2 + J^2}$	$1 - \dfrac{J}{\sqrt{(\nu_0\delta)^2 + J^2}}$

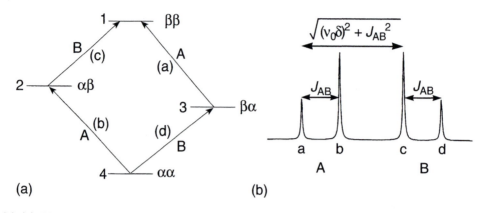

(a)   (b)

Figure 3.24 (a) The energy level diagram for a system of two spins related to the resulting AB quartet. Note that the vertical scale is misleading. The energy gaps represented by the arrows are in MHz, while the energy gap between energy levels 2 and 3 is in Hz. (b) The spectrum was calculated for $J = 10$ Hz, $v_0\delta = 25$ Hz, i.e. $J/v_0\delta = 0.4$.

The energy levels are marked with the appropriate spin state and the transitions, which in the first-order case can be regarded as arising from transitions of the A or B nucleus, form opposite sides of the figure. The three line separations are $a - b = c - d = J$ and $b - c = \sqrt{(v_0\delta)^2 + J^2} - |J|$. The separation $a - c$ or $b - d$, which in the first-order case is the same as the separation between the doublet centres, and is therefore the chemical shift in hertz, $v_0\delta$, is now simply $\sqrt{(v_0\delta)^2 + J^2}$ and is larger than the true chemical shift. In other words, although $v_0\delta$ is reduced to zero, the doublet centres never coincide and are separated by J Hz. The outer lines, however, have intensity zero at this point, while the inner lines are coincident, i.e. we predict a singlet spectrum as is observed (cf. Fig. 3.23(h)). Thus arises our rule, 'isochronous coupled protons resonate as a unit'.

We can calculate some simple rules for analysing an AB spectrum:

1. The spectrum contains two intervals equal to J, $a - b$ and $c - d$.
2. The true AB chemical shift $v_0\delta$ is found as

$$(a - d)(b - c) = \left\{ \sqrt{(v_0\delta)^2 + J^2} + J \right\} \left\{ \sqrt{(v_0\delta)^2 + J^2} - J \right\}$$

$$= (v_0\delta)^2 + J^2 - J^2$$

$$= (v_0\delta)^2$$

so

$$v_0\delta = \sqrt{(a - d)(b - c)} \qquad (3.4)$$

where $a - d$ is the separation between the outermost lines and $b - c$ is the separation between the innermost pair of lines.

3. The assignment can be checked against the intensity ratios of the larger and smaller lines. The intensity ratio, stronger/weaker, is

$$\dfrac{1 + \dfrac{J}{\sqrt{(\nu_0\delta)^2 + J^2}}}{1 - \dfrac{J}{\sqrt{(\nu_0\delta)^2 + J^2}}}$$

which gives

$$\frac{\sqrt{(\nu_0\delta)^2 + J^2} + J}{\sqrt{(\nu_0\delta)^2 + J^2} - J}$$

for the ratio of the line separations $(a - d):(b - c)$. Note that changing the sign of J does not alter the pattern.

Figure 3.23 shows the form of the AB quartet for several values of $J/\nu_0\delta$. It is important to remember that if a multiplet shows signs of being highly second order then both intensities *and* resonance positions are perturbed from their first-order values. A spacing corresponding to J_{AB} remains in AB-type spectra since only one coupling interaction exists, but in more complex systems the spacings are combinations of coupling constants. On the other hand, if the intensity perturbation is only slight {Fig. 3.23(a)} then the line positions are not detectably perturbed.

Examination of the spectra in Fig. 3.23 show several features. The first visible onset of second-order behaviour is the inner lines becoming more intense. The two doublets are said to 'lean' towards each other. This feature can be very valuable in deciding which multiplets are coupled but care is necessary as the heights of signals can be distorted by insufficient digitisation, see section 5.6.3. There are two features which are traps waiting for the unwary and have led to the misinterpretation of spectra.

1. When the separation of the two signals is $J\sqrt{3}$, the spectrum has the appearance of a $1:3:3:1$ quartet as if it arises from coupling to a methyl group (Fig. 3.23(e)).
2. When $J/\nu_0\delta$ is large, the outer lines are weak and can be lost in the noise leaving an apparent doublet.

Four examples of actual spectra that contain an AB multiplet are given in Fig. 3.25. We have however reached a point in the history of NMR where, in many cases, the analysis of second-order spectra can be avoided by operating at a high enough frequency. In those cases where this will not work, computer programs are available that permit the full analysis of virtually any system.

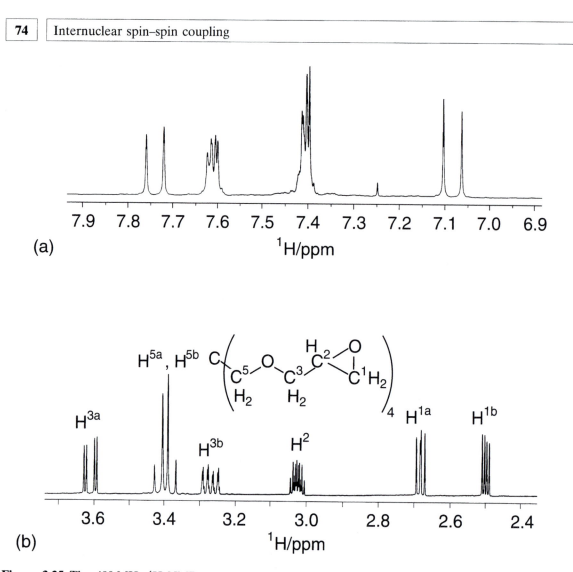

Figure 3.25 The 400 MHz ^{1}H NMR spectra of some compounds in CDCl$_3$ containing AB groupings of protons, in order of decreasing $v_0\delta/J$. (a) (PhCH=CH)$_2$CO. Note the AB pattern arising from the CH=CH group at δ 7.08 and 7.74. The signal at δ 7.61 is due to the *ortho*-hydrogen atoms of the phenyl groups, while the *meta*-and *para*-hydrogen atoms are at δ 7.4. (b) The 400 MHz ^{1}H NMR spectrum of C(CH$_2$OCH$_2$CHCH$_2$O)$_4$ in CDCl$_3$. The inequivalence of the CH$_2$ protons is caused by the presence of a chiral CH group in the epoxide ring. This produces a substantial chemical shift difference for the CH$_2$ protons on both C^1 and C^3, but is also observed at C^{5}H$_2$, where a tightly coupled AB pattern is produced centred at δ 3.4. (c) The 101.2 MHz ^{31}P NMR spectrum with the ^{1}H coupling removed of [Pt(η^2-CH$_2$=CHCO$_2$Me)(PPh$_3$)$_2$] in d_8-toluene. Note the three AB patterns centred at 13, 32, and 51 ppm due to ^{31}P–^{31}P coupling. The AB patterns at 13 and 51 ppm arise from coupling to ^{195}Pt, $I = 1/2$, 34% abundant, and the AB pattern at 32 ppm rises from the ^{31}P attached to NMR inactive platinum. (Reproduced from Chaloner *et al.* (1997) *J. Organomet. Chem.*, **527**, 145, copyright (1997), with permission from Elsevier Science.) (d) 109 MHz ^{31}P NMR spectrum with ^{1}H coupling removed of [Pt$_2$Cl$_2$(CO)$_2${P(naphthyl)Ph$_2$}$_2$] in CDCl$_3$. The signal at 30 ppm is due to molecules containing no NMR active platinum nuclei. There are two AB patterns, one at 32.2 and 40 ppm, and the other at 20 and 27.6 ppm, due to [Pt^{195}PtCl$_2$(CO)$_2${P(naphthyl)Ph$_2$}$_2$]. There are additional weak signals due to [^{195}Pt$_2$Cl$_2$(CO)$_2${P(naphthyl)Ph$_2$}$_2$]. (Reproduced from Mingos *et al.* (1997) *J. Organomet. Chem.*, **528**, 163, copyright (1997), with permission from Elsevier Science.)

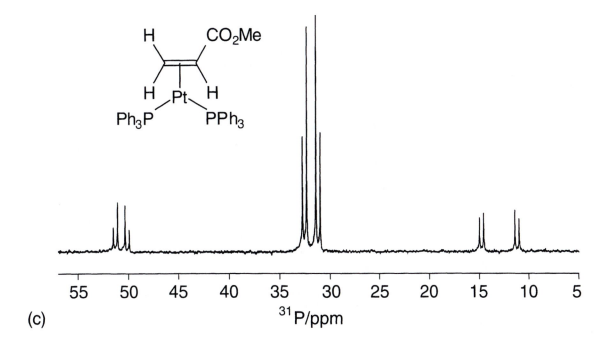

(c)

^{31}P/ppm

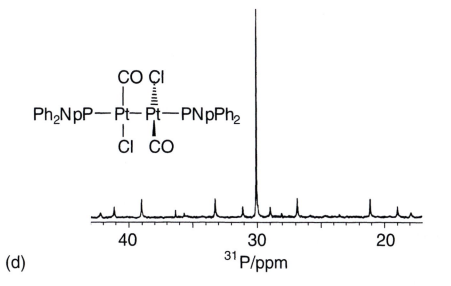

(d)

^{31}P/ppm

3.5.2 The AB$_2$ second-order system

The complete analysis of the AB$_2$ spin system will not be performed here, but an examination of the spin system is instructive, as on going from the AX$_2$ spin system to the AB$_2$ one, not only do the positions of lines change, but lines which were degenerate singlets become non-degenerate doublets. This is a common feature of second-order spectra, with, in some cases, what would be sharp multiplets in a first-order spectrum becoming broad due to the overlap of many lines, degenerate in the first-order case but moving apart in the second-order case, though not enough to be resolved.

A calculated AB$_2$ spectrum is given in Fig. 3.26. When $J/v_0\delta$ is small, a first-order spectrum is observed with a $1:2:1$ triplet for A and a $1:1$ doublet for B. When $J/v_0\delta$ is significant, the central line of the triplet and both lines of the doublet split, resulting in eight lines. This can be misleading causing the erroneous interpretation that two different coupling constants are operating in the A and B parts of the spectrum. The derivation of the chemical shifts and coupling constant is trivial. If the lines are lettered a to h from the weaker A to the stronger B signals, then

$$v_A = c$$

$$v_B = \frac{e+g}{2}$$

$$J_{AB} = \frac{|a-d+f-h|}{3}$$

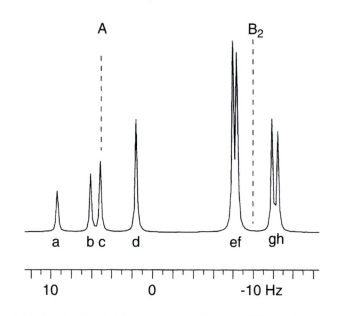

Figure 3.26 A calculated AB$_2$ spectrum with $v_A = 5$ Hz and $v_B = -10$ Hz, and $J_{AB} = 4$ Hz. Dotted lines are included to mark the chemical shifts of A and B.

3.5.3 The ABX second-order system

It is to be expected that the positions of the lines of nuclei involved in second-order spectra are moved from their ideal or first-order position, but less evident is the fact that nuclei distant in frequency which couple with the second-order group of nuclei are also affected. The simplest example of this occurs for the ABX spin system. In addition, the form of the spectrum depends on the relative signs of the coupling constants $J(AX)$ and $J(BX)$. This behaviour is frequently encountered in second-order spectra.

The ABX spin system also introduces the concept of sub-spectral analysis. As the frequency difference between the two AB and the X nuclei is large compared with $J(AX)$ and $J(BX)$, the spin states of X are not perturbed, and it is possible to analyse the AB part of the system as two separate problems for each of the two possible signs of the magnetic quantum number, m, of X. Hence the AB part of the spectrum consists of two AB sub-spectra. See Fig. 3.27, which has been calculated by computer simulation using input data supposing that A, B, and X are protons. The two AB sub-spectra can readily be

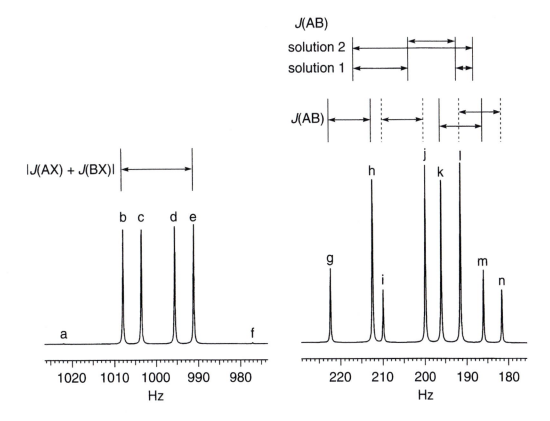

Figure 3.27 Simulated ABX spectrum using $\delta(A)$ = 190 Hz, $\delta(B)$ = 210 Hz, $\delta(X)$ = 1000 Hz, $J(AB)$ = 10 Hz, $J(AX)$ = 4 Hz and $J(BX)$ = 13 Hz.

identified as they are centro-symmetric and have the same $J(AB)$ (10.0 Hz in this case). Examination of Fig. 3.27 shows the AB groups to be g, h, k, m and i, j, l, n.

Taking the first sub-spectrum, its centre of gravity is 204.23 Hz, half way between lines h, k, or better, a quarter the sum of the shifts of all four lines. From equation (3.1) in section 3.5.1 we have:

$$\text{Separation} = \sqrt{(g - m)(h - k)} = 24.46 \text{ Hz}$$

between the A and B resonances in this multiplet.

On the scale of the spectrum, these points are at 204.23 ± 12.43 Hz or 216.46 Hz and 192.00 Hz.

For the second AB sub-spectrum, we find that the centre of gravity is 195.72 Hz, whence the AB shift separation is 15.42 Hz, placing them on the spectral scale at 195.72 ± 7.41 Hz or 204.43 Hz and 188.01 Hz.

These four points at 216.46, 204.43, 192.00 and 188.01 Hz are the frequencies of the A and B signals in the absence of AB coupling but still with coupling to X. There are two solutions which give the same AB sub-spectra and we need to find both then decide which is correct. The two solutions are:

1. A frequencies 192.00, 188.01; difference 4 Hz; average A shift 190 Hz, B frequencies 216.46, 203.43; difference 13 Hz; average B shift 210 Hz;
2. A frequencies 203.43, 192.00; difference 11.4 Hz; average A shift 197.72 Hz, B frequencies 216.46, 188.01; difference 28.5 Hz; average B shift 202.24 Hz.

The *differences* are the coupling constants of A or B to X and the averages of the doublet A or B frequencies is then the actual A and B shifts. The difference between the two solutions is that for 1, $J(AX)$ and $J(BX)$ have the same sign whereas in 2 they have opposite signs. It is evident that the first solution is the correct one from the constants fed into the simulation. In practice, however, we would not know this and would have to make a logical choice, for which there are two bits of information to help us in the present case. One is that our experience should tell us that 28.5 Hz is unreasonably large for a proton–proton coupling constant so that 1 is the correct solution. The second is to realize that the X part of the spectrum depends on the relative signs of $J(AX)$ and $J(BX)$, even though the AB part does not. A simulated spectrum will show which is the correct solution, as is demonstrated by Figs 3.27 and 3.28, which latter shows a simulated ABX spectrum using the alternative parameters. Comparison shows that all the lines are in the same positions, but there are marked changes in the intensities of lines a, c, d, and f. It is generally easiest to compare the two solutions by simulating the spectra.

It is very tempting to analyse the X part of the spectrum to obtain $J(AX)$ and $J(BX)$. In many cases it has the appearance of a doublet of doublets, especially when the outer weak lines, a and f, are lost in the noise. The separation of lines b and e does give $|J(AX) + J(BX)|$,

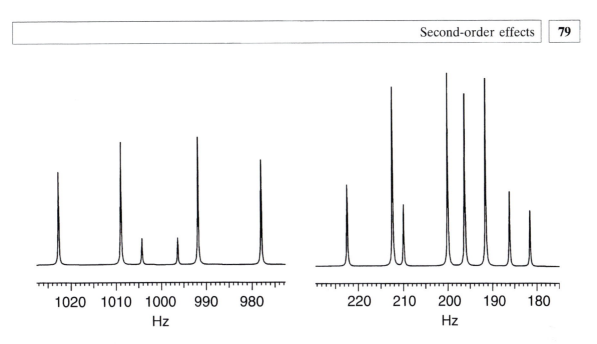

1020 1010 1000 990 980

Hz

220 210 200 190 180

Hz

Figure 3.28 Simulated ABX spectra using $\delta(A) = 197.72$ Hz, $\delta(B) = 202.24$ Hz, $\delta(X) = 1000$ Hz, $J(AB) = 10$ Hz, $J(AX) = 11.4$ Hz and $J(BX) = -28.5$ Hz.

but the lines c and d move together as a result of the second-order nature of the AB part of the spectrum and true values of $J(AX)$ and $J(BX)$ are not derived from a simple first-order analysis. In the extreme, the lines c and d can merge to give a deceptively simple triplet, as is discussed later.

To emphasize the sensitivity of the spectra to the signs of $J(AX)$ and $J(BX)$, we calculate a spectrum in Fig. 3.29 using the data for Fig. 3.27 in which only $J(BX)$ has been changed from $+13$ to -13 Hz. Note that the spectrum differs substantially from that given in Fig. 3.27 as a result of the change of sign of $J(BX)$. The sensitivity of second-order coupling patterns to sign matters. For example, in molecules containing sp^3 carbon atoms $^2J(^1HC^1H)$ is negative, while $^3J(^1HCC^1H)$ is positive and this influences the nature of the second-order spectra obtained from such groups.

Note that there are four states of the AB nuclei ($\alpha\alpha$, $\alpha\beta$, $\beta\alpha$, $\beta\beta$) and there should, on a simple view, be four X lines, but six may be observed. This is due to combined transitions of the type $\alpha\beta\beta$–$\beta\alpha\alpha$ for which $\Delta m = 1$ even though all three spins flip.

The ABX spin system is often observed in ^{13}C NMR spectra where the AB nuclei are abundant $I = 1/2$ nuclei, such as ^{19}F, ^{31}P, or ^{103}Rh. This occurs because ^{13}C is only 1.1% abundant so that there is effectively only one such atom present per molecule. An example is shown in Fig. 3.30, for $[(\eta^5\text{-}C_5Me_5)_2Rh_2(\mu\text{-}CH_2)_2\{\mu\text{-}CH_2CH(CH_2CH=CH_2)CH_2\}]$, (**3.11**). The ^{13}C NMR signal shown arises from those molecules with a $^{13}CH_2$ group in the hexylene bridging ligand. The chemical shifts of the two ^{103}Rh nuclei in this isotopomer are different due to a secondary isotope shift of 0.36 ppm due to one ^{103}Rh nucleus being

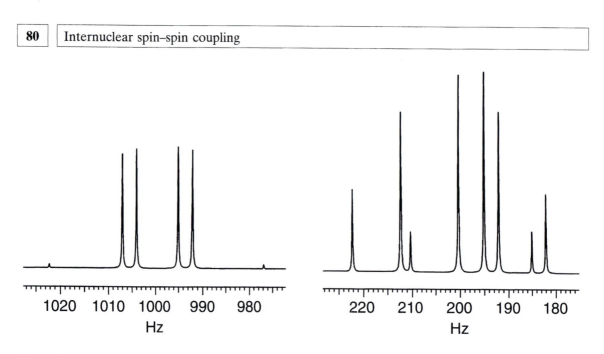

Figure 3.29 Simulated ABX spectra using $\delta(A) = 190$ Hz, $\delta(B) = 210$ Hz, $\delta(X) = 1000$ Hz, $J(AB) = 10$ Hz, $J(AX) = 4$ Hz and $J(BX) = -13$ Hz.

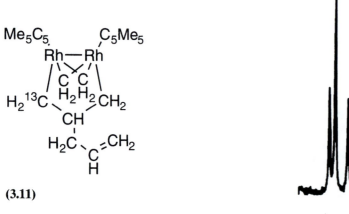

(3.11)

Figure 3.30 The 100.62 MHz ^{13}C NMR spectrum of the ^{13}CH$_2$ group of $[(\eta^5\text{-}C_5Me_5)_2Rh_2(\mu\text{-}CH_2)_2\{\mu\text{-}CH_2CH(CH_2CH{=}CH_2)CH_2\}]$, **(3.11)**. The spectrum is ABX with the two ^{103}Rh nuclei giving the AB part of the spectrum, not shown, and the ^{13}CH$_2$ giving the X part. This spectrum has $^1J(^{103}Rh^{103}Rh) = 13.5$ Hz, $^1J(^{103}Rh^{13}C) = 30.0$ Hz, $^2J(^{103}Rh^{13}C) = 0.0$ Hz, and $\Delta\delta(^{103}Rh) = 0.36$ ppm or $\nu_0\delta = 4.5$ Hz. (Reproduced with permission from Mann *et al.* (1985) *J. Chem. Soc., Dalton Trans.*, 1555.)

attached to ^{12}C and the other to ^{13}C. A double irradiation process is used to eliminate any effect of the hydrogen nuclei on the spectra.

When $J(AB)$ is much larger than the chemical shift difference between the A and B nuclei, the X part of the spin system collapses to a $1:2:1$ triplet with the two **outer** lines separated by

$|J(AX) + J(BX)|$ and the rest of the intensity concentrated in the centre of the multiplet. This is often observed in the ^{13}C NMR spectra of ligands of metal complexes with two identical phosphorus ligands. Frequently these spectra are referred to as AA'X rather than as ABX, to indicate that no $^{12}C/^{13}C$ isotope shift of the phosphorus has been detected. An example of this occurs with the *meta*-carbon in the phenyl groups of $Ph_2PCH_2CH_2PPh_2$. The spectrum is a $1:2:1$ triplet (Fig. 3.31). It is tempting, but wrong, to attribute the occurrence of the triplet to equal coupling between the *meta*-carbon and both ^{31}P nuclei. It arises because $^3J(^{31}P^{31}P)$ is large compared with $^2J(^{31}P^{13}C)$ and the spectrum is AA'X. Such multiplets are referred as being 'deceptively simple'. The only coupling information that can be derived from such a spectrum is $|^3J(^{31}P^{13}C) + {}^6J(^{31}P^{13}C)|$ from the separation of the outer lines.

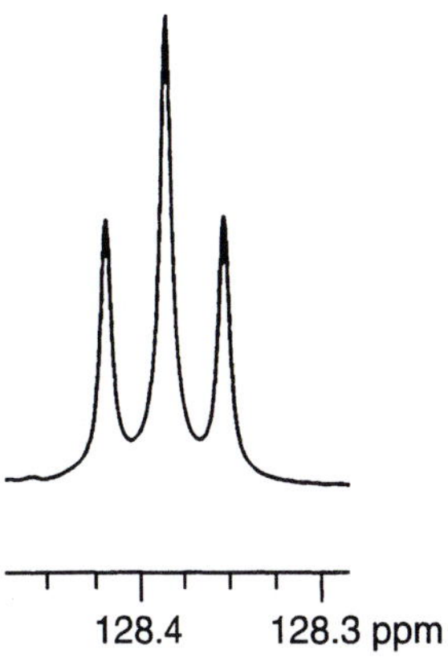

3.5.4 A four-spin [AX]₂ system

As discussed in sections 3.4.1 and 3.4.2, the $[AX]_2$ spin system frequently arises for molecules such as (**3.3**) to (**3.5**). Both the A and X parts of the spectrum are identical in appearance. The spectra can be readily analysed. 1,1-difluoroethene, (**3.3**), will be taken as an example. A simulated spectrum of either the ^1H or ^{19}F NMR signal is shown in Fig. 3.32. Sub-spectral analysis indicates that when both spins of one nucleus are in the same state ($\alpha\alpha$ or $\beta\beta$), the other two nuclei will resonate as a unit and produce a singlet. There are two such singlets separated by the sum $|J(A^1X^1) + J(A^1X^2)|$. The analysis is

Figure 3.31 The 100.62 MHz ^{13}C NMR spectrum of the *meta* carbon of the phenyl groups of $Ph_2PCH_2CH_2PPh_2$ in $CDCl_3$ with coupling to hydrogen removed by double resonance.

128.4   128.3 ppm

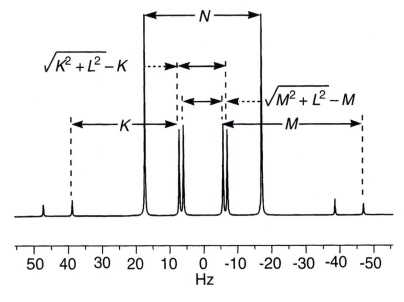

Figure 3.32 A simulated spectrum for either the ^1H or ^{19}F NMR spectrum of 1,1-difluoroethene, (**3.3**).

straightforward. It is usual to define some terms.

$$K = J_{A^1A^2} + J_{X^1X^2}$$

$$L = J_{A^1X^1} - J_{A^1X^2}$$

$$M = J_{A^1A^2} - J_{X^1X^2}$$

$$N = J_{A^1X^1} + J_{A^1X^2}$$

Each multiplet consists of an intense doublet, separation N, and two AB patterns, apparent coupling constants K and M. The inner lines of each AB pattern are separated by $\sqrt{K^2 + L^2} - K$ and $\sqrt{M^2 + L^2} - M$ respectively.

N can be easily determined, 34.6 Hz. K and M are 31.6 and 41.2 Hz. The corresponding values of $\sqrt{K^2 + L^2} - K$ and $\sqrt{M^2 + L^2} - M$ are 14.2 and 11.6 Hz. Substituting values of K and M yields $L = 33.2$ Hz. The coupling constants are then obtained from combinations of K, L, M, and N as follows.

$$J_{A^1A^2} = \frac{K + M}{2} = 39.4 \text{ Hz}$$

$$J_{X^1X^2} = \frac{K - M}{2} = 4.8 \text{ Hz}$$

$$J_{A^1X^1} = \frac{L + N}{2} = 33.9 \text{ Hz}$$

$$J_{A^1X^2} = \frac{L - N}{2} = 0.7 \text{ Hz}$$

Typically, in inorganic chemistry, $[AX]_2$ systems are found in metal phosphine complexes such as *cis-* and *trans-*$[PdI_2\{PF(OPh)_2\}_2]$ where the nuclei A are the ^{31}P and the nuclei X are the ^{19}F (Fig. 3.33). The remaining nuclei either have no influence on the ^{31}P or ^{19}F spectra, or their effect is removed by double irradiation. There need be no coupling between the X nuclei, though in fact there will be a small, not necessarily observable, X–X coupling. There will be strong coupling between the A nuclei via the metal atom. It should be noted that the metal atom need not be present from the point of view of the NMR analysis of the system.

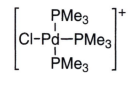

(3.12)

Were we to investigate more complex spin systems, for example *bis* complexes of methyl phosphines, we would find the same general features. An example is shown in Fig. 3.34. This is a ^1H NMR spectrum of $[PdCl(PMe_3)_3]^+$, (3.12). The PMe_3 *trans* to Cl is a doublet of triplets. The doublet coupling is $^2J(^{31}P^1H)$, i.e. proton–phosphorus coupling in the same ligand, while the triplet coupling, $^4J(^{31}P^1H)$ is to the phosphorus in the two equivalent PMe_3 ligands. These latter two PMe_3 ligands give rise to an $[AX_9]_2$ spin system giving the triplet to low frequency. The structure of the spectrum is very similar to that

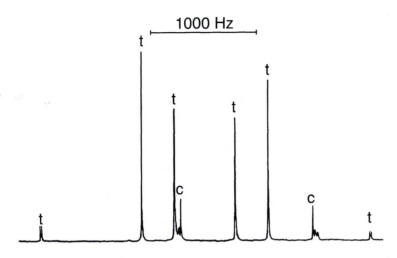

Figure 3.33 An example of magnetic inequivalence in a chemically equivalent system. The ^{31}P spectrum, with the effect of the protons removed by double irradiation, of a mixture of cis-[PdCl$_2$\{PF(OPh)$_2$\}$_2$], marked with a c and trans-[PdCl$_2$\{PF(OPh)$_2$\}$_2$], marked with a t. Both complexes give rise to [AX]$_2$ spin systems, the two phosphorus nuclei being inequivalent because the two P – F couplings are unequal. The spectra of the two complexes are drastically different from the simple 1 : 2 : 1 triplet that would have been observed for an A$_2$X$_2$ spin system. (Reproduced from Crocker and Goodfellow (1981) *J. Chem Res(M)*, 742 with permission.)

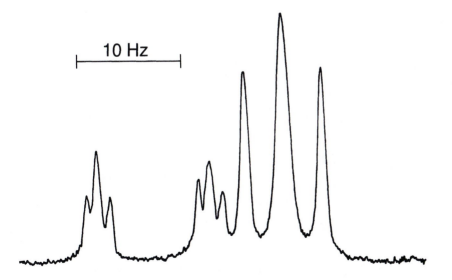

Figure 3.34 The 100 MHz ^1H NMR spectrum of [PdCl(PMe$_3$)$_3$]$^+$ in CH$_2$Cl$_2$. (Reproduced with permission from Goggin *et al.* (1973) *J. Chem. Soc. Dalton Trans.*, 2220.)

of the $[AX]_2$ spin system, with the intense N doublet and a series of AB patterns, but on account of the large value of $^2J(^{31}P^{31}P)$, which means that K, M, etc. are large, the central lines merge into a broad central singlet and the outer lines vanish into the noise. The result is a deceptively simple triplet. On the other hand, when two phosphorus ligands are equivalent and are mutually *cis*, $^2J(^{31}P^{31}P)$ is small for Pd^{II} and other complexes, K, M, etc. are small and now the lines crowd around the N doublet lines and only a doublet is observed, see Fig. 3.33, complex c. This behaviour has proved to be a valuable tool in determining the stereochemistry of phosphorus complexes of Os^{II}, Ru^{II}, Rh^I, Ir^I, Rh^{III}, Ir^{III}, Pd^{II}, Pt^{II}, Pd^{IV}, and Pd^{IV}.

Figure 3.35 shows the proton spectrum of $[PtMe(PMe_2Ph)_3]^+$, already encountered in Fig. 3.16, where the effect of the ^{31}P atoms was removed by double irradiation. The full spectrum contains a triplet at δ 1.68 due to the mutually *trans* PMe_2 protons, a doublet at δ 1.29 due to the PMe_2 protons *trans* to the PtMe group, and a doublet of triplets due to the Pt–Me group at δ 0.54. In each case, the signals are flanked by ^{195}Pt satellites. The methyl groups of the phosphine ligands cannot in any way couple equally to the two phosphorus atoms and their triplet is the deceptively simple triplet described above. Such triplets are taken to indicate *trans* structures for such complexes, whereas doublets arise from *cis* configurations.

An example of an organic molecule that has a four-spin $[AX]_2$ spectrum is furan, whose proton spectrum is shown in Fig. 3.36 together

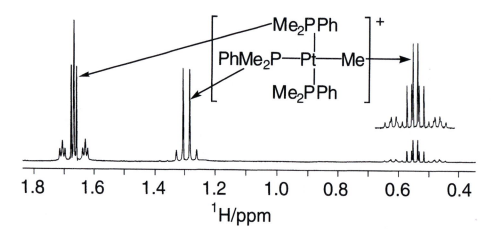

Figure 3.35 The 400 MHz ^1H NMR spectrum of the complex $[PtMe(PMe_2Ph)_3][PF_6]$ in $CDCl_3$ showing the three methyl resonances in the intensity ratio 4 : 2 : 1. The flanking resonances are ^{195}Pt satellites. The small triplet splittings of the signal at δ 1.68 are due to coupling to the two magnetically inequivalent but chemically equivalent ^{31}P atoms giving rise to a deceptively simple triplet. The Pt–Me at δ 0.55 is a doublet of triplets due to coupling to the ^{31}P atoms.

with a formula provided with the various proton spin–spin coupling constants. The resonances of the α and β protons appear at first glance to be triplets, but close inspection at high resolution shows the central line to be composite. A full analysis of the system gives a doublet spectrum plus a pair of AB sub-spectra with most intensity in the central resonances. Interestingly, the ^{13}C spin satellites of this molecule are first order in appearance. The presence of a ^{13}C atom in a molecule splits the resonance of the attached proton into a widely separated doublet, so that it is no longer isochronous with its twin. This introduces an effective chemical shift, which permits these hydrogen atoms to couple with each of the other three and give simplified spectra that can aid interpretation of the main spectrum.

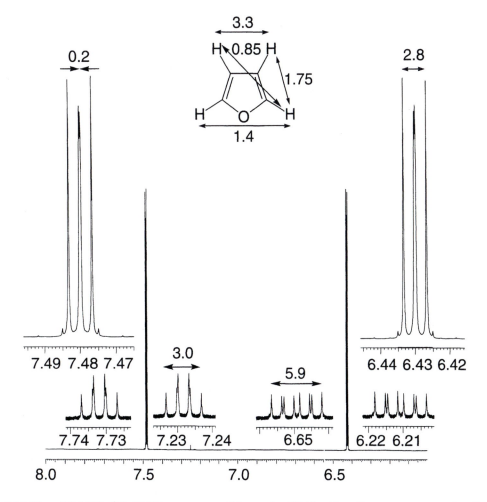

Figure 3.36 The 400 MHz ^1H NMR spectrum of furan in $CDCl_3$. Coupling constants and splittings are given in hertz. $^1J(^1H - {}^{13}C) = 201.4$ Hz and $^1J(^1H - {}^{13}C) = 175.3$ Hz respectively.

3.6 QUESTIONS

3.1. Figure 3.6 shows the ^{1}H spectrum of an ethyl group. Measure the chemical shift of the two multiplets and their coupling constant.

3.2. Verify that the ratio between the coupling constants to ^{1}H of the nuclei ^{10}B and ^{11}B in the $[BH_4]^-$ are in the ratio of the magnetogyric ratios of the two boron isotopes (Fig. 3.8). What do you expect the ^{11}B NMR spectrum of $[BH_4]^-$ to look like?

3.3. Figure 3.9 shows the proton spectrum of a CD_2H group in isotopically substituted acetone. $^2J(^2H^1H)=2.2$ Hz. Calculate the coupling constant $^2J(^1H^1H)$ between the protons in the CH_3 groups of normal acetone, ignoring isotope effects on J.

3.4. Doublet satellite signals are observed in the ^{1}H spectrum of acetone, close to the main singlet signal, with a spacing of 5.9 Hz, which is the value of the coupling $^2J(^1H-^{13}C)$. What will be the pattern in the ^{13}C spectrum of this carbonyl group, i.e. number of lines and their relative intensities?

3.5. The 60 MHz ^{1}H spectrum of ascaridole features a tightly coupled AB quartet at about $\delta = 6.45$ ppm. The line positions, starting at the highest frequency one, are 395.5, 386.9, 385.5 and 376.9 Hz from TMS. Calculate $^3J(H_a-H_b)$ and the chemical shift between H_a and H_b in Hz and ppm. Then calculate the relative line intensities and positions expected when the same sample is observed at 500 MHz. What is the exact chemical shift of the centre of the quartet?

3.6. Explain why, in Fig. 3.25(a), the two halves of the AB quartet have different intensities and so different linewidths.

3.7. Assign the spin systems, e.g. AB, to the following compounds,
- a. ethyl chloride
- b. CH_2F_2

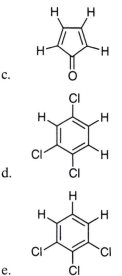

- c.

- d.

- e.

f. $CH_2=CHBr$

g. $[PdCl_2(PMe_3)_2]$

h.   *fac*-$[IrH_3(PR_3)_3]$ (ignore any spins in the R groups).

3.8. Figure 3.37 shows the ^{19}F resonance of the CF_2H group of 1,1,1,2,2,3,3-heptafluoropropane, $CF_3CF_2CF_2H$. The two fluorines are equivalent and are coupled to all the other magnetically active nuclei in the sample. Pick out the various multiplet patterns and measure the coupling constants $^2J(F-H)$, $^3J(F-F)$ and $^4J(F-F)$.

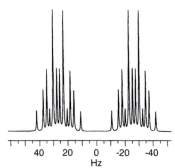

Figure 3.37 The ^{19}F resonance of the CF_2H group of $CF_3CF_2CF_2H$. The spectrum has been simulated using experimental coupling constants.

3.9. When $IrCl_3(PEt_2Ph)_3$ is treated with $LiAlH_4$, then a hydride of formulation $IrH_m(PEt_2Ph)_n$ is formed. The hydride ^1H NMR spectrum is shown in Fig. 3.38(a). Use the pattern produced by ^{31}P coupling to the hydride to deduce n. The ^{31}P NMR spectrum, with coupling from the ethyl and phenyl protons removed, is shown in Fig 3.38(b). Deduce the value of m from the coupling of the ^{31}P to hydride. Also deduce the stereochemistry of the compound.

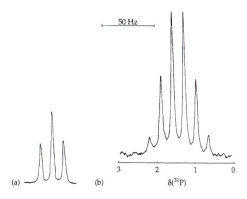

Figure 3.38 (a) 90 MHz ^1H NMR spectrum of the hydride signal of $IrH_m(PEt_2Ph)_n$ in C_6D_6. The coupling is with ^{31}P. (b) 36.43 MHz ^{31}P NMR spectrum of $IrH_m(PEt_2Ph)_n$ in C_6D_6. The coupling is with ^1H of the hydride. Coupling due to the ethyl and phenyl protons has been removed by decoupling. (Reproduced with permission from Mann *et al.* (1971) *J. Inorg. Nucl. Chem.*, **33**, 2195, copyright (1971) Elsevier Science.)

When $IrH_m(PEt_2Ph)_n$ is warmed in benzene with $AsMe_2Ph$, then one mole of H_2 is evolved per mole of compound and $IrH_x(PEt_2Ph)_y(AsMe_2Ph)$ is formed, which has the hydride ^1H spectrum in Fig. 3.39. Deduce the values of x and y.

You are given one additional piece of information: it has been shown empirically for iridium complexes that, when a hydride is *trans* to a tertiary phosphine, $^2J(^1H–Ir–^{31}P)$ is large (about 120 Hz), but, when a hydride is *cis* to a tertiary phosphine, $^2J(^1H–Ir–^{31}P)$ is small (about 15 Hz).

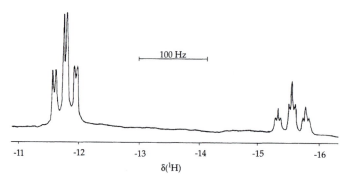

Figure 3.39 90 MHz ^1H NMR spectrum of the hydride region of $IrH_x(PEt_2Ph)_y(AsMe_2Ph)$ in C_6D_6. Note that the relative intensities of the two signals are 2:1. (Reproduced with permission from Mann *et al.* (1971) *J. Inorg. Nucl. Chem.*, **33**, 2195, copyright (1971) Elsevier Science.)

3.10. In d_8-THF, the ^{13}C NMR spectrum with the effect of the protons removed by double irradiation, of the lithiated carbon atom of $[(CH_2{=}^{13}CH)^6Li]$ shows two sets of signals. Fig. 3.40 shows the ^{13}C NMR spectrum of the lithiated carbon atom of the major isomer. The coupling is to ^6Li. How many ^6Li atoms are attached to each carbon atom?

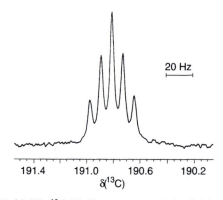

Figure 3.40 The 100.6 MHz ^{13}C NMR spectrum of the lithiated carbon atom of the major isomer of $[(CH_2{=}^{13}CH)^6Li]_n$ in d_8-THF at –100°C. (Reproduced with permission from Brauer and Griesinger (1993) *J. Am. Chem. Soc.*, **115**, 10871, copyright (1993) American Chemical Society.)

The signal due to the lithiated carbon atom of the minor isomer is shown in Fig. 3.41 at both –90°C when the molecule is rigid and at –60°C when the atoms of the molecule are mobile and the lithium moves between the vinylic CH carbon atoms in the molecule. Use the coupling pattern to determine how many lithium atoms are attached to a lithiated carbon atom in the rigid structure, and how many lithium atoms visit each vinyl CH carbon atom.

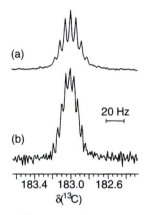

Figure 3.41 The 100.6 MHz ^{13}C NMR spectrum of the lithiated carbon atom of the minor isomer of $[(CH_2={}^{13}CH)^6Li]_n$ in d_8-THF at (a) –90°C and (b) –60°C. (Reproduced with permission from Brauer and Griesinger (1993) *J. Am. Chem. Soc.*, **115**, 10871, copyright (1993) American Chemical Society.)

3.11. Figure 3.42 shows the ^{119}Sn NMR spectrum of the mixture of isotopomers, $[SnH_nD_{3-n}]^-$, $n = 0$ to 3, although there is only a trace of $[SnD_3]^-$. Identify all the signals present. Account for the multiplicity of each signal. Analyse the coupling patterns and indicate which splittings are due to $^1J(^{119}Sn^1H)$ and $^1J(^{119}Sn^2H)$. Derive the secondary isotope shift which occurs on the replacement of ^1H by ^2H.

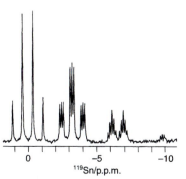

Figure 3.42 The 134 MHz ^{119}Sn NMR spectrum of the mixture of isotopomers, $[SnH_nD_{3-n}]^-$, $n = 0$ to 3, in liquid ammonia at 20°C. (Reproduced with permission from Wasylishen and Burford (1987) *J. Chem. Soc., Chem. Commun.*, 1414.)

3.12. Figure 3.43 shows the 100.62 MHz ^{13}C NMR spectrum of the mixture of isotopomers, $CH_nD_{3-n}I$, $n = 0$ to 3. Identify all the signals present. Analyse the coupling patterns and derive $^1J(^{13}C^1H)$ and $^1J(^{13}C^2H)$. Derive the secondary isotope shift which occurs on the replacement of ^1H by ^2H.

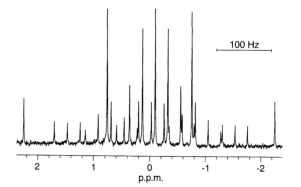

Figure 3.43 The 100.62 MHz ^{13}C NMR spectrum of the mixture of isotopomers, $CH_nD_{3-n}I$, $n = 0$ to 3. (Reproduced with permission from Sergeyev *et al.* (1994) *Chem.Phys.Lett.*, **221**, 385, copyright (1994), with permission from Elsevier Science.)

3.13. Figure 3.44 shows the hydride signals from $[Ru(CO)H_2(PPh_3)_3]$. The multiplet structure arises from coupling between the hydrides and ^{31}P nuclei. Analyse the multiplets to derive the coupling constants. Identify which coupling constant is for $^2J(^1H^1H)$ and which coupling constants are for $^2J(^{31}P^1H)$. Given that $^2J(^{31}P^1H)$ is large when the stereochemistry is *trans* and smaller when it is *cis*, propose a structure for the compound.

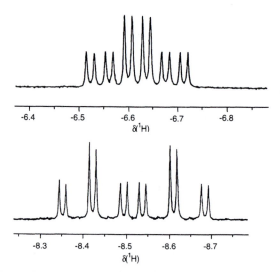

Figure 3.44 Expansions of the 400 MHz ^1H NMR signals from the hydride atoms of $[Ru(CO)H_2(PPh_3)_3]$ in $CD_3C_6D_5$.

3.14. Figure 3.45 shows the ^1H NMR signals for the CH_2 protons of *trans*-$[PdCl_2\{P(CH_2Ph)_2Ph\}_2]$. Account for the appearance of the spectrum. Predict what signal would be observed for the CH_2 protons of *trans*-$[PdCl_2\{P(CH_2Ph)Ph_2\}_2]$.

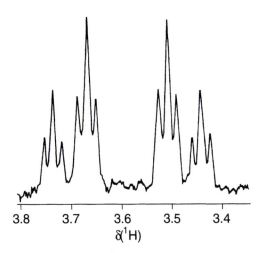

Figure 3.45 The 220 MHz ^1H NMR spectrum of the CH_2 protons of *trans*-$[PdCl_2\{P(CH_2Ph)_2Ph\}_2]$. (Reproduced with permission from Nelson and Redfield (1973) *Inorg. Nucl. Chem. Letts.*, **9**, 807, copyright (1973) Elsevier Science.)

3.15. Figure 3.46 shows the ^6Li and ^{15}N NMR spectra of the product obtained when $Me_2^{15}NCH_2CH_2^{15}NMe_2$ is added to a solution of $^6LiBu^n$ in d_8-toluene. Suggest a structure for the product. Account for the multiplicity of each signal.

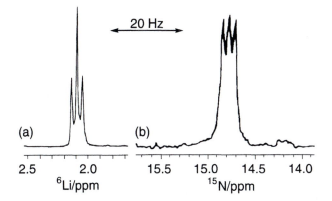

Figure 3.46 44.15 MHz ^6Li and 30.408 MHz ^{15}N NMR spectra of the product obtained when $Me_2^{15}NCH_2CH_2^{15}NMe_2$ is added to a solution of $^6LiBu^n$ in d_8-toluene. (Reproduced with permission from Nicholls *et al.* (1997) *J. Am. Chem. Soc.*, **119**, 5479, copyright (1997) American Chemical Society.)

3.16. Figure 3.47 shows the 235.36 MHz ^{19}F NMR spectrum of *cis*-[Ag(CF$_3$)$_2$Cl(CN)]$^-$ in THF. Account for the multiplicity of each signal.

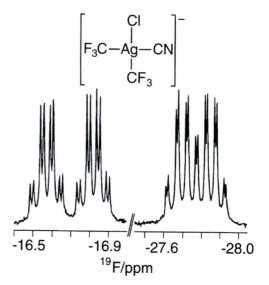

Figure 3.47 The 235.36 MHz ^{19}F NMR spectrum of *cis*-[Ag(CF$_3$)$_2$Cl(CN)]$^-$ in THF. Note that silver has two NMR active nuclei, ^{107}Ag, 51.82% abundant, $\gamma = -1.0828 \times 10^7$ rad T^{-1} s^{-1}, and ^{109}Ag, 48.18% abundant, $\gamma = -1.2449 \times 10^7$ rad T^{-1} s^{-1}. Both nuclei have $I = 1/2$. (Reproduced with permission from Eujen *et al.* (1997) *Inorg. Chem.*, **36**, 1464, copyright (1997) American Chemical Society.)

3.17. Assign the following proton signals from menthol, Fig. 3.2, using the multiplicities and coupling constants.
  δ 2.18, septet (7 Hz) of doublets (3 Hz).
  δ 1.12, doublet (12 Hz) of doublets (10 Hz) of triplets (3 Hz).

SPECTRAL INTERPRETATION

The proton spectra of simple molecules are often sufficient to provide a full structural description of the molecule. The task is made even easier if an infrared (IR) spectrum is also to hand, since the two often give complementary information. Since we are, however, concentrating on NMR spectra here, we will ignore the IR and attempt to obtain the maximum from the NMR spectra.

An NMR spectrum contains several pieces of usable information. First of all is the chemical shift. Thus the position of a resonance indicates the type of group in which the protons reside, sometimes with remarkable clarity, though often there will be ambiguities. The chemical shift ranges within which several types of proton are found are given in the accompanying chemical shift chart (Fig. 3.48). It is only necessary to add that hydrogen-bonded protons are found to high

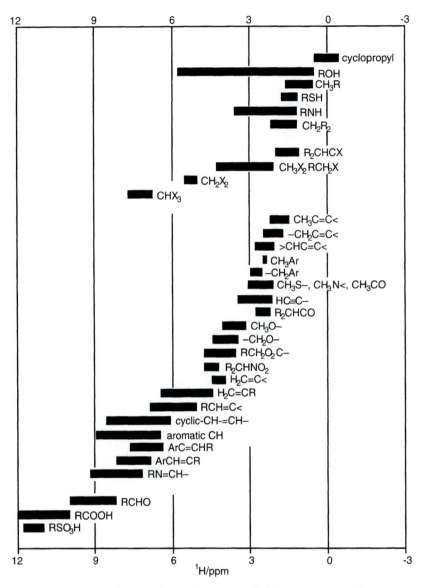

Figure 3.48 Chart of approximate ^{1}H chemical shift ranges of different types of protons in organic compounds. X = halogen, R = organic substituent, Ar = aryl. (From Akitt (1987) in *Multinuclear NMR*, (ed. Mason), Plenum, New York, with permission.)

frequency, often below 10 ppm, and that metal alkyls are found to low frequency around and above TMS. Hydrogen bonded to metals occurs in a wide range also at low frequency of TMS and varies from about –3 ppm, e.g. [HRe(PPh$_3$)$_3$(CO)$_2$], to –50 ppm, e.g. [HIrCl$_2$(phosphine)$_2$], though there are one or two exceptions with shifts to high frequency of TMS. The second piece of information is the integral trace of the spectrum, which gives the area under each resonance and so the relative numbers of protons contributing to each resonance. This, coupled with the empirical formula, will enable the hydrogen in the molecule to be split up into chemically different subgroups. Spin–spin coupling patterns also give this type of quantitative information, with the difference that an ethyl quartet–triplet pattern is diagnostic but does not tell us how many ethyl groups are present. The existence of coupling also tells us that the coupled groups are proximate.

The interpretation of spectra is thus a deductive process in which one attempts to account for all the spectral features in terms of a single molecular structure. We will work through some examples and will start with a few simple spectra where the only information is the chemical shift and the formula:

1. The compound has a singlet at 7.27 ppm and formula C_6H_6. Evidently we have a sample of benzene.
2. The compound produces a singlet at 2.09 ppm and has a formula C_3H_6O. The chemical shift is typical of methyl or methylene in unsaturated molecules or of CH_3CO. Again, it is a short step to acetone, $(CH_3)_2CO$.
3. The compound gives two singlets of equal intensities at 2.01 and 3.67 ppm and has a formula $C_3H_6O_2$. Clearly, the two singlets arise from three protons each, which form two equivalent sets and which are not coupled. There are thus two CH_3 and by difference we have CO_2. The chemical shifts are typical of CH_3O and CH_3CO. Thus we have the ester CH_3COOCH_3, with four bonds between the hydrogen atoms in the two groups and so no detectable coupling.

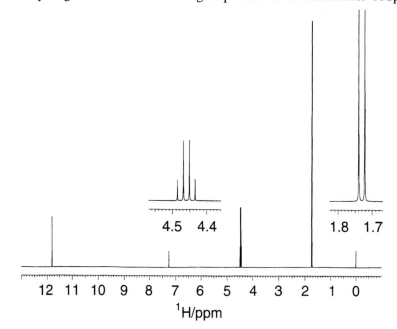

Figure 3.49 The 400 MHz ^{1}H NMR spectrum of $C_3H_5O_2Br$ in $CDCl_3$.

Now we will consider an example where there is spin–spin coupling. The spectrum shown in Fig. 3.49 is that of the substance $C_3H_5O_2Br$. It consists of nine resonances, of which we can discount two as arising from the reference TMS (0 ppm) and the remnant $CHCl_3$ in the deuterated $CDCl_3$ solvent (7.25 ppm). Two of the resonances are split into regular multiplets with identical splitting of 7.3 Hz. This informs us that there is spin–spin coupling between two of the groups of protons and the value of the coupling constant is typical of a vicinal three-bond coupling pathway, though not exclusively so. The doublet–quartet pattern must arise from a $H–H_3$ interaction. The integrals are in the ratio 3(doublet) : 1(quartet) : 1(singlet), and since there are five protons in the molecule the hydrogen atoms must also be split up in this ratio. The chemical shift of the doublet is typical of CH_3 alkyl; that of the quartet is rather high frequency for CHX but we need also to keep in mind the influence of the oxygen-containing part of the molecule, which is also to high frequency. Finally, the singlet to high frequency is typical for acidic protons or aldehydes. We can now attempt to work out the structure, and it helps to do this if we settle upon one structural unit at a time and subtract this from the formula. Thus we have $C_3H_5O_2Br$. The doublet is evidently a CH_3 unit. This leaves $C_2H_2O_2Br$. It is coupled to an alkyl proton, i.e. a CH, to give CH_3CH, leaving CHO_2Br. Two valencies on the CH carbon need to be filled. Bromide must take one, leaving CHO_2, and this carboxyl

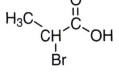

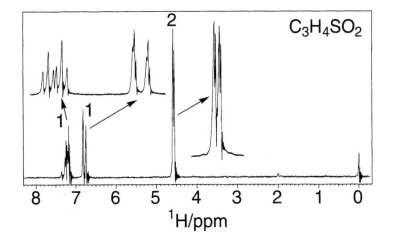

Exercise 1 The 60 MHz ^{1}H NMR spectrum of $C_3H_4SO_2$ in $CDCl_3$. Deduce a structure which is consistent with this spectrum. (Reproduced with permission of Varian, Inc.)

Possible answers: $CH_3SO_2C\equiv CH$, $HOCH_2SOC\equiv CH$,

$$
\begin{array}{ccc}
H & & H \\
C & = & C \\
| & & | \\
C & - & SO_2 \\
H_2 & &
\end{array}
$$

group must then take the other. The molecule is as shown in the margin.

Now apply the same approach to Exercises 1 to 3. Bear in mind the following: (a) The fluorine resonances of the compounds containing fluorine are not visible but the proton resonances may be coupled to the ^{19}F and so split into multiplets. The same comments apply also to the phosphorus compound. (b) Alcoholic protons are variable in position because of differing hydrogen bonding effects. The numbers on the spectra give the relative numbers of protons contributing to each resonance and have been obtained from integral traces. Exercises 1 and 2 show expanded regions of the spectra so as to allow fine structure to be distinguished. The answers are given below, though several structures are given, only one being correct. *All* the features of a spectrum should be explicable from the structure. It is instructive to also consider what differences in the spectrum would be obtained from the incorrect structures given.

A spectrum is shown in Fig. 3.50 with numbers corresponding to the integrals of each group of resonances. The empirical formula is $C_6H_{14}O_2$. Evidently there are five types of hydrogen and all show spin–spin coupling. The doublet and triplet at 1.37 and 1.25 ppm are in the chemical shift region typical of CH_3R and the integrals indicate that they are due to a CH_3 and two identical CH_3 groups. The former are coupled to a single proton which should have a quartet resonance. This is seen at 4.7 ppm, the line spacings being equal, and the shift

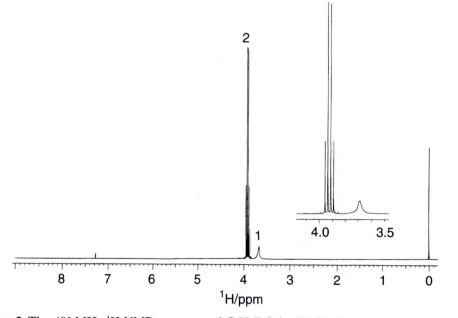

Exercise 2 The 400 MHz ^{1}H NMR spectrum of $C_2H_3F_3O$ in CDCl$_3$. Draw a structure which is consistent with this NMR spectrum.
Possible answers: CF$_2$HCHFOH, CF$_3$CH$_2$OH, CH$_2$FCF$_2$OH.

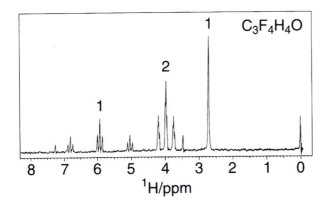

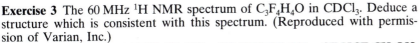

Exercise 3 The 60 MHz ¹H NMR spectrum of $C_3F_4H_4O$ in $CDCl_3$. Deduce a structure which is consistent with this spectrum. (Reproduced with permission of Varian, Inc.)
Possible answers: $CF_2HCH_2CF_2OH$, CF_3CHFCH_2OH, $CF_2HCF_2CH_2OH$, $CH_3CF_2CF_2OH$.

being typical of a RCH_2O type of situation. We thus have a fragment CH_3CH with oxygen close by. The other CH_3 groups are coupled to two hydrogen atoms and are exactly equivalent. The conclusion is that we have two equivalent ethyl groups but we do not see any other quartets in the spectrum though the four hydrogen atoms to which they must be coupled give a slightly second-order group of lines around 3.6 ppm. We have to conclude provisionally that we have two equiv-

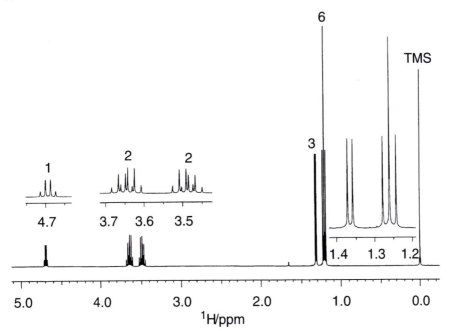

Figure 3.50 The 400 MHz ¹H NMR spectrum of $MeCH(OCH_2CH_3)_2$ in $CDCl_3$. The intensities of the signals are given above each one.

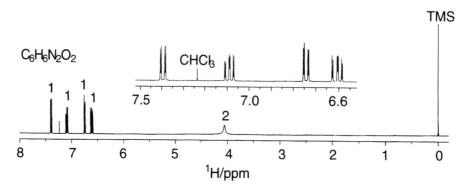

Figure 3.51 The 400 MHz ^1H NMR spectrum of 2-nitroaniline in $CDCl_3$. The intensities of the signals are given above each one.

alent ethyl groups CH_3CH_2-. This accounts for all the atoms in the molecule except the two oxygen atoms. Because the methyl groups are identical, the oxygen atoms have to be distributed symmetrically and this leads inevitably to the structure $CH_3CH(OCH_2CH_3)_2$. The CH_2 pairs of hydrogen atoms in this molecule are diastereotopic and so non-equivalent which accounts for their complex spectrum.

In Fig. 3.51, we see four groups of resonances in the aromatic/alkenic region of shift which exhibit a larger and a smaller coupling. There are also two uncoupled protons in the region typical of RNH and, in view of the empirical formula given on the figure, this suggests an NH_2 group which leaves us with a C_6H_4 aromatic nucleus and an NO_2 group. We thus have a nitro-aniline and need only to deduce which isomer we have. This can be done using the coupling patterns. Were it the *para* isomer, the two pairs of protons would be equivalent and an $[AB]_2$ pattern with two groups of resonances would result. Were it the *meta* isomer then there would be an isolated hydrogen (H_1) which would show only the small coupling. This is plainly not the case. However, in the *ortho* isomer all protons have near neighbours and

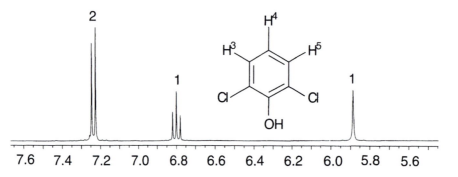

Figure 3.52 A 400 MHz ^1H NMR spectrum and its integral of 2,6-dichlorophenol in $CDCl_3$. The intensities of the signals are given above each one.

so the larger coupling interaction. H_1 and H_4 are close to one neighbour and are doublets while the other two are near to two neighbours and appear to be doublets of doublets though the spectra are complicated by the small longer range couplings. A very similar example will be found in Fig. 8.7 for the R groups of a phosphine PR_3.

In Fig. 3.52 we see the ^1H NMR spectrum of $C_6H_4OCl_2$. There are three resonances with a triplet, intensity 1 and a doublet, intensity 2 in the aromatic region and a singlet at 5.9 ppm still on the edge of the aromatic region. The coupled protons have the typical pattern for two chemically equivalent protons coupled equally to one proton, which is in accord with the intensities. The fourth proton is isolated from the others. With 6 carbon atoms we then have a benzene nucleus and 7 substituent atoms and we need to decide how they are arranged. The coupling constant is quite large at 8.5 Hz so the coupled protons are likely to be on adjacent carbon atoms. For two such protons to be equivalent, they have to be the outer ones and the molecule has to be symmetrical around an axis passing through the central one. The two chlorine substituents have then to be placed next to the two outer hydrogens which leaves us an OH group as sixth substituent on the ring to give the formula 2,6-dichlorophenol.

Nuclear magnetic relaxation

4

4.1 RELAXATION PROCESSES IN ASSEMBLIES OF NUCLEAR SPINS

If we perturb a physical system from its equilibrium condition and then remove the perturbing influence, the system will return to its original equilibrium condition. It does not return instantaneously, however, but takes a finite time to readjust to the changed conditions. The system is said to relax. Relaxation to equilibrium usually occurs exponentially, following a law of the form

$$(n–n_e)_t = (n–n_e)_0 \exp(- t/T)$$

where $(n—n_e)_t$, is the displacement from the equilibrium value n_e at time t and $(n—n_e)_0$ that at time zero. The relaxation can be characterized by a characteristic time T. If T is small, relaxation is fast; whereas if T is long, relaxation is slow. An alternative description is to use rates of relaxation, which are given the symbol R:

$$R = \frac{1}{T}$$

R has the advantages that its value increases as the relaxation becomes faster and that, if a system is subject to several parallel relaxation processes, then the overall rate is the simple sum of the rates of all the processes, i.e.

$$R = R_a + R_b + R_c + \cdots$$

The relaxation behaviour of assemblies of nuclear spins shows up directly in their NMR spectra and is related to the molecular dynamics of the system. For these reasons, it is of considerable importance to NMR spectroscopists, on the one hand allowing them to optimize experimental conditions, even to eliminate some undesirable spectral feature, or on the other hand allowing close study of the physical and chemical properties of the motion of a system.

We have already seen that in an assembly of $I = 1/2$ nuclei immersed in a strong magnetic field the spins are polarized into two populations with opposite senses and with a small excess number in the lower energy state. The nuclei precess around the magnetic field direction with a net magnetization M_z and no detectable transverse magnetization in the xy plane. It turns out that this system can be perturbed in two ways, and that we have to expect that there may be two relaxation processes with different relaxation times, which are called T_1 and T_2, or rates of relaxation R_1 and R_2. We have already seen in Chapter 1 (Fig. 1.6) that a B_1 radiofrequency pulse can swing the total nuclear magnetization away from its equilibrium position in the z direction, and this is essentially a perturbation of the system. If we apply a rather long pulse, we can swing the magnetization back into the z direction but pointing in the opposite direction. Such a pulse is called a 180° pulse, the reason for this name being evident in Fig. 4.1. The magnetization has been inverted and, immediately following the end of the pulse, relaxation processes start to return the magnetization to its normal state. Thus the magnetization decays from $-M_z$ via zero to M_z via first-order rate processes. The characteristic time for this process is T_1. The process is also called longitudinal relaxation since it takes place in the direction of B_0, and it is also called spin–lattice relaxation. In all cases it must be emphasized that the inverted magnetization has higher energy than the normal magnetization and that the return to equilibrium involves an exchange of magnetic energy with the surroundings.

If instead we use a 90° pulse to perturb the spin system, we move the magnetization into the xy plane as in Fig. 4.2. Now the magnetization in the z (B_0) direction is zero, and this returns to its normal

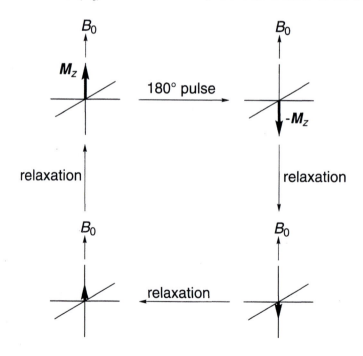

Figure 4.1 The T_1 relaxation process. If the magnetization is inverted, then it has to return to its equilibrium state and does so by decaying to zero and then increasing again in the normal B_0 direction. This process involves an exchange of energy between the spins and their environment.

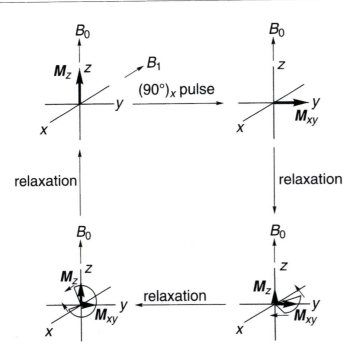

Figure 4.2 The T_2 relaxation process. A 90° pulse swings the magnetization into the xy plane around the B_1 vector. This is shown as stationary in the figure as if the observer were rotating in the same direction at the Larmor frequency, thus giving a static picture. There is a spread of nuclear frequencies, which causes the spins to fan out and reduce the resultant M_{xy}. M_z increases at the same time due to the T_1 process.

value by the mechanism just discussed. However, we have also created transverse magnetization in the xy plane, which rotates at the nuclear Larmor frequency. This has to decay to zero at equilibrium and does so because the frequency of each spin differs slightly from that of its companions, each varying randomly around the mean precession frequency. This means that, on average, some spins are slower and some are faster than the mean, so that the xy magnetization starts to fan out, to lose coherence and the resultant to become less in magnitude. Eventually the spins take all directions in the xy plane and M_{xy} is zero. The characteristic time of this process is T_2. This is also called transverse relaxation, and in solid materials it is known as spin–spin relaxation. Since the process is related to the spread of frequencies of a nuclear resonance, it is evident in the spectral linewidth. T_1 and T_2 may be equal or they may differ by orders of magnitude. The T_2 process involves no energy change. Evidently M_z increases as M_{xy} decreases, so that T_1 and T_2 can be equal, but T_2 cannot be longer than T_1. On the other hand, T_2 can be less than T_1, and in this case M_{xy} decays more rapidly than M_z is re-established and the signal (derived from M_{xy}) disappears well before equilibrium of the spins is attained.

T_2 is most commonly encountered in the linewidth of a signal. The linewidth at half height, $W_{1/2}$, is given by

$$W_{1/2} = \frac{1}{\pi T_2} \tag{4.1}$$

This is true for a singlet in a perfectly homogeneous magnetic field, but in practice there are often extra contributions to the linewidth due to inhomogeneity of the magnetic field and unresolved coupling.

The nuclei of atoms are extremely well isolated from their surroundings and, because the energy of NMR transitions is small, the chance of a spin transition occurring spontaneously is negligibly small. The fact that relaxation times can be quite short indicates that transitions are stimulated, and we must thus consider the various ways that this can happen.

4.2 DIPOLE–DIPOLE RELAXATION

We have already noted in section 2.1 that the magnetic field at a given nucleus due to the magnetic moment of a near-neighbour nucleus is very high but is averaged to zero by the random rotational diffusion of the molecule in which the nuclei reside. The magnitude of this field is such that the instantaneous chemical shift displacement of one ^1H nucleus due to the other in a methylene group can be as high as 150 000 Hz. As the group rotates, this field varies by such an amount on each side of zero. Thus the nuclei have instantaneously different precession frequencies since all possible orientations of the molecules will exist at any one instant. Randomization of the frequencies means not that all will have the same frequency in the long term but rather that once out of step a nucleus is just as probable to move further away from its companion's frequency as to reconverge to it. This dipole–dipole fluctuating field then is the cause of the loss of coherence between spins and so the source of the T_2 relaxation process.

The chaotic random motion of a solute in a solvent is called Brownian motion. This has a timescale that depends upon a number of factors such as mass of solute, solution viscosity and temperature. Because the motion is random, this timescale is characterized by a somewhat loosely defined term, the rotational correlation time τ_c. This is the time taken on average for a solute molecule to rotate by one radian or, more precisely, the time interval after which the molecular motion contains no vestige of its earlier angular momentum, i.e. has lost all memory of its previous behaviour. Not only the overall rotation of the molecule contributes to τ_c, but also local motion, such as the rotation of a methyl group. The time τ_c is typically 10^{-11} s for small molecules in liquids of low viscosity, which converts on the frequency scale to 10^{11} rad s^{-1} or 15 920 MHz. This is around the maximum rate of motion in the system, and all slower rates of motion can exist. The frequency spectrum of such random motion and associated magnetic fields is simple and is essentially white noise at all frequencies less than $1/\tau_c$. Since the maximum NMR frequency that is likely to be encountered is at present 800 MHz, it is clear that there is a component at all possible NMR frequencies. The relaxation field thus provides a B_1 component that varies in intensity and direction and causes random precession of the nuclei also: hence the dephasing of spins and also the possibility of energy transfer needed for the T_1 mechanism to operate.

The field intensity at any frequency, $K(\nu)$, is given by

$$K(\nu) \propto \frac{2\tau_c}{1 + 4\pi^2 \nu^2 \tau_c^2} \qquad (4.2)$$

This function is plotted in Fig. 4.3, where it will be seen that, when $\tau_c = 10^{-11}$ s, then over the NMR frequency band the intensity of the relaxation field is constant. However, if τ_c is rather longer, this may not be true. The Debye theory of electric dispersion shows that, for a spherical molecule rotating in a liquid, the correlation time is given by

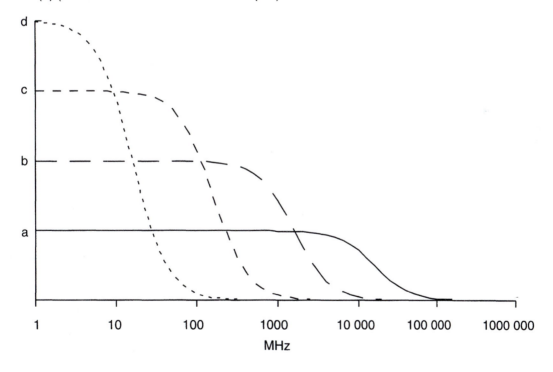

Figure 4.3 Intensity of fluctuations in magnetic fields in a liquid sample due to Brownian motion, as a function of frequency. (a) $\tau_c = 10^{-11}$ s⁻¹. (b) $\tau_c = 10^{-10}$ s⁻¹. (c) $\tau_c = 10^{-9}$ s⁻¹. (d) $\tau_c = 10^{-8}$ s⁻¹. Note that a different vertical scale is used for each plot as the intensity of fluctuations in the flat region of the plot is proportional to τ_c. $\tau_c \sim 10^{-11}$ s⁻¹ is found for very small molecules, $\tau_c \sim 10^{-9}$ s⁻¹ is commonly encountered with molecules of molecular mass, 1000 to 3000 D, while $\tau_c > 10^{-9}$ s⁻¹ is normally found for molecules of molecular mass > 5000 D.

$$\tau_c = \frac{4\pi a^3}{3k} \frac{\eta}{T}$$

where η is the viscosity of the liquid, T is the temperature and a is the radius of the molecule. Thus if we vary the viscosity of a sample, or its temperature, or the mass of the solute molecule, we will change τ_c. Examination of equation (4.2) shows that, when $4\pi^2\nu^2\tau_c^2 \ll 1$, the intensity of the relaxation field is proportional to τ_c. For the higher NMR frequencies and the longer correlation times, we start to leave the flat portion of the noise spectrum and the relaxation field starts to decrease. Provided we limit our range to the portion of these curves where $4\pi^2\nu^2\tau_c^2 \ll 1$, then the relaxation field and so the relaxation rate increases with τ_c. Long relaxation times thus occur for low viscosity, high temperature and small molecular mass.

It will be seen from equation (4.1) that the relaxation field intensity has its flat frequency response when the quantity $4\pi^2\nu^2\tau_c^2$ is very much less than unity, i.e.

$$4\pi^2\nu^2 \ll \frac{1}{\tau_c^2}$$

This is known as the region of extreme narrowing, where the correlation time is much shorter than one Larmor period of the nuclei.

We can now make some qualitative predictions about how the relaxation times will vary with τ_c. In the extreme narrowing limit, T_1 and T_2 are determined by the same relaxation field and so are equal. Increasing τ_c reduces the relaxation times until we reach the point at which the frequency spectrum of the field is no longer flat. Its intensity begins to decrease at the higher NMR frequencies, and so the chance for energy exchange is decreased and T_1 increases with further increase in τ_c. There is thus a minimum in T_1. The behaviour of T_2 is quite different since the correlation time is now similar in length to a Larmor period, and superimposed on the random field we see for short periods the rotating vector of the neighbouring nuclear magnetic moment. This is at the nuclear frequency and provides a second means for loss of coherence in the xy plane. T_2 continues to decrease as τ_c increases. In the limit of infinite τ_c we have the solid state. Here there is no dipole–dipole relaxation field and T_1 is very long and is determined by the presence of ferromagnetic impurities in the lattice. Hence the name spin–lattice relaxation. T_2 is determined by the now very strong interaction between spins via the rotating field generated by the Larmor precession of the spins. Rapid exchange of spin states is stimulated, the lifetime of the individual spin state is short and the uncertainty principle dictates a short T_2. Hence also the name spin–spin relaxation.

A full analysis of the spectral density of the relaxation field and the way this influences the spins gives the following equations for T_1 and T_2, expressed as the rates. We will also introduce the notation R_{1DD} and R_{2DD} or T_{1DD} and T_{2DD} to indicate that the mechanism discussed is dipole–dipole interaction.

For $I = 1/2$ nuclei of the same isotope situated in the same molecule, the intramolecular dipole–dipole relaxation rates for the jth type of nuclei, being relaxed by n different types of nuclei of the same isotope are

$$\frac{1}{T_{1DD}} = R_{1DD} = 2a\gamma^4 \left(\frac{\tau_c}{1 + \omega^2\tau_c^2} + \frac{4\tau_c}{1 + 4\omega^2\tau_c^2} \right) \sum_{i=1(i \neq j)}^{n} \frac{1}{r_{ij}^6}$$

$$\frac{1}{T_{2DD}} = R_{2DD} = a\gamma^4 \left(3\tau_c + \frac{\tau_c}{5 + \omega^2\tau_c^2} + \frac{2\tau_c}{1 + 4\omega^2\tau_c^2} \right) \sum_{i=1(i \neq j)}^{n} \frac{1}{r_{ij}^6}$$

where $a = 3\mu_0^2\hbar^2/320\pi^2$, μ_0 is the permeability of a vacuum, r is the distance between the nuclei, γ is their magnetogyric ratio, ω is the NMR frequency in rad s^{-1}, and $\hbar$ is (Planck's constant)/2π. The way these rates of relaxation vary with τ_c is shown in Fig. 4.4 for two protons at two different spectrometer frequencies, ω. The main feature to note is the T_1 minimum, which marks the limit of the extreme narrowing region and the way this moves with spectrometer frequency. The higher this frequency, the shorter becomes the maximum permitted τ_c, the result being that, for large complex molecules where the highest frequencies may be needed to give the necessary degree of resolution, the increased correlation times of such molecules may result in reduced T_2 and so increased linewidths. The expression for T_{1DD} and T_{2DD} is much simpler in the extreme narrowing region

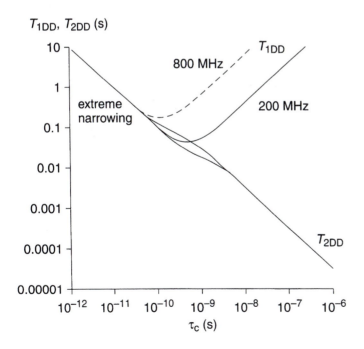

Figure 4.4 Variation of T_{1DD} and T_{2DD} with τ_c for two different spectrometer frequencies. The figures given apply to two protons separated by 160 pm.

$$\frac{1}{T_{1DD}} = \frac{1}{T_{2DD}} = R_{1DD} = R_{2DD} = 10a\gamma^4\tau_c \sum_{i=1(i \neq j)}^{n} \frac{1}{r_{ij}^6} \tag{4.3}$$

Two further points should be emphasized. In the first place, the rate of dipole–dipole relaxation depends upon the fourth power of the magnetogyric ratio, and nuclei with large magnetic moments will be most strongly subject to such relaxation, for example, ^{1}H or ^{19}F. The mechanism will be of lesser importance for nuclei with smaller magnetogyric ratios. Secondly, the efficiency of relaxation depends strongly upon the inverse distance between the spins. The effect of more distant spins will be almost negligible.

For an $I = 1/2$ nucleus in the same molecule as n different $I = 1/2$ nuclei with magnetogyric ratios γ_I and γ_S and frequencies ω_I and ω_S rad s^{-1}, respectively, the intramolecular dipole–dipole relaxation rates of one (species I) due to interaction with other (species S) are

$$\frac{1}{T_{1DD}} = R_{1DD} = \frac{a}{3}\gamma_I^2\gamma_S^2 \left\{ \frac{2\tau_c}{1 + (\omega_I - \omega_S)^2\tau_c^2} + \frac{6\tau_c}{1 + \omega_I^2\tau_c^2} \right.$$

$$\left. + \frac{12\tau_c}{1 + (\omega_I + \omega_S)^2\tau_c^2} \right\} \sum_{i=1(i \neq j)}^{n} \frac{1}{r_{ij}^6}$$

$$\frac{1}{T_{2DD}} = R_{2DD} = \frac{R_{1DD}}{2} + \frac{a}{3}\gamma_I^2\gamma_S^2 \left(4\tau_c + \frac{6\tau_c}{1 + \omega_S^2\tau_c^2} \right) \sum_{i=1(i \neq j)}^{n} \frac{1}{r_{ij}^6}$$

These equations, though more complex, give plots of form very similar to those depicted in Fig. 4.4. In the extreme narrowing limit

$$R_{1DD} = R_{2DD} = \frac{20a}{3}\gamma_I^2\gamma_S^2\tau_c \sum_{i=1(i \neq j)}^{n} \frac{1}{r_{ij}^6} \tag{4.4}$$

The most common example of relaxation of one spin by a different species occurs in ^{13}C spectroscopy, where the carbon nucleus is relaxed by the moments of directly bonded hydrogen atoms. In addition, because the relaxation field is stronger if more hydrogen atoms are attached to the ^{13}C, the rates of relaxation are proportional to the number of directly bonded hydrogens. Thus for relaxation of ^{13}C in a CH$_3$ group, the above expression has to be multiplied by three. Provided that relaxation occurs solely by the dipole–dipole mechanism, and the C–H bond length is known, then the correlation time may be calculated from the measured relaxation time. Relaxation of tertiary or carbonyl ^{13}C is much slower than for CH$_n$ groups since the C–H distance is much greater, and in this case other relaxation mechanisms may be important. The relaxation of protons on ^{13}C is dominated by interaction with other protons in the molecule.

An unpaired electron also generates a magnetic field. It influences relaxation in exactly the same way as a nucleus with $I = 1/2$ relaxes a second nucleus, but $\gamma_e = 658\gamma_H$. The result is that unpaired electrons can be very effective in relaxing other nuclei. The effectiveness depends on T_1^e, the spin–lattice relaxation time of the electron. When T_1^e is long, the unpaired electron is very effective at causing relaxation, and equation (4.3) applies, with γ_e replacing γ_S. When T_1^e is short, the unpaired electron is ineffective at causing relaxation. The result is that metal ions such as Cr^{III}, where T_1^e is long, cause substantial broadening of ^1H NMR signals (Fig. 2.10(a)), while Pr^{III} and Eu^{III}, where T_1^e is short, cause little broadening (Fig. 2.12). When ^2H is observed, the line is sharper as $\gamma_D = 0.1535\gamma_H$ (Fig. 2.10(b)).

In the case of ligands attached to metals, the unpaired electron is frequently delocalized onto the ligand. It then becomes meaningless to use equation (4.4) as the distance r is ill-defined. T_1 and T_2 are then given by

$$\frac{1}{T_1} = \frac{1}{T_2} = \frac{\mu_0^2\gamma_S^2 a_N^2 S(S + 1)\tau_c}{24\pi^2} \tag{4.5}$$

where a_N is the nuclear electron hyperfine interaction constant.

4.3 QUADRUPOLAR RELAXATION

Relaxation times for protons in organic molecules, as will be seen from Fig. 4.4, are of the order of seconds. For $I = 1/2$ nuclei with small magnetogyric ratios, the times are in general much longer, particularly in the absence of neighbouring protons. If the nucleus in question, however, has $I > 1/2$, it has a quadrupole moment, and this introduces a second and very efficient relaxation mechanism, which results in a generally much reduced relaxation time, a factor of as high as 10^8 over the expected dipole–dipole relaxation time being possible.

The quadrupole moment of a nucleus arises because the distribution of charge is not spherical, as is the case for $I = 1/2$ nuclei, but is ellipsoidal, i.e. the charge distribution within the nucleus is either slightly flattened (oblate, like the Earth at its poles) or slightly elongated (prolate, like a rugby ball or an American football). Cross-sections of the charge of two such nuclei are shown in Fig. 4.5 with the departure from spheroidal form much exaggerated. If the electric fields due to external charges vary across the nucleus, then the torque on each dipole component of the quadrupole is different, and a net torque is exerted on the nucleus by the electric field as well as by the magnetic fields present. Quadrupolar electric field gradients, that is field gradients which are shaped like d-orbitals, exist at atomic nuclei due to asymmetries in the spatial arrangement of the bonding electrons. The Brownian motion of the molecule causes the direction of the resulting electric quadrupole torque to vary randomly around the nucleus in exactly the same way as does the torque due to the magnetic relaxation field. Rotating electric torque components thus exist at the nuclear resonant frequency, which can also cause interchange of energy between the nucleus and the rest of the system (T_1 mechanism) and randomization of nuclear phase (T_2 mechanism).

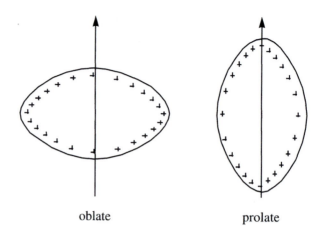

oblate   prolate

Figure 4.5 A representation of the charge distributions in quadrupolar nuclei. Note that there is a higher concentration of positive charge as the curvature of the surface increases. The magnetic axis is also marked.

The quadrupolar electric field gradient at a point in a molecule arises from the effect of the surrounding electric charges. The quadrupolar electric field can be described by three components in a Cartesian system of axes, but the field magnitude will change as one moves along an axis and in general all three of its components will change. Further, the changes may be different depending on which axis one chooses. The quadrupolar electric field gradient (EFG) thus requires nine numbers in order to describe it fully, and is a tensor quantity. Fortunately, by choosing an appropriate system of axes relative to the molecular geometry, this number can be reduced to three, V_{xx}, V_{yy} and V_{zz}, the diagonal components of the tensor, which have the further property that their sum is zero. This permits a further simplification by introducing the asymmetry factor, η, such that

$$\eta = \frac{V_{yy} - V_{xx}}{V_{zz}}$$

V_{zz} is chosen to be the largest component of the EFG tensor, and this can then be described by two quantities, V_{zz} and η. If the system is axially symmetric so that $V_{yy} = V_{xx}$ then $\eta = 0$. The V_{xx}, etc. are calculated as the sum of contributions from all charges i

$$V_{xx} = \sum_i q_i r^{-5}(3x_i^2 - r_i^2) \tag{4.6}$$

where r_i is the distance of charge q_i from the nucleus and x_i is the coordinate of the charge in the axis system chosen. Similar expressions give V_{yy} and V_{zz}.

The equation describing the quadrupole relaxation times (T_{1Q}, T_{2Q}) of a quadrupolar nucleus with spin I situated in a molecule with an isotropic correlation time τ_c and in the extreme narrowing region is

$$\frac{1}{T_{1Q}} = \frac{1}{T_{2Q}} = R_{1Q} = R_{2Q} = \frac{3\pi^2}{10} \frac{(2I+3)}{I^2(2I-1)} \left(\frac{e^2Q}{h}\right)^2 V_{zz}^2 \left(1 + \frac{1}{3}\eta^2\right)\tau_c$$

(4.7)

where Q is the quadrupole moment of the nucleus of spin, I, e is the electronic charge and h is Planck's constant. Outside the region of extreme narrowing, the behaviour is reminiscent of that of dipole–dipole relaxation but is complicated by the fact that several nuclear transitions are possible. Thus for a nucleus with spin 3/2 there are three transitions, $3/2 \leftrightarrow 1/2$, $1/2 \leftrightarrow -1/2$ and $-1/2 \leftrightarrow -3/2$, and the relaxation rate for nuclei undergoing the first and last transitions differs from that of those undergoing the $1/2 \leftrightarrow -1/2$ transition. This produces non-exponential relaxation.

Quadrupolar relaxation depends strongly upon both nuclear properties (Q, I) and molecular properties (V_{zz}, η, τ_c). Its effectiveness increases rapidly as Q is increased, though this is to some extent offset by the fact that large Q tends to be associated with large I and the function of I in the expression for T_{1Q} decreases rapidly as I increases (Table 4.1). As a result, almost all quadrupolar nuclei are capable of being detected, only a few having such short relaxation times that they are unobservable. Of the molecular properties, the correlation time has a particularly strong influence in that changes in τ_c brought about by changes in temperature alter the relaxation times considerably. Increasing temperature reduces τ_c and so increases the relaxation time and reduces the resonance linewidth. Since the resonances of quadrupolar nuclei are usually quite broad, these changes are evident. The same effect occurs with dipole–dipole relaxation of $I = 1/2$ nuclei, but here the relaxation times are long, the linewidths are small, and the linewidth changes are not so immediately obvious. However, it is frequently possible to differentiate the narrow signals from a small impurity molecule from the broader signals from a higher molecular weight compound by the different linewidths in their ^1H NMR spectra.

For a given nuclear species, the quadrupolar relaxation time is determined mainly by the EFG and this can vary from almost zero to very large values, so that relaxation times can vary by several orders of magnitude depending upon the situation of the nucleus. In principle, the EFG can be calculated at any point in a molecule so that one could obtain τ_c, or if τ_c were known (e.g. from the ^{13}C relaxation time of a CH moiety in the molecule) the EFG could be determined and compared with that calculated so as to verify the accuracy of the calculation. Unfortunately, it proves in practice difficult to calculate the

Table 4.1 The dependence of $F(I) = (2I + 3)/\{I^2(2I - 1)\}$ on I

I	1	$\frac{3}{2}$	$\frac{5}{2}$	3	$\frac{7}{2}$	$\frac{9}{2}$
$F(I)$	5	$\frac{4}{3}$	$\frac{8}{25}$	$\frac{1}{5}$	$\frac{20}{147}$	$\frac{2}{27}$

EFG with sufficient accuracy. Equation (4.4) shows that the EFG is proportional to the inverse cube of the distance of the charge from the nucleus. Thus charge close to the nucleus has the predominating effect, and it is here that the calculations are least accurate. In fact, there is some confusion in the literature on this point. In the solid state, the EFG arises from quite distant charges as well as those close by, in the same way that a Madelung constant is calculated. In a liquid, the movement of the molecules reduces the distant effects to zero and the EFG arises quite locally around the nucleus. For instance, in the anion $[AlCl_3(NCS)]^-$ both the quadrupolar nuclei ^{27}Al and ^{14}N have long relaxation times and show spin–spin coupling. There are two points of low EFG in the molecule, and this can only arise if the EFG arises in the region quite close to the nucleus. The question is, of course, how close has the charge to be, to be effective?

The cases of greatest interest are those in which the relaxation times are relatively long, since this gives the best resolution of resonances and the possibility of seeing coupling effects. This means that we should understand the requirements for obtaining a low EFG at a nucleus. We can do this easily if we remember that for a traceless EFG tensor the sum of the diagonal elements is zero

$$V_{xx} + V_{yy} + V_{zz} = 0$$

If the system is axially symmetric (i.e. $\eta = 0$), $V_{yy} = V_{xx}$, and it follows that if V_{zz} is zero then all three terms must be zero and the EFG vanishes. Thus in an axially symmetric system we need only calculate V_{zz}. We take for the model in Fig. 4.6 a tripod of bonds to a nucleus N disposed regularly around the z axis with effective, equal charges q called q_1, q_2 and q_3 at a distance r from N.

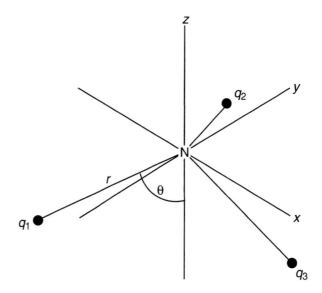

Figure 4.6 Calculation of the EFG at N due to three equal charges q.

The z coordinate of each charge is $r\cos\theta$. According to equation (4.6) then

$$V_{zz} = \frac{3q(3r^2\cos^2\theta - r^2)}{r^5} = \frac{3q(3\cos^2\theta - 1)}{r^3}$$

In order that V_{zz} be zero, $(3\cos^2\theta - 1)$ must be zero, which requires that

$\theta = 54.7356°$

As we shall see, the quantity $(3\cos^2\theta - 1)$ appears in many expressions describing NMR phenomena, and the value of θ at which it becomes zero has become known as 'the magic angle'. It may seem astonishing that such a one-sided arrangement of charge should give zero EFG. If, however, we remember that the electric field is zero where the z axis intersects the plane defined by the three charges and is also zero at $-\infty$, we will see that it has to have a maximum at some point on the z axis, a point where its gradient is zero. We should also note that V_{zz} is zero for any number greater than three of equal regularly disposed charges that form the magic angle with the axis.

It has been customary to state as the necessary condition for low EFG a cubic arrangement of equal charges (or identical ligands) around a nucleus. This is correct, and indeed a nucleus at the centre of a regular tetrahedron, octahedron or cube has near-zero EFG. It is, however, too limiting a statement. The octahedron is in fact made up of two back-to-back tripods as depicted in Fig. 4.6 with all six charges equal and pairs of charges collinear with N. Equally, an octahedral complex in which there are two tripods of differing charge will still have zero EFG, i.e. an all-*cis* or *fac* structure MA_3B_3 where A and B are ligands. Such structures, while non-cubic, still have long relaxation times, and we can begin to see that long relaxation times for quadrupolar nuclei are a source of structural information.

A specific example is found with the molybdenum tricarbonyl arene complexes, see formula in margin.

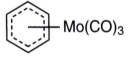

The nucleus ^{95}Mo has a significant quadrupole moment and its relaxation times vary typically from about 1 s in the regular tetrahedron $[MoO_4]^{2-}$ up to about 0.15 ms in less symmetric compounds. In the arene carbonyl complexes, the times are near 0.07 s and are relatively long for a non-cubic symmetry. A to-scale sketch of the molecule is given in Fig. 4.7. The crystal structure of the mesitylene complex has been obtained and this shows that the three carbonyl ligands are orthogonal, which means that they each make the magic angle with the symmetry axis of the molecule. Following Fig. 4.6 then, the carbonyl ligands give zero net contribution to the EFG at the molybdenum. This means that the remaining bonding electrons to the arene ring must also produce a very small EFG at the metal. So the metal–arene bonds must lie on a conical surface with a half-apex angle equal to the magic angle also. This argument does not depend upon our knowing how many such bonds there may be; they have simply

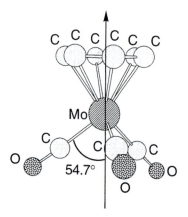

Figure 4.7 A drawing taken from the crystal structure of $[(\eta^6\text{-}1,3,5\text{-}Me_3C_6H_3)$ $Mo(CO)_3]$ with the methyl groups and hydrogen atoms omitted. The carbonyl ligands are approximately orthogonal and form half a regular octahedron. The arene ring carbon atoms give an average angle of 36.4° to the axis.

to be disposed regularly on the conical surface. It is clear that such bonding orbitals do not intersect the arene ring, that the conventional picture of overlap of metal and arene orbitals is correct, and that the bonds are bent. It is also clear that the electron density on the arene cannot influence the EFG and that this is determined by the electron density nearest to the metal.

In general, however, many molecules either have some distortion of their symmetry or their symmetry is non-cubic around the nucleus to be observed, and the relaxation times and linewidths observed are very variable. For octahedral complexes, the point-charge model makes some predictions which are useful in distinguishing between *cis*- and *trans*- and between *mer*- and *fac*-isomers, see Table 4.2.

Where terminal atoms are concerned, the lines are often very broad and may not even be detectable. This missing intensity problem has to be kept in mind when working with quadrupolar nuclei, and it is

Table 4.2 The predicted quadrupolar electric field gradients and observed ^{59}Co linewidths for some cobalt(III) amine complexes

Structure	Relative magnitude of quadrupolar electric field gradient	Complex	Linewidth (Hz)
MA_6	0	$[Co(NH_3)_6]^{3+}$	87
MA_5B	2	$[Co(NH_3)_5(N_3)]^{2+}$	350
cis-MA_4B_2	2	*cis*-$[Co(NH_3)_4(N_3)_2]^+$	370
trans-MA_4B_2	4	*trans*-$[Co(NH_3)_4(N_3)_2]^+$	520
fac-MA_3B_3	0	*fac*-$[Co(NH_3)_3(N_3)_3]$	300
mer-MA_3B_3	3	*mer*-$[Co(NH_3)_3(N_3)_3]$	1240

often necessary to check the resonance intensity against that of a standard sample in order to ensure that something is not being missed. Even if the resonance is visible, its intensity will be incorrect if its width is more than one-sixth that of the Fourier transform spectral width. On the other hand, if the resonance is broad, but is resolved from other resonances, then structural information is present in the spectrum. A few examples of quadrupolar nuclei in different environments are given in Table 4.3 to illustrate the possibilities. Linewidths are quoted rather than relaxation times, since it is the linewidths that have been measured in most cases and these are proportional to R_{2Q} or $1/T_{2Q}$.

Solid-state NMR spectroscopy of quadrupolar nuclei or the zero-field technique of nuclear quadrupole resonance can give values for the interaction between the nuclear quadrupole and the EFG, and this is called the quadrupole coupling constant, χ, where in terms of equation (4.4)

$$\chi = \frac{eQV_{zz}}{h} = \frac{e^2qQ}{h}$$

If one can be certain that the EFG in solid and liquid are equal, and this is likely to be so where the EFG is large and determined

Table 4.3 Linewidths of the resonances of some compounds of quadrupolar nuclei

Nucleus	Molecule	Linewidth (Hz)	Comments
^{14}N	$[Me_4N]^+$	0.1	Regular tetrahedron, small $J(^{14}N^1H)$ resolvable
^{14}N	$[PhNH_3]^+$	100	Non-regular tetrahedron
^{14}N	MeCN	80	Terminal position
^{14}N	MeNC	0.26	Apparently linear and low EFG not possible. Requires annular electronic distribution around CN bond. See Fig. 4.6.
^{27}Al	$[Al(OH_2)_6]^{3+}$	2	Regular octahedron, linewidth influenced by proton exchange.
^{27}Al	$[AlBu^i_3]$	3990	Planar trigonal monomer with high EFG. Linewidth at ∞ dilution.
^{14}N	$[NO_3]^-$	3.7	Note contrast with previous example. The difference is explained by there being electron density above and below the plane of this trigonal ion.
^{35}Cl	$[ClO_4]^-$	1.2	Regular tetrahedron, aqueous solution.
^{35}Cl	PCl_3	7600	Terminal atom. Figure based on relaxation time measurement and is typical of many chlorides, e.g. CH_2Cl_2, $SiCl_4$.

primarily by the bonding electrons rather than by long-distance effects in the solid, then it is possible to obtain correlation times from the relaxation times of quadrupolar nuclei in molecules in solution.

4.3.1 Spin–spin coupling to quadrupolar relaxed nuclei

Longitudinal relaxation involves changes in spin orientations relative to B_0 and so we might expect relaxation processes to modify the spin coupling patterns that we observe. In fact, the relaxation times of $I = 1/2$ nuclei are sufficiently long that relaxation normally has no effect on the analysis of the spectra, but the much shorter relaxation times of the quadrupolar nuclei do lead to considerable modification of the observed patterns.

If T_{1Q} is long, then the normal spin coupling effects are observed, such as the coupling of protons to both ^{10}B and ^{11}B shown in Fig. 3.8.

At the other extreme, if T_{1Q} is very short, the nuclear spins interchange energy and change orientation so rapidly that a coupled nucleus interacts with all possible spin states in a short time. It can distinguish only an average value of the interaction and a singlet resonance results. This explains, for instance, why the chlorinated hydrocarbons show no evidence of proton spin coupling to the chlorine nuclei ^{35}Cl and ^{37}Cl, both with $I = 3/2$. At intermediate relaxation rates, the coupling interaction is indeterminate and a broad line is observed. The line shapes calculated for the resonance of a $I = 1/2$ nucleus coupled to a quadrupole relaxed $I = 1$ nucleus such as ^{14}N are shown in Fig. 4.8. The shape of the spectrum observed depends upon the product $T_1 J$ where T_1 is the relaxation time of the quadrupole nucleus, since if the frequency defined by $1/2\pi T_1$ is comparable with the coupling constant in hertz, then the coupled nucleus cannot distinguish the separate spin states. The situation is equivalent to attempting to measure the frequency of a periodic wave by observing only a fraction of a cycle. The resonance of the quadrupolar nucleus will, of course, be split into a multiplet by the $I = 1/2$ nuclei, but each component line will be broadened by its relaxation.

A common example of lines broadened by coupling to quadrupolar nuclei is found in amino compounds. The protons on the nitrogen are usually observed as a broad singlet, a good example appearing in Fig. 3.51. It is important to remember in this case that the spin–lattice relaxation time of the amino proton is unaffected and can cause normal splitting in vicinal protons bonded to carbon. In contrast, the protons in the highly symmetric ammonium ion give narrow resonances because the nitrogen quadrupole relaxation is slow (Fig. 3.8).

Since quadrupole relaxation is sensitive to temperature and viscosity, the line shapes observed for coupled nuclei are altered by viscosity and temperature changes, an increase in temperature leading to *slower* relaxation and a better resolved multiplet. This fact is stressed, since on a first encounter it seems contrary to one's expectation. Alternatively, lowering the temperature increases the rate of quadru-

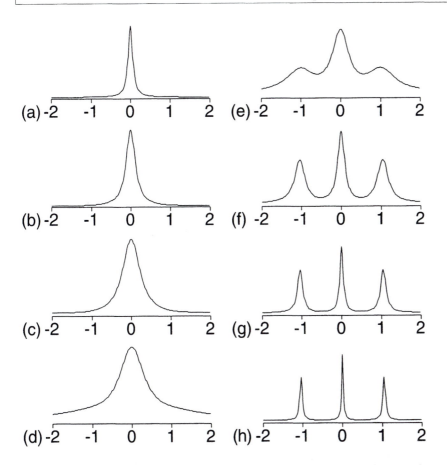

Figure 4.8 The resonance line shape of a spin-1/2 nucleus coupled to a spin-1 nucleus having various rates of quadrupole relaxation. (a) $T_1 \approx 0.05/\pi J$. (b) $T_1 \approx 0.1/\pi J$. (c) $T_1 \approx 0.2/\pi J$. (d) $T_1 \approx 0.4/\pi J$. (e) $T_1 \approx 0.8/\pi J$. (f) $T_1 \approx 2.0/\pi J$. (g) $T_1 \approx 4.0/\pi J$. (h) $T_1 \approx 8.0/\pi J$. The intensities are not to scale. The horizontal axis is in units of $\Delta v/J$.

pole relaxation and, if the resonance of the coupled $I = 1/2$ nucleus was already broad, this may well narrow as the relaxation becomes too fast for any coupling to be exhibited. This technique can sometimes be used to advantage to simplify the spectra of $I = 1/2$ nuclei coupled to quadrupolar nuclei.

4.4 SPIN ROTATION RELAXATION: DETAILED MOLECULAR MOTION

Theories of molecular motion differentiate two linked types of motion, which are given two correlation times τ_2 and τ_{SR}. The time τ_2 is the orientational correlation time or the time between significant changes in orientation. This is clearly equivalent to the correlation time τ_c used here, which can be obtained from measurements of T_{1DD} or T_{1Q}, both of which measure the reorientation of internuclear vectors, which may or may not correspond to bonds. The time τ_{SR} is the angular momentum correlation time or the time between significant changes in angular momentum. τ_c and τ_{SR} are not independent since, if the angular velocity is maintained for a longer time and τ_{SR} is long, the orientation must change more rapidly and τ_c is short. Alternatively, rapid random changes in angular velocity leave the orientation more or less unchanged. Thus τ_c and τ_{SR} are related by Hubbard's equation

$$\tau_c \tau_{SR} = \frac{I}{6kT}$$

where T is the absolute temperature and I is the moment of inertia.

This molecular rotation has a further effect upon nuclear relaxation, which can be detected if the dipole–dipole mechanism is weak. If the rotation is particularly fast, the system of bonding electrons is subject to some displacement relative to the atomic nuclei, which gives rise to a small magnetic moment proportional to the angular momentum of the molecule. Changes in the direction of the magnetic moment provoke relaxation in the same way as with other processes, but with the difference, which is unique to spin rotation relaxation, that the faster the rotation, the longer the corresponding correlation time τ_{SR}. The time T_{1SR} is given by

$$\frac{1}{T_{1SR}} = R_{1SR} = \frac{2IkTC^2 \tau_{SR}}{3\hbar^2}$$

τ_{SR} can be replaced with τ_c by using Hubbard's equation to give

$$\frac{1}{T_{1SR}} = R_{1SR} = \frac{2I^2 kC^2}{18k\hbar^2 \tau_c}$$

Thus increasing the temperature reduces τ_c and increases the efficiency of this relaxation process so reducing T_{1SR}. A study of the temperature dependence of the nuclear relaxation time can thus distinguish the spin rotation mechanism from the other mechanisms. T_{1SR} is significant for small molecules, or rapidly rotating side groups in larger molecules. The mechanism is particularly efficient in the gas phase, and thus resonances of gases are much broader than observed in solutions of the same molecules.

Two examples will illustrate these points. In the first, the relaxation time of the nitrogen nuclei is measured in liquid nitrogen under pressure and over a wide range of temperature. The two types of motion were separated by using $^{14}N_2$, which relaxes by the quadrupolar mechanism and gives T_{1Q} and so τ_c and nitrogen enriched in the $I = 1/2$ nucleus ^{15}N, which, because of its small magnetogyric ratio, relaxes by the spin rotation mechanism and gives T_{1SR} and so τ_{SR}. The results are shown in Fig. 4.9, where the opposite temperature dependences are obvious.

The second example concerns the relaxation behaviour of the quadrupolar nucleus ^{9}Be in the hydrated cation $[Be(H_2O)_4]^{2+}$. The quadrupole moment of this nucleus is small, so that the quadrupole relaxation mechanism does not necessarily dominate its relaxation. Interaction with the protons of the water ligands causes dipole–dipole relaxation, but this can be eliminated by using deuteriated water as solvent. The results are shown in Fig. 4.10, where it will be seen that the relaxation shows three different types of behaviour with change in temperature: T_1 increases with temperature at low temperatures, then has a maximum where there is little change and then decreases with further increase in temperature. Replacing H_2O by D_2O increases the relaxation time at low temperature but has no effect at high temperature. In the low-temperature region, then, for $[Be(H_2O)_4]^{2+}$ we have both quadrupolar and dipole–dipole relaxation, and for $[Be(D_2O)_4]^{2+}$ we have effectively only quadrupole relaxation, so that the measurements in the two solvents allow us to separate the two processes. Evidently, even for this tetrahedral molecule with low EFG, the quadrupolar mechanism is the most effective. At high temperature the changed temperature dependence indicates that the spin

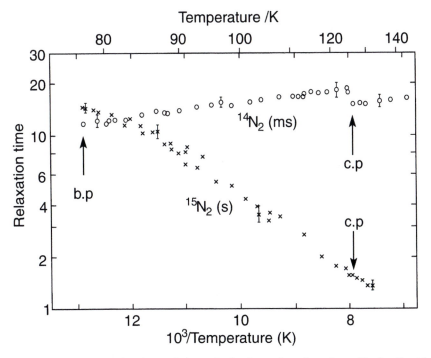

Figure 4.9 Measured values of the spin–lattice relaxation time, T_1, for liquid $^{14}N_2$ (in ms), o, and liquid $^{15}N_2$ (in s), x, on the liquid vapour coexistence line *versus* temperature (on a scale of 10^3 K/T, where T is the absolute temperature); c.p. indicates the critical point, and b.p. the boiling point at 1 atm. (From Powles *et al.* (1975) *Mol. Phys.* **29**, 539, reprinted with permission, copyright (1975) Taylor & Francis.)

rotation mechanism now predominates. The various relaxation rates at 80°C are approximately $R_{1DD} = 7.7 \times 10^{-3}$, $R_{1Q} = 4.3 \times 10^{-2}$ and $R_{1SR} = 0.24$ s^{-1} respectively, the first two processes accounting for less than 18% of the overall relaxation rate.

4.5 CHEMICAL SHIFT ANISOTROPY RELAXATION

We have already discussed in Chapter 2 how the chemical shift is a tensor quantity and varies as a molecule tumbles relative to the magnetic field direction. This averages to an isotropic value but nevertheless means that there is effectively a fluctuation in magnetic field strength at the nucleus that can also cause relaxation. The mechanism is not very efficient and, in the extreme narrowing region, depends upon

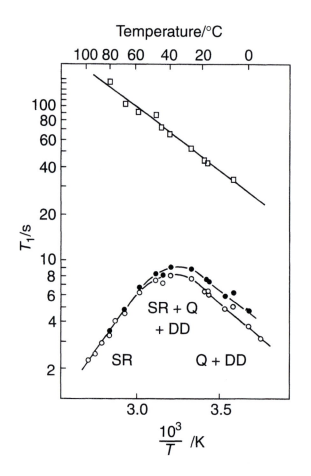

Figure 4.10 The ^{9}Be spin–lattice relaxation time in 1 M aqueous [Be(NO$_3$)$_2$] as a function of temperature: (○) H$_2$O solution; (●) D$_2$O solution; (□) T_{1DD} calculated from the difference between the rates of relaxation in both solvents. (From Wehrli (1976) *J. Magn. Reson.*, **23**, 181, with permission.)

$$\frac{1}{T_{1CSA}} = \frac{2\gamma_I^2 \boldsymbol{B}_0^2 (\sigma_{||} - \sigma_{\perp})^2 \tau_c}{15} \tag{4.8}$$

where γ_I is the magnetogyric ratio of the nucleus whose relaxation time is required and the σ are defined in Chapter 2. This mechanism can be distinguished from others by the fact that it depends upon $\boldsymbol{B}_0^2$. It is important for nuclei with a rather high screening anisotropy and high chemical shift ranges and at high magnetic fields. In some cases, it is an advantage since it reduces otherwise long relaxation times at high field, making easier the observation of ^{13}C signals from non-proton bearing carbon atoms which are *sp*² or *sp* hybridized, with $\sigma_{||} - \sigma_{\perp}$ of the order of 200 and 400 ppm, respectively. In other cases such as for ^{199}Hg in $[Me_2Hg]$ where $\sigma_{||} - \sigma_{\perp}$ is of the order of 7500 ppm, signals can be substantially broadened.

An example has recently been reported that contrasts the relaxation behaviour of the $I = 1/2$ nucleus ^{195}Pt in the two anions $[Pt(CN)_4]^{2-}$ and $[Pt(CN)_6]^{2-}$. The former has square planar geometry and so a large chemical shift anisotropy (CSA), which has been measured to be a $\sigma_{||} - \sigma_{\perp} = 2500$ ppm in the solid state. The octahedral anion has no such anisotropy. The relaxation times were measured as functions of both temperature and of magnetic field. Two spectrometers were used, one operating at 4.70 T (^{195}Pt at 42.8 MHz) and one at 8.48 T (^{195}Pt at 77.04 MHz). The results are plotted in Fig. 4.11. Only one set is shown for $[Pt(CN)_6]^{2-}$ since the relaxation rates were the same at both magnetic fields, the rate of motion being fast enough to ensure that the extreme narrowing condition was met. However, the rate of relaxation increases with temperature and so the mechanism of relaxation must be that of spin rotation, any dipolar contribution from the ^{14}N nuclei or from the solvent (D_2O) being small. The behaviour of the relaxation of $[Pt(CN)_4]^{2-}$ is quite different; the rate is increased at the higher magnetic field and the plots are curved, indicating the presence of two competing relaxation mechanisms, spin rotation at the highest temperatures and CSA at the lowest, with opposite temperature dependences.

It is not only the nucleus itself that is affected, but nuclei to which it is coupled. This is illustrated in Fig. 4.12 for $^3J(^{195}Pt^1H)$. At 90 MHz, the ^{195}Pt satellites on either side of the proton signal for H^2 are sharp, but as the magnetic field is increased, so that ^1H resonates at 250 and then 400 MHz, the ^{195}Pt satellites progressively broaden, reflecting the shortening of $T_1(^{195}Pt)$.

4.6 SCALAR RELAXATION

We have already mentioned that in many cases quadrupolar nuclei relax so rapidly that any effects of spin–spin coupling to other nuclei are completely lost. If, however, the rate of relaxation of the quadrupolar nucleus is very fast indeed, and if the coupling constant in the

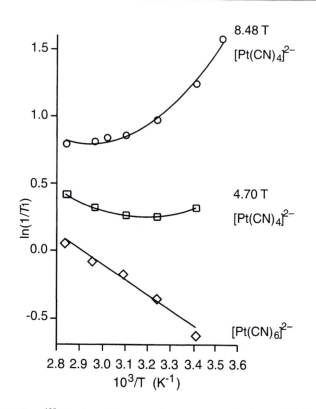

Figure 4.11 The ^{195}Pt relaxation rates for aqueous solutions of $K_2Pt(CN)_4$ and $K_2Pt(CN)_6$ at $\boldsymbol{B}_0 = 4.70$ and 8.48 T as functions of inverse temperature. The ^{195}Pt relaxation rates of $K_2Pt(CN)_6$ are independent of $\boldsymbol{B}_0$. (From Wasylishen and Britten (1988) *Magn. Reson. Chem.*, **26**, 1075, copyright John Wiley and Sons Ltd, New York, reprinted with permission.)

absence of the quadrupolar relaxation is significant, then the relaxation time of the coupled nucleus may be influenced by the quadrupolar nucleus. For T_1, the effect of scalar coupling is rarely significant, especially at high field. The contribution of scalar coupling to T_1 of one nucleus, I, by a second rapidly relaxing nucleus, S, is given by

$$\frac{1}{T_{1SC}} = \frac{8\pi^2 J^2 S(S+1) T_1^S}{3\{1 + (\omega_I - \omega_S)^2 (T_1^S)^2\}}$$

where J is the coupling constant between I and S, S is the nuclear spin quantum number of nucleus S, T_1^S is the spin–lattice relaxation time of nucleus S, ω_I and ω_S are the frequencies in rad s^{-1} of the nuclei I and S. Clearly this term is only significant when $(\omega_I - \omega_S)^2$ is small and J is large. This condition is rarely met. T_{1SC} has been reported as being significant for ^{13}C attached to ^{79}Br and to ^{185,187}Re.

In contrast, T_{2SC} can be dominant. It follows the equation

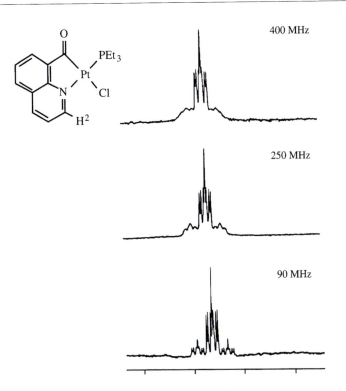

Figure 4.12 ^{1}H NMR spectra for H^2 in [PtCl(COC$_9$H$_6$N)(PEt$_3$)] at magnetic field strengths corresponding to 90, 250, and 400 MHz for ^{1}H. $^3J(^{195}$Pt^{1}H) = 27.1 Hz. Note that the ^{195}Pt satellites are sharpest at 90 MHz. (Reproduced with permission from Anklin and Pregosin (1985) *Magn. Reson. Chem.*, **23**, 671, John Wiley and Sons Ltd, reprinted with permission.)

$$\frac{1}{T_{2SC}} = \frac{1}{2}\frac{1}{T_{1SC}} + \frac{4}{3}\,\pi^2 J^2 S(S+1)T_1^S$$

where J is the coupling constant between the observed nucleus and the quadrupolar one, and T_1^S is the relaxation time of the quadrupolar nucleus.

There can be substantial contributions to T_2 and hence the linewidth when J is comparable with $1/T_1^S$. This condition is commonly met for nuclei attached to nuclei such as ^{11}B, ^{14}N, ^{27}Al, ^{51}V, ^{55}Mn, ^{59}Co, ^{63}Cu, and ^{65}Cu, often resulting in substantial line broadening. This is illustrated in Fig. 4.13 for [Mn(CH$_2$Ph)(CO)$_5$]. At 30°C, the carbonyl signals are very broad due to broadening by the ^{55}Mn via T_{2SC}, as a result of the large $^1J(^{55}$Mn^{13}C) and moderate $T_1^S(^{55}$Mn). On cooling to −87°C, the T_1 for the ^{55}Mn ($I = 5/2$, 100% abundant) shortens due to the lengthening of τ_c, and the broadening of the ^{13}CO signals due to the ^{55}Mn is considerably reduced because $^1J(^{55}$Mn^{13}C) has not changed in value, but $T_1^S(^{55}$Mn) is much shorter producing a smaller $1/T_{2SC}$ and sharper ^{13}C signals.

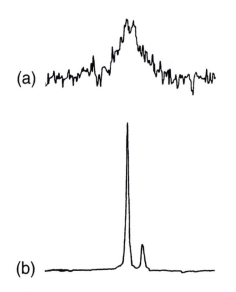

Figure 4.13 The ^{13}C NMR spectra of $[Mn(CH_2Ph)(CO)_5]$ in CH_2Cl_2 at (a) 30°C, and (b) –87°C. Note the improvement in both the linewidth and signal:noise at low temperature. (Reproduced from Todd and Wilkinson (1974) *J. Organomet. Chem.*, **80**, C31, copyright (1974), with permission from Elsevier Science.)

In the case of $[^{195}Pt(CN)_6]^{2-}$, for instance, if the T_2 value of the ^{195}Pt equalled T_1, the linewidth would be around 1 Hz. In fact, it is 25 to 60 Hz, depending upon temperature, and this is due to coupling with the rapidly relaxing ^{14}N nuclei in the anion. In this case the frequency difference ^{195}Pt–^{14}N is too great for there to be any influence on T_1.

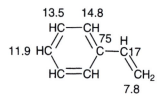

(4.1)

4.7 EXAMPLES OF ^{13}C RELAXATION TIMES

When we are setting up an NMR experiment, it is important that we have some idea of T_1 for the observed nuclei so that an appropriate pulse angle and relaxation delay can be selected. In small molecules T_1 is generally long. For example, in styrene, **4.1**, the T_1 values for ^{13}C range from 7.8 to 75 s. When it is remembered that ^{13}C relaxation is dominated by dipole–dipole relaxation, the variations in T_1 can be explained in terms of the C–H distance and the correlation time. The ipso, non-proton bearing carbon has the longest T_1 as there are no close protons to relax it. The CH_2 group has the shortest as the carbon has two attached protons. The variation in T_1 for the remaining CH carbons is due to differences in molecular motion. The molecule can tumble anisotropically, but it is simplest to attribute most of the differences to the differences in τ_c due to rotation about the vinyl–phenyl bond. The *para*-CH group lies on this axis and is not moved by the rotation and has the shortest T_1 value of the CH carbon atoms. The *ortho*- and *meta*-CH and the vinylic CH groups lie off the axis and the hydrogen atoms move with respect to the carbon atoms due to the rotation. The result is that τ_c is shortened and T_1 becomes longer.

Thiamine hydrochloride provides a more representative example. The T_1 values are given in Fig 4.14 for the compound in CH_3OH and CD_3OD. Note that certain non-proton bearing carbon atoms have much longer T_1 values than the proton bearing ones, but the T_1 is reduced for these non-proton bearing carbons in the protio solvent due to the proximity of the solvent protons. The T_1 of the methyl group is lengthened by rotation relative to the CH carbon atoms. By

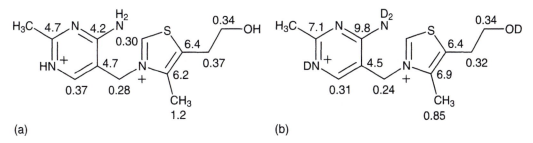

(a)   (b)

Figure 4.14 The ^{13}C relaxation times of thiamine hydrochloride in (a) CH_3OH and (b) CD_3OD, measured at 25.15 MHz.

extrapolation, one might expect the relaxation time to be about 0.1 s in the absence of CH_3 rotation.

The final example is a compound with a long chain, phytol, **4.2**. The central CH_2 carbon atoms have T_1 values around 0.3 s, while the CH carbon atoms have T_1 values around 0.5 s as they only have one proton to provide the relaxation. The carbon atoms at the ends of the chain have longer T_1 values. This is particularly noticeable at the $CHMe_2$ end of the chain. This arises because rotation around the C–C bonds provides an extra mode of motion, increasing T_1. The effect is not so marked at the $HOCH_2$ end of the chain as hydrogen bonding to the OH group tends to anchor this end of the chain. Despite having three protons to provide relaxation, the methyl carbon atoms have considerably longer T_1 values due to rotation of the methyl groups.

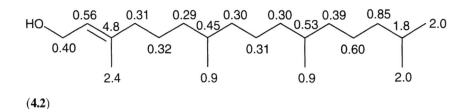

(4.2)

4.8 QUESTIONS

4.1. A $I = 1/2$ nucleus is relaxed by three different mechanisms and has a measured relaxation time of 1 s. The three contributions are spin rotation, whose characteristic time is 2.5 s, chemical shift anisotropy, with characteristic time 1.8 s and a long-range dipole–dipole interaction. Calculate the characteristic time of this latter contribution.

4.2. Given an isolated $^{13}C^1H_2$ fragment in which the C–H distance is 109 pm and the H–H distance is 177 pm, use equations (4.2) and (4.3) to calculate the ratio of the rates of dipolar relaxation of the two nuclear species. The magnetogyric ratios should be taken as proportional to the nuclear frequencies given in Chapter 1.

4.3. It is pointed out in Table 4.1 that the electric field gradient at ^{14}N in the nitrate ion, $[NO_3]^-$, is almost zero, and that this is surprising. Use equation (4.4) to obtain an expression for the EFG at ^{14}N given that there are three charges q producing the field gradient and that these are situated in the N–O bonds at a distance r from the nitrogen atom. It will be found most convenient to choose the z axis as that perpendicular to the NO_3 plane and passing through the nitrogen atom, and to calculate V_{zz}. How far do we have to displace the charges from the NO_3 plane keeping r constant in order that the EFG will be zero at N? Does it make any difference as to which side we make the displacement?

4.4. In metal carbonyls, T_1 of ^{13}CO is normally dominated by chemical shift anisotropy. Take $\sigma_{||} - \sigma_{\perp} = 400$ ppm and $\tau_c = 10^{-11}$ s. Calculate T_{1CSA} for ^{13}C in such a carbonyl at 4.7 and 9.4 T. Note that $\gamma_C = 6.7263 \times 10^7$ rad T^{-1} s^{-1}.

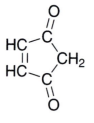

4.5. In the molecule, shown in the margin, use equation (4.2) to calculate the relative T_1 values of the CH and CH_2 protons, assuming that only dipole–dipole relaxation contributes to T_1 and the same τ_c applies to both types of protons. Take the HC=CH and CH_2 interproton distances as 270 pm and 185 pm respectively. Ignore any dipole–dipole relaxation arising between the HC=CH and CH_2 groups.

4.6. For a C–D bond, the ^2H nuclear quadrupole coupling constant, χ, is usually 180 ± 20 kHz. Given that $T_1(^2H)$ of CD_3I is 6 s, use equation (4.5) to calculate τ_c for the molecule. Assume that η is negligible.

4.7. The ^{13}C NMR signal of $C^{2,6}$ of pyridine, with the effect of the protons removed by double irradiation, is much broader than the signals due to $C^{3,5}$ and C^4. Explain why this occurs. On cooling, the signal sharpens. Explain why this happens.

4.8. The 25.2 MHz ^{13}C T_1 values for PhC≡CC≡CPh are given overleaf.

    (a) Account for why the T_1 of the *para*-carbons are much shorter than for the *ortho*- and *meta*-carbons.

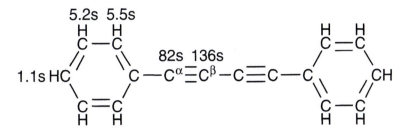

(b) When the ^{13}C T_1 values are determined at 63.1 MHz, values of 15 s and 30 s were found for C^α and C^β. Explain why the T_1 values have decreased by so much.

4.9. The ^{13}C T_1 values have been determined at 25.15 MHz for the *ipso* carbon in bromobenzene as 3.6 s for $C_6H_5{}^{79}Br$ and 16 s for $C_6H_5{}^{81}Br$. Explain why the two values are different. What is going to be the effect on these T_1 values of increasing the magnetic field so that the ^{13}C frequency becomes 50.3 MHz?

4.10. Although the relaxation of proton bearing ^{13}C nuclei is normally dominated by the dipole–dipole relaxation mechanism, no significant dipole–dipole relaxation has been found for ^{103}Rh in rhodium hydrides. Explain this observation.

4.11. In Fig. 4.11, it is shown that for $B_0 = 8.48$ T and $10^3/T = 2.85$, the ^{195}Pt relaxation rate in $[Pt(CN)_4]^{2-}$ is given by $\ln R_1 = 0.8$. Calculate the ^{195}Pt linewidth of this anion under these conditions assuming $T_1 = T_2$ and a perfectly homogeneous magnetic field. The measured value of the linewidth is, in fact, 90 Hz. Calculate T_2 and estimate the contribution of scalar relaxation to this value.

The spectrometer 5

We have seen in the last three chapters that there are three principal parameters in an NMR spectrum: the chemical shift, the coupling and, less evidently, the relaxation. The present chapter is devoted to showing in more detail how the spectrum is obtained and how these parameters may be derived from the data collected.

Chapter 1 discussed how a 90° $\boldsymbol{B}_1$ pulse at the nuclear frequency can swing the magnetization from the direction of the magnetic field into the xy plane, and showed that this magnetization, which is now rotating at the nuclear frequency, can be detected in a suitable coil. In Chapter 4 we discussed how this magnetization is affected by a variety of relaxation processes, in particular, how T_2 leads to an exponential reduction with time of the xy magnetization. The spectrometer output thus consists of a signal at the nuclear frequency, which decays with time until it is no longer detectable. We need to know how this can be translated into typical spectral form. The equations governing the behaviour of the transverse and longitudinal magnetization $\boldsymbol{M}_{xy}$ and $\boldsymbol{M}_z$ and their return to equilibrium following a 90° $\boldsymbol{B}_1$ pulse are

$$(\boldsymbol{M}_z)_t = (\boldsymbol{M}_z)_\infty \left[1 - \exp\left(-\frac{t}{T_1}\right) \right]$$

i.e. $\boldsymbol{M}_z$ increases from zero to its equilibrium value, and

$$(\boldsymbol{M}_{xy})_t = (\boldsymbol{M}_{xy})_0 \exp\left(-\frac{t}{T_2}\right)$$

i.e. the transverse magnetization falls from its maximum value (equal to $\boldsymbol{M}_z$) to zero after sufficient time has elapsed. This behaviour is summarized diagrammatically in Figs 4.1 and 4.2.

In general, the rate of decay of the spectrometer output is faster than predicted from the value of T_2 in a given system. This occurs because of inhomogeneities in the magnetic field throughout the sample due principally to imperfections in the magnet system, which means that the nuclear frequencies are slightly different in different parts of the sample volume and this increases the rate at which the spins throughout the sample lose phase and coherence. We write that the apparent relaxation time, T_2^*, is equal to

$$\frac{1}{T_2^*} = \frac{1}{T_2} + \frac{1}{T_{\text{inhomo}}}$$

where T_{inhomo} is the decay in intensity due to the field inhomogeneities alone. This decaying output is known as the free induction decay or FID. We now need to know how the FID is produced in practice, how it is collected and how it is processed to produce a spectrum. A block diagram of an NMR spectrometer is shown in Fig. 5.1 and should be referred to in the discussion which follows.

5.1 THE MAGNET AND FIELD HOMOGENEITY

A modern NMR spectrometer is built around a superconducting solenoid magnet which, indeed, is the heart of the system. The magnet consists of a coil of superconducting niobium–tin alloy sitting in a bath of liquid helium at 4.2 K. In order to reduce liquid helium loss by boiling off, the liquid helium is protected by a vacuum jacket, which

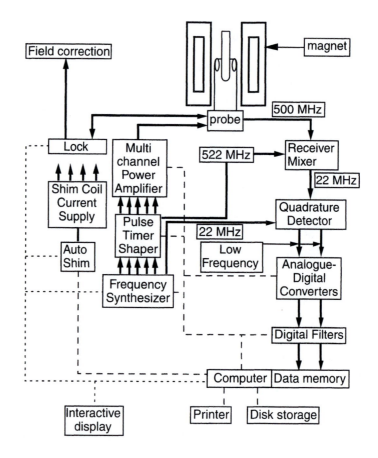

Figure 5.1 A block diagram of an NMR spectrometer.

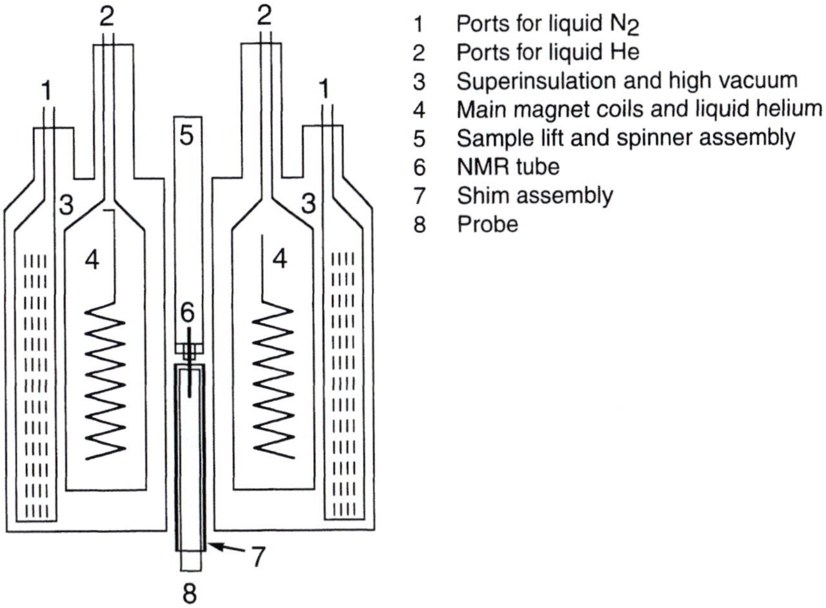

1	Ports for liquid N_2
2	Ports for liquid He
3	Superinsulation and high vacuum
4	Main magnet coils and liquid helium
5	Sample lift and spinner assembly
6	NMR tube
7	Shim assembly
8	Probe

Figure 5.2 A representative cross-section through a superconducting magnet. (Reproduced with permission from Braun *et al.* (1996) *100 and More Basic NMR Experiments*, VCH, Weinheim.)

itself is then cooled with liquid nitrogen, which in turn is also protected by the vacuum jacket (Fig. 5.2). Modern systems are very efficient with 25 l of liquid helium lasting for at least three months. The liquid nitrogen needs topping up regularly, typically once a week. Once the current has been established in the superconducting coil, it will continue to give an acceptable magnetic field for at least ten years, before the current has decayed sufficiently to need to be topped up again. The current in the superconducting coil is quantized, as is the magnetic field. If an NMR tube containing $CHCl_3$ is placed in the magnet, and a single pulse spectrum taken every hour, then it will be found that the position of the signal remains constant for a while, and then jumps one or two hertz to lower frequency, and again remains constant for a while. This experiment monitors the quantized jumps of the magnetic field, which of course have to be counteracted to give a permanently stable field.

Although there are no reasons to believe that a static magnetic field can cause harm to personnel, it is usual to mark a 5 gauss (0.0005 T) contour line around the magnet, and to try to stay outside this region as far as is possible. The major recognized hazard associated with the stray magnetic field is its ability to capture large metallic objects such as tools and gas cylinders which accelerate rather rapidly towards the magnet! Nor should those with heart pacemakers go anywhere near. Anything, including floppy discs and credit cards, containing magnetic records can be damaged by the stray magnetic field.

The sample sits in the middle of the solenoid. It is vital to obtain a uniform magnetic field over the sample. A modern superconducting magnet is capable of producing a magnetic field which gives a sample linewidth of better than 0.1 Hz in, say, 400 MHz. The magnetic field then has to be uniform to better than 1 part in 4×10^9. This is achieved by attention to three techniques:

1. shimming;
2. sample spinning;
3. the use of very uniform samples.

5.1.1 Sample spinning

For high resolution NMR spectroscopy, the solution of the sample is placed in an NMR tube, frequently of 5 mm outside diameter. In order to partially average inhomogeneity at right angles to the main magnetic field direction, the sample is rotated around its axis, typically at 20 Hz, using an air turbine which consists of a cylindrical spinner in which the sample tube fits and a static outer fixed in the centre of the solenoid and to which the spinning air is fed. The spinner/tube combination is removed from, or put into, the outer by means of an air lift, which gently manipulates the spinner up or down.

5.1.2 Shimming

Two sets of coils are wound to produce magnetic field gradients to cancel those inherent in the main magnetic field. The first set consists of superconducting coils, which are wound within the liquid helium bath and are adjusted as part of the magnet installation. The currents in these coils are not subsequently changed by the operator, but the operator has to be aware of their existence as they can quench, i.e. lose their current, and the resolution deteriorate. The second set of shim coils is mounted around the probe and the currents in them are adjusted to cancel any remaining field gradients and give optimum resolution. This second set of coils comprises:

1. Zero-order coil. There is a single zero-order coil with its axis along the direction of the superconducting solenoid and this is used to adjust the main magnetic field strength within narrow limits.
2. First-order coils. There are three first-order coils called $X1$, $Y1$, and $Z1$, which produce magnetic fields as the functions x, y, and z. They produce field gradients shaped like p-orbitals. The x and y coils are aligned at right angles to the main magnetic field, while the z coil is coaxial with the main field.
3. Second-order coils. There are five second-order coils called XY, $X^2 - Y^2$, ZX. ZY, and Z^2, which produce magnetic fields as the functions xy, $x^2 - y^2$, zx. zy, and z^2. They produce field gradients shaped like d-orbitals.

4. Third-order coils. There are seven third-order coils called X^3, Y^3, Z^2X. Z^2Y, ZXY, $Z(X^2 - Y^2)$, and Z^3, which produce magnetic fields as the functions x^3, y^3, z^2x. z^2y, zxy, $z(x^2 - y^2)$, and z^3. They produce field gradients shaped like f-orbitals.
5. The magnets designed for the highest magnetic fields are often also equipped with fourth- and fifth-order z-shims.

It has to be emphasized that the quality of the spectra depends absolutely on minimizing the field gradients with the shims. Adjusting all the shims, however, is a highly skilled operation and beyond the scope of this book, but some excellent texts are available, and listed in the Bibliography. The inexperienced user is best advised to restrict changes in the shim currents to those in the Z and Z^2 coils when the sample is spinning. The shim currents are adjusted by optimizing the height of the lock signal for maximum, see section 5.3. This may need to be accompanied by adjustment of the lock phase for maximum height. If the X and/or Y shim currents are mis-set, then the magnetic field at a given point in the sample fluctuates as the sample spins, the nuclear frequencies are modulated at the spinning frequency and the NMR signals will be flanked by spinning side-bands (Fig. 5.3). These shim currents then have to be adjusted with a non-spinning sample. Shimming is critically important in obtaining a high-quality NMR spectrum. A good quality control of the shim settings is to routinely examine the signal from the TMS reference. It should be a singlet with well-resolved ^{29}Si satellites, $^2J(^{29}Si^1H) = 6.6$ Hz. It is also useful to watch the FID as it accumulates. It is often possible to identify a sample that is giving poor resolution after only one FID has been acquired. For ^1H, the FID should last for several seconds, and the presence of a beat after one or two seconds while it may be due to coupling, often indicates poor resolution.

The position of the bottom of the NMR tube and the top of the solution in the tube distort the magnetic field and so contribute to the field gradient. Provided a gauge is used to position the spinner on the NMR tube, so that the bottom of the NMR tube is always in the same position, and a relatively long sample length is used, then, on change of sample, it should only be necessary to adjust the shim currents in the Z and Z^2 coils when the sample is spinning.

Manual shimming is becoming replaced by computer shimming. For many years, it has been possible to use iterative routines to adjust the shims automatically, but these routines are slow, and are normally only used to compensate for drift during long data accumulations. It is becoming common for probes to be equipped with a z-magnetic field gradient coil, see section 5.8.6. This coil can be used to map magnetic field inhomogeneity in the z-direction, and the Z, Z^2, Z^3, Z^4 and Z^5 corrections are then calculated and applied by the computer. This only takes a few minutes and uses the same methods as will be described in Chapter 11 for magnetic resonance imaging. Less common is for probes to be equipped with x-, y- and z-magnetic field gradient coils,

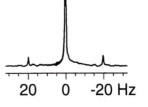

20   0   -20 Hz

Figure 5.3 A ^1H NMR spectrum of $CHCl_3$ showing spinning side bands due to mis-setting of the current in the X shim coil. The sample tube was spun at 20 Hz and the spinning side bands are symmetrically situated at ± 20 Hz of the main signal. (Reproduced with permission from Braun *et al.* (1996) *100 and More Basic NMR Experiments*, VCH, Weinheim.)

where the complete magnetic field inhomogeneity can be mapped and corrected, but this is a lengthy procedure.

5.1.3 Sample preparation for high resolution NMR spectroscopy

Sample preparation is extremely important to obtain well-resolved spectra. The sample must be dissolved, usually in a deuterated solvent. The resulting solution must be homogeneous and free from particles and paramagnetic impurities. This is usually achieved by making up the solution in a sample tube or flask and then filtering the solution or passing it through a short alumina plug in a Pasteur pipette into the NMR tube. If it is then necessary to add more solvent to the tube, the resulting solution must be well mixed to avoid concentration gradients which imply changes in magnetic susceptibility and which destroy the resolution. In order to obtain good quality NMR spectra, the NMR tube must be accurately cylindrical and of uniform wall thickness to avoid distortion of the field by irregularities in the glass, which is of different magnetic susceptibility to the sample. Thus, good quality NMR tubes must be used, with the specification becoming tighter for increasing magnetic field strength.

5.2 THE PROBE

A cross-section through a typical multinuclear NMR probehead is shown in Fig. 5.4. There are two coils. In a conventional probe, the inner coil provides X-nucleus observation, usually covering the range of frequencies from ^{109}Ag to ^{31}P. This coil is double-tuned so that it is also used for the ^{2}H lock channel, see section 5.3. The outer coil is used for ^{1}H observation and decoupling. The sensitivity of such a probe is optimized for the X-nucleus, and the 90° pulse for the X-nucleus will be short, but the ^{1}H 90° pulse will be long. An alternative configuration is found in the inverse probe, where ^{1}H and the ^{2}H lock are on the inner coil and the X-nucleus is on the outer coil. This type of probe is ideal for ^{1}H sensitivity and inverse detection of X using techniques such as HMQC, HMBC, and HSQC, see Chapter 9. There are many variations on this design.

There are several glass tubes present which support the coils and direct the variable temperature gas flow. Immediately below the sample is a thermocouple. Around the whole assembly is a Dewar that stabilizes the temperature. Warmed or cooled gases can be passed through the probe in order to vary the temperature of the sample, which is held constant at a given value by a control circuit based on the thermocouple. Temperature stability is very important since temperature gradients in the sample cause changes in magnetic susceptibility and chemical shifts which can degrade the resolution. If the spectrum is temperature sensitive (see Chapter 6) then temperature gradients can also broaden the lines. The presence of glass tubes in

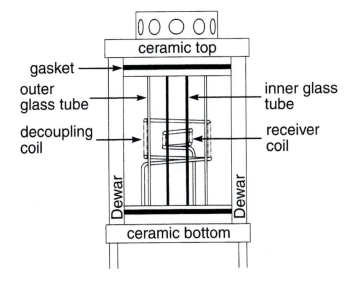

Figure 5.4 A cross-section through a typical NMR probehead. Note all the glass in the probehead which makes it fragile. The coils are wound as saddles on the glass tubes (Helmholtz double coils) to give the RF field normal to B_0. (Reproduced with permission of Bruker Spectrospin.)

the probehead means that it is fragile and failure to insert the NMR tube on a cushion of air can result in breakage.

5.3 FIELD-FREQUENCY LOCK

These devices, which are relatively simple, are used in conjunction with FT spectrometers to provide a final stage of stabilization using a nuclear resonance in the sample other than the one to be observed, often the strong deuterium signal from a deuteriated solvent. They are based on a time sharing principle with transmitter and receiver on alternately (Fig. 5.5). The response which is detected during receiver on-periods is strong enough to give an analogue output. In fact, the modulation of the transmitter gives sidebands at the pulse modulation frequency which are continuous and which provoke slight precession of the magnetization of the lock nucleus which therefore gives a continuous output which can be detected as either absorption or dispersion signal, see section 5.6. The device is used in two ways. A small triangular repetitive sweep is applied temporarily to the magnetic field and the deuterium is observed on the monitor in the absorption mode as its resonance is traversed repeatedly. We observe a signal, and can use this, maximizing its height with shim adjustment, in order to optimize the homogeneity of the magnetic field. We then switch off the field sweep and turn our attention to the dispersion signal produced by the solvent. This has the property that, at resonance, the output is zero,

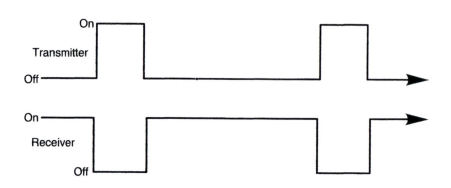

Figure 5.5 Receiver/transmitter time sharing of the field-frequency lock. The receiver is switched off just before the transmitter is turned on, and remains off for a while after the transmitter is turned off.

but if drift of field or frequency occurs, either a negative or a positive output is obtained, the sign depending upon the direction of drift, and we can use this output to provide a correction voltage to alter the magnetic field until the output is again zero (Fig. 5.6). Thus the frequency and field are locked together indefinitely and we can in principle record as many FIDs as we wish and collect them in memory, 100 000 being quite feasible, though the number is always kept as small as possible since there is very often a queue of people waiting to use the spectrometer. It is, of course, necessary that the frequency of the lock device is derived from the FT drive frequency, otherwise it could vary independently. The lock then keeps the ratio B_0/ν_0 constant and is known as a field-frequency lock. The whole system is shown in the block diagram of Fig. 5.7.

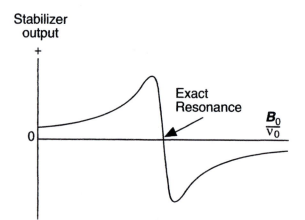

Figure 5.6 A resonance in the dispersion mode can be used to stabilize an NMR spectrometer. At exact resonance, the output is zero, but if drift occurs, it becomes either positive or negative. The output can be used to provide a correction to the magnetic field, which reverses the drift until the exact resonance position is regained.

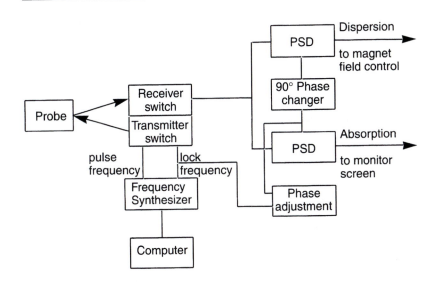

Figure 5.7 A block diagram of the field-frequency lock system. The frequency synthesizer provides both the pulsing frequency and the lock frequency. The pulsing produces sidebands of the lock frequency and the lock operates on one of these. The continuous output of the nuclear lock signal is fed to two phase sensitive detectors (PSD, to be described in section 5.5.1) operating in quadrature which produce the dispersion and absorption signals to correct the magnetic field and enable the signal to be monitored. Means are provided to adjust the phase to ensure that the spectrometer is operating at its optimum position.

It is possible to use a number of deuteriated solvents for the lock, such as $CDCl_3$, C_6D_6, $(CD_3)_2CO$, etc., and these all have different chemical shifts and so lock frequencies. It is arranged that each lock substance has its own frequency stored in memory and that the magnetic field shall be identical whatever lock substance is used. This means that all standards are referenced to an absolute value.

5.4 THE TRANSMITTER

The transmitter provides the pulses to stimulate the FID and also to manipulate the nuclear spins in a variety of ways. The centre of the transmitter system is a quartz controlled frequency synthesizer which generates the frequencies needed for a given experiment and also that required for the lock system described above. The frequencies produced are locked together so that they always have the same relationship and are continuous and coherent, which is to say that they are pure, continuous sine waves with no irregularities. Some of the outputs are also used in the receiver, as we shall see. The transmitter frequencies are used to drive the power amplifiers which produce the pulses but have first to be processed. Pulses are produced by a switch which applies the signal to the amplifier for the required short time. Such pulses may be rectangular or they may be shaped by a computer controlled device to give pulses with either a broad bandwidth or a narrow bandwidth. These pulses are effectively bits of the input continuous wave. Means are also provided to shift the pulse oscillations relative to the drive signal, in other words to phase shift them. The processed signals are then amplified in the power amplifiers, one for each frequency, and provision is made for the output to be varied in intensity, the degree of attenuation usually being expressed in a dB (decibel) scale, the more dB, the less the intensity. The decibel scale is a log scale, being related to power, P, by

$$dB = 10\log_{10}\left(\frac{P_1}{P_2}\right) = 20\log_{10}\left(\frac{V_1}{V_2}\right)$$

where P_1 and P_2 are the powers and V_1 and V_2 are the applied voltages being compared. Hence increasing the dB value by 3 units approximately halves the power being used. However, as the radiofrequency field strength is proportional to voltage, to halve it, an approximate increase in the dB value of 6 units is required.

It is particularly important to be able to control the relative phases of the output signals. This can be understood by referring to Fig. 5.8. The pulse gives the field $\boldsymbol{B}_1$ in the xy plane, rotating with the nuclei on, say, the x' axis in the rotating plane and the nuclear magnetization rotates around $\boldsymbol{B}_1$ into the y' axis. If we use a 90° pulse then this is called a $(90°)_x$ pulse. If we phase shift the transmitter output by 90° (do not confuse pulse lengths with phase shifts even though both are expressed in the same units), this moves $\boldsymbol{B}_1$ 90° in the $x'y'$ plane to the y' axis and the nuclear magnetization now rotates into the $-x'$ axis. This is known as a $(90°)_y$ pulse. It is equally possible to place the magnetization along the x' or $-y'$ axes using a $(90°)_{-y}$ or $(90°)_{-x}$ phase shifted pulse. We shall find this facility extremely useful below.

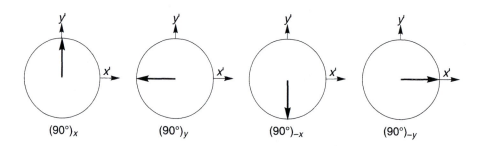

$(90°)_x \qquad\qquad (90°)_y \qquad\qquad (90°)_{-x} \qquad\qquad (90°)_{-y}$

Figure 5.8 The effect of the phase of the 90° pulse on the magnetization with respect to the $x'y'$ rotating frame. The direction of the arrow within the circle represents the direction of the magnetization in the $x'y'$ plane after applying the 90° pulse about the x, y, $-x$, or $-y$ axes as indicated by the subscripts after the bracket.

5.5 THE DETECTION SYSTEM

For further processing the FID needs to be detected and stored in a computer memory. The nuclear frequency is high, tens or hundreds of megahertz (MHz), and so is difficult to handle with normal data storage devices. For this reason it is necessary to reduce its frequency, and this is done in two stages. First, the superheterodyne principle is used and the nuclear signal is mixed with another a little different in frequency and locked to the spectrometer frequency, to give the much lower difference frequency which is called the intermediate frequency or IF. Then this is detected using a device called a phase-sensitive detector, in which the modified nuclear signal is compared with the frequency source used to generate B_1, also modified by mixing to reduce its frequency. The frequency of the nuclear signal thus is reduced to the difference between the two, which is zero to a few kilohertz. The intensity of each signal present is preserved and, in addition, the phase of the signal is determined relative to the phase of B_1, the reference signal. Knowledge of all three factors is essential in order that a spectrum can be extracted from the data, and knowledge of phase is particularly useful in some of the more complex experiments that we will encounter. It is useful, perhaps essential, to understand what the phase-sensitive detector does.

5.5.1 The phase-sensitive detector

This device is essentially a switch driven by the B_1 signal and which reverses its polarity in step with each half cycle of B_1. A simplified circuit is shown in Fig. 5.9. The input nuclear signal is thus directed in one direction during the positive B_1 half-cycle and in the opposite sense during the other half-cycle. If the B_1 and nuclear frequencies

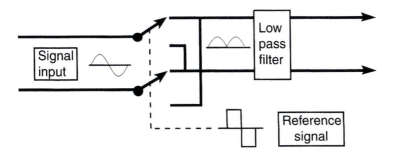

Figure 5.9 Schematic diagram of a phase-sensitive detector. The reference signal operates the switches, one way for the positive half cycle and the other way for the negative part of the wave. This regularly reverses the output polarity relative to the input polarity. The input sine wave is thus rectified and the fast fluctuations in voltage removed by the filter to give a smoothed, constant or slowly varying output. The polarity of the output depends on the relative phases of signal and reference and is zero if these are shifted by 90°. The switching is done digitally using fast gallium arsenide transistors.

are exactly the same, then the nuclear signal is rectified with, say, the negative half cycles reversed in polarity, and a direct current output is obtained after removal of the high frequency components by a suitable low pass filter. If the two signals are in phase, then the output is positive; if they are 180° out of phase, the output is negative; and if they are of intermediate phase, then the output is reduced and is zero when the phase difference is 90°. The intensity of the output thus depends on both the intensity and phase of the nuclear signal input. If the nuclear and B_1 frequencies are not the same, then their relative phase fluctuates with time and the output becomes a wave oscillating at the difference frequency. The phase information is still retained since the output will have some phase angle relative to B_1 immediately after the 90° pulse, i.e. at the start of the decaying output. The spectrometer output after phase-sensitive detection is thus oscillatory, referred to B_1 and with a much reduced frequency. It decays with time characterized by T_2^* (Fig. 5.10). We should also note that this result is obtained if the spectral frequencies are all higher than the B_1 frequency or all lower, though the order of the spectral lines is reversed between the two cases. Evidently, we cannot have some lower and some higher as this would mix the resonance positions up so as to prevent sensible analysis of the spectrum.

5.5.2 Analogue filters

The output of the phase-sensitive detector consists of wanted, low frequency signals, electronic noise of all frequencies and the higher

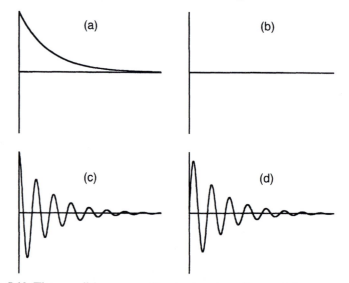

Figure 5.10 The possible range of outputs when the nuclei are exactly on resonance with B_1 and are in-phase (a) or 90° out-of-phase (b). If the nuclear and B_1 frequencies are different, then the output is at the difference frequency, but starts on a maximum if they start in phase (c) or a minimum if they start 90° out-of-phase. Any intermediate phase is possible.

frequency IF inputs. To separate the signal we want from the rest, we use an analogue filter, based on a simple resistance–capacitor network. Such filters are designed to pass the low frequencies and attenuate the high frequencies. Thus there is no attenuation up to what is called the cut off frequency and then the attenuation increases steadily with frequency. Such filters remove the IF satisfactorily but leave much noise at frequencies above the pass band and so reduce the signal to noise ratio in the eventual spectrum, thus reducing our ability to detect weak signals. Unwanted NMR signals outside the pass band also get through, and as we shall see this gives rise to problems. Unfortunately, there is a stability limit to the sharpness of cut off of an analogue filter and we can do no more than remove IF and some of the noise intensity outside the pass band. There are, fortunately, ways around these problems but before we describe them it is necessary to discuss the needs of the computer and how this handles the input and what restraints it introduces.

5.5.3 Collection of data in the time domain

The output of the detector is a decaying wave which is entirely a function of time, and is said to exist in the time domain. In order to carry out numerical processing of the output of an NMR spectrometer, we need to convert the analogue electrical output into digital information using an analogue–digital converter. This is characterized by the maximum number of bits by which it can represent a voltage. Typically this is 16 bits (though this can be varied at will) and the system gain is arranged to avoid overflow of the converter output with the maximum voltage present at the start of the FID. Sixteen bits gives an enormous dynamic range for the input so that weak signals are also well digitized. The signal is sampled at regular intervals, the voltage registered converted into a binary number, and this number then stored sequentially in one of a series of computer memory locations, one location being a word of 32 bits (Fig. 5.11). The magnitude of the sampling interval is very important, and is often known as the dwell time, DW. The maximum frequency or minimum period that sampled information stored in a computer memory can represent is limited to that where alternate numbers are positive and negative, assuming a cosine wave input varying symmetrically about zero. The period of this wave is then $2(DW)$ and its frequency is $1/2(DW) = F$. A lower frequency, $F - f$, or a higher frequency, $F + f$, give the same pattern in memory, so that there is an ambiguity in the digital representation of the signals. This is avoided by ensuring that all the nuclear frequencies lie in the range 0 to F Hz for single phase detection or $\pm F$ Hz for quadrature detection. This is often known as the sweep width or as the spectral width. The frequency F is known generally as the Nyquist frequency and the low pass filter should cut off above this frequency.

Given that the spectrometer is highly stable, a series of isolated 90° pulses will each give an almost identical nuclear response, so that these can be added together in the computer memory to give a much

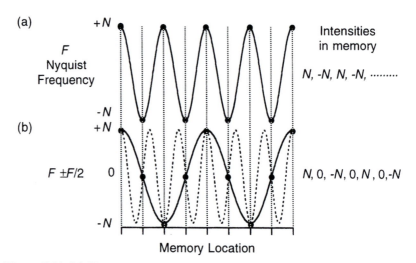

Figure 5.11 (a) How a waveform at the Nyquist frequency, F, gives alternate positive and negative values of the number N corresponding to peak voltage. This assumes that the wave and the computer memory sweep are in a certain timing relationship. (b) A waveform of lower frequency, $F - F/2$, gives the same pattern as a waveform of frequency higher than the Nyquist frequency by the same amount, $F + F/2$.

stronger total response. Each signal is, however, accompanied by unwanted random noise; indeed, weak signals may not be visible because of the noise. While the noise intensity also increases as more responses are added, the noise is incoherent, i.e. its intensity at a given memory location is sometimes + and sometimes –, and so it adds up relatively slowly. A series of N FIDs when added together in this way have a signal-to-noise ratio that is $\sqrt{N}$ times better than that of a single FID. It is this feature that renders Fourier transform spectroscopy so useful for the less receptive nuclei, allowing a FID to emerge from an apparently hopeless morass of noise (Fig. 5.12).

We should also remember that the FID appears immediately following the transmitter pulse. In practice, the pulse breaks through with the FID and distorts the first point, which gives considerable distortion of the final spectrum. This is avoided by leaving a short delay between the end of the pulse and the start of data collection so that the pulse break-through has time to die out. This has to be allowed for in subsequent treatment of the spectrum. It is a great disadvantage when the FID is very short as important initial intensity is lost and in this case special electronic circuitry has to be used to eliminate break-through.

5.5.3.1 Folding

There are two sorts of folding, or aliasing, of signals and noise. One is due to the ambiguity described above where the computer cannot distinguish between signals higher or lower than the Nyquist frequency. This means that the spectrometer has to be set up with all nuclear

1 acquisition

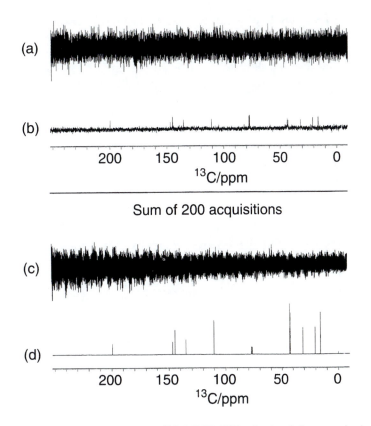

Sum of 200 acquisitions

Figure 5.12 (a) An 100.62 MHz ^{13}C NMR FID obtained from a single pulse applied to carvone in CDCl$_3$. (b) The resulting Fourier transform. (c) The resulting FID after adding together 200 FIDs. (d) The resulting Fourier transform of the sum of 200 FIDs.

signals appearing between 0 and the Nyquist limit, otherwise signals above this limit appear out of their proper place. The high frequency noise, of course, is always folded in and is only partially limited by the analogue filters. The other type of folding occurs with the simple phase-sensitive detector described above which cannot discriminate between signals higher or lower in frequency than the reference frequency. All nuclear signals have to be placed on one side or the other of reference but noise is always collected from both sides, the effective noise bandwidth being twice that of the filter. Fortunately, both these types of folding can be avoided in a modern spectrometer.

5.5.3.2 Quadrature detection

Signals higher or lower in frequency than the B_1 frequency fold into the same spectral space with the phase-sensitive detector described above. There is, however, a difference between them even if they are equally separated from B_1 in that the higher frequency component reaches its first null beat with B_1 earlier than the lower frequency signal. In other words, they are phase shifted and it is possible to discriminate between them using two phase-sensitive detectors (Fig. 5.13). This quadrature detector as it is called, then gives two outputs with phase in quadrature which are collected separately in memory and then Fourier transformed together so that all the signals appear in their correct place relative to B_1 and without phase anomalies. It will be clear from Fig. 5.13 that this eliminates one kind of folding and also allows us to place B_1 in the centre of the spectral range. The maximum frequency which we need now is only half that needed for a single phase-sensitive detector for the same spectral range. This means that the cut off frequency of the low pass filter can be halved which in turn means that less noise is collected, halving the bandwidth giving an increase in signal to noise ratio of $\sqrt{2}$ for an equal number of FIDs. This is the normal method of operation.

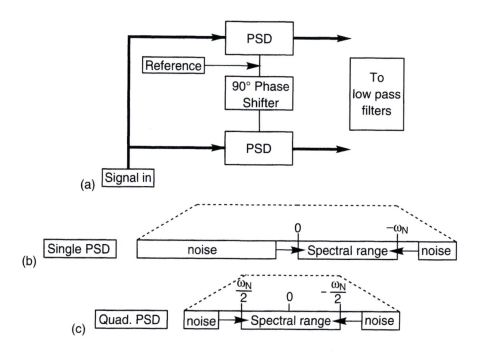

Figure 5.13 A quadrature detection system. The two boxes marked PSD are phase-sensitive detectors of the type described in Fig. 5.9. The same reference is applied to both but one is made to be 90° out of phase with the other. The two low frequency outputs are filtered and then converted to numbers and fed sequentially to the computer memory. The lower part of the figure demonstrates the difference between operating with a single or quadrature phase-sensitive detector. In the first case the spectrum range is 0 to the Nyquist frequency $-\omega_N$. Noise folds in from both ends of the spectrum but is particularly troublesome on the low frequency side because the low pass filter allows noise through without attenuation from frequencies of 0 to $+\omega_N$ Hz. With the quadrature system the filter can cut off at half the frequency and both the high and low frequency noise is attenuated. The dotted lines represent the fraction of signal with a given frequency which pass through the filter into the receiver.

A problem that occurs with the quadrature detector is that of quadrature images in which an intense signal to one side of the reference frequency is reflected through to appear, albeit attenuated, on the other side of reference. These artefacts are eliminated by phase cycling, a process which involves changing the receiver phases and switching the two outputs in concert between the two different computer locations where the two FIDs are collected. Different phase cycling sequences are used for different NMR experiments and it is useful to know what is going on during a spectral accumulation even if it is all set up by the computer. A commonly used sequence is called CYCLOPS.

We should perhaps note at this point that NMR spectroscopists put considerable effort into creating these acronyms which help to give the subject a human face. We will meet many such as we delve deeper!

5.5.3.3 Digital filtering

This still leaves the problem of folding around the Nyquist frequency because of the slow cut off of the analogue filters. This situation is improved by adding devices called digital filters which can be designed to have nearly infinitely sharp cut off. The first step necessary to achieving this is to digitize the detector outputs not once every dwell time but many times for each dwell period. This does not affect the Nyquist frequency since this is determined by the intervals at which the computer accepts data. If there are twenty points produced for each dwell period then these are averaged using an appropriate weighting and a single data point produced which forms part of the FID to be processed. This operation can be carried out either in the computer memory, i.e. all points collected then averaged or in a separate digital averaging device before being transmitted to the computer. This technique first of all cuts out the folded noise but also the averaging process reduces the remaining noise further, the gain in signal to noise ratio being $\sqrt{}$ (samples per dwell period). A similar gain in dynamic ratio is also realized.

These digital filters are indeed so effective that they can be used to select only a part of a spectrum and study it in isolation without any folding from the rest of the non-observed resonances present. This multiple sampling technique is known as over sampling.

Modern spectrometers use digital technology to the maximum, both for the convenience of interaction with the computer and for the increased stability of digital circuits which are much less disturbed by drift in component values than are the equvalent analogue circuits. Thus throughout, analogue signals are converted to digital signals by an analogue to digital converter at the earliest possible moment, are processed and then converted back to analogue form.

5.5.3.4 Phases of the pulses and receiver

The radiofrequency used in an NMR experiment is coherent. That means that the sine wave oscillation of the electromagnetic radiation continues without any discontinuity for ever. This is the basis of many NMR experiments. This means that it is possible to phase shift the frequency. In other words, shift the sine waves by any angle, frequently 90°, 180°, or 270°. In terms of the rotating frame, the maximum can then be at y', $-x'$, or $-y'$ instead of x, see Fig. 5.8.

It is not only the transmitter that has variable phases. As a direct consequence of quadrature detection, the receiver has phase encoding also. The quadrature detector produces two outputs, 1 and 2. These outputs are stored at two locations in the computer, A and B. The way that the outputs are transferred to the computer provides a mechanism for receiver phase cycling. This is summarized in Table 5.1. For example a y receiver phase means that output 2 is subtracted from location A and output 1 is added to location B.

Extensive use is made of phase changing in producing the required changes in magnetization. We will meet many examples of this in Chapters 7 and 8. Phase is also cycled, i.e. changed systematically over a group of 4, 8, 16, or 32 spectra in order to minimize artefacts. Even the routine observation of an NMR spectrum involves the use of the CYCLOPS phase cycle to remove images due to errors in the quadrature detection by averaging the response from imperfectly balanced receivers.

The CYCLOPS phase cycle is given in Table 5.2. The phase of the transmitter pulse is cycled as is that of the receiver. The result is to considerably reduce the size of the quadrature image.

5.5.4 The analogue–digital converter and dynamic range

Sixteen bit analogue to digital converters or digitizers are typical but they can be bigger or smaller. The speed at which they operate is inversely proportional to the number of bits and a 16 bit device can cope with a digitization rate of some 200 kHz. A sixteen bit digitizer

Table 5.1 The switching of the quadrature receiver outputs, 1 and 2, between the two computer memory locations A and B

Receiver phase	Location A	Location B
x	+1	+2
y	−2	+1
$-x$	−1	−2
$-y$	+2	−1

Table 5.2 The CYCLOPS phase cycle for reducing quadrature images.

Acquisition	Pulse phase	Receiver phase
1	x	x
2	y	y
3	$-x$	$-x$
4	$-y$	$-y$

produces a number within the range $\pm(2^{15}-1)$ or $\pm 65\,535$ with the sixteenth bit being used to indicate the sign. This gives a very large dynamic range, permitting weak signals to be detected in the presence of very strong signals. However, this can be compromised by incompetent spectrometer operation. The receiver gain has to be set to use most of this digitization range, but not so high that the signal ever lies even slightly outside the range. For example, an analogue signal of strength $+65\,536$ turns on the sign bit as well and produces a number which the computer reads as $-65\,535$. The resulting distortion of the FID produces a wave in the baseline in mild cases, and extra signals in more extreme cases (Fig. 5.14).

5.6 PRODUCTION OF THE SPECTRUM

5.6.1 Exponential relaxation and line shape

The T_2 relaxation process means that the nuclear frequency is not precisely defined and that the spins can be thought of as having a frequency distribution around their resonance frequency ω_0. A plot of this frequency distribution against frequency is the line shape $f(\omega)$ (Fig. 5.15). Individual nuclei have quite random frequencies within this range, but over a short enough time interval we can regard them as having this quite regular behaviour.

If we have an assembly of N nuclei, we can imagine that a small proportion n_k will have a particular angular velocity, ω_k. Each n is given by the line shape function

$$n_k \propto f(\omega_k - \omega_0)$$

which we can normalize by writing

$$n_k = Nf(\omega_k - \omega_0)$$

i.e. we have chosen n_0 so that $\int_k f(\omega_k - \omega_0) = 1$.

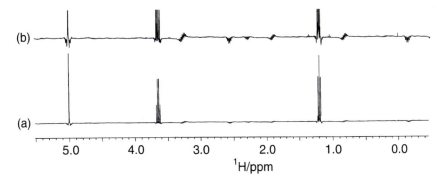

Figure 5.14 A 400 MHz ^{1}H NMR spectrum of ethanol in CDCl$_3$ recorded with the receiver gain set too high. (a) Normal gain. (b) Increased amplitude to show artefacts.

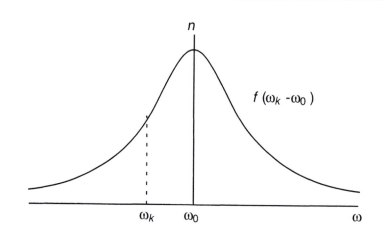

Figure 5.15 A plot of the number of nuclei with angular frequency ω_k different from the resonant frequency ω_0 for a small time interval t. The function $n_k = f(\omega_k - \omega_0)$ is the line shape function.

At a small time t after the $\boldsymbol{B}_1$ pulse, those nuclei with angular velocity ω_k will have moved an angle θ_k from those nuclei exactly at resonance, ω_0 (Fig. 5.16). We can resolve the magnetic moments of the n_k nuclei along the ω_0 direction and at right angles to it, and, following standard alternating current theory, we shall differentiate the normal component by the prefix i, where i = $\sqrt{-1}$. We have that

$$\theta_k = (\omega_k - \omega_0)t$$

and the two components are

$$n_k\cos[(\omega_k - \omega_0)t] \text{ and } in_k\sin[(\omega_k - \omega_0)t]$$

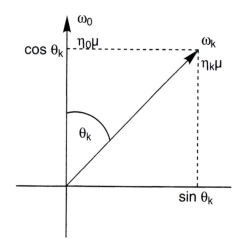

Figure 5.16 How the total magnetic moment of the group of nuclei n_k with angular velocity ω_k is displaced an angle θ_k from the on-resonance nuclei time t after a $\boldsymbol{B}_1$ pulse has produced magnetization in the xy plane.

The next step is to replace the n_k by the line shape function. However, Fig. 5.10 indicates that these may not be the same for the two components, and so we shall denote them as two separate functions by the letters v and u. The total intensity of the two components is then proportional to

$$Nv(\omega_k - \omega_0)\cos[(\omega_k - \omega_0)t] \text{ and } iNu(\omega_k - \omega_0)\sin[(\omega_k - \omega_0)t]$$

Summing over all possible values of k gives us the total intensity of each component. We normally would wish to compare this with the initial intensity at $t = 0$. This is proportional to the sum of all the nuclei, N. Thus N cancels from the formulae. The intensity of the normal component is, of course, only significant if $\omega_0 \neq \omega_{B_1}$. If N is large we can use the integral form of the equations, which, remembering that ω is the variable and t is constant, gives

$$\int_0^\infty v(\omega_k - \omega_0)\cos[(\omega_k - \omega_0)t]d\omega = e^{-t/T_2} \tag{5.1}$$

and

$$\int_0^\infty u(\omega_k - \omega_0)\sin[(\omega_k - \omega_0)t]d\omega = e^{-t/T_2} \tag{5.2}$$

It remains to find the form of the functions v and u. Inspection of tables of definite integrals will show that these are

$$u = \frac{(\omega_k - \omega_0)T_2^2}{1 + T_2^2(\omega_k - \omega_0)^2} \tag{5.3}$$

$$v = \frac{T_2}{1 + T_2^2(\omega_k - \omega_0)^2} \tag{5.4}$$

These two functions are plotted in Fig. 5.17; u is the dispersion mode and v is the absorption mode of the Lorentzian line shape. The absorption mode has its maximum at ω_0 and its intensity is proportional to the number of nuclei producing the signal. The dispersion mode is of zero intensity at resonance and of different sign above and below resonance. Spectra are displayed in the absorption mode, though, as we have already seen, the dispersion mode has an important use in spectrometer locking systems.

The problem remains of how to obtain the plots of Fig. 5.17 from the actual spectrometer output of Fig. 5.10. This latter is known as the free induction decay (FID) and is stored digitally in computer memory where it can easily be mathematically processed. Time and frequency domains are related through the Fourier relationship

$$F(\omega) = \int_{-\infty}^\infty f(t)e^{-i\omega t}dt$$

This can be written as

$$F(\omega) = \int_{-\infty}^\infty f(t)[\cos(\omega t) - i\sin(\omega t)]dt \tag{5.5}$$

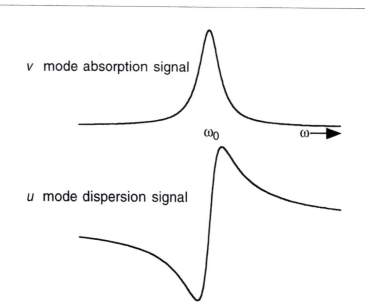

Figure 5.17 The absorption and dispersion mode signals available at the output of the phase-sensitive detector

The Fourier transform of the output data thus contains two components, often called real (R) and imaginary (I), which correspond to the u and v components. The inverse transform is also possible, i.e.

$$f(t) = \frac{1}{2\pi} \int_{-\infty}^{\infty} F(\omega) e^{-i\omega t} dt$$

and this should be compared with the relations (5.1) and (5.2) derived above.

5.6.2 Time and frequency domains

The Fourier transform (FT) is a mathematical relationship that relates a function of time to one of frequency, i.e. the time and frequency domains. The output of an NMR spectrometer is a sinusoidal wave that decays with time, varies entirely as a function of time and so exists in the time domain. Its initial intensity is proportional to M_z and so to the number of nuclei giving rise to the signal. Its frequency is a measure of its chemical shift, and its rate of decay is related to T_2 and the quality of the magnetic field. Fourier transformation of this FID gives a function whose intensity varies as a function of frequency and is said to exist in the frequency domain. The parameters of the absorption curve contain all those of the FID: the position reproduces the frequency and so the chemical shift. Equation (5.4) can be used to predict the linewidth, which we define as the width at half-intensity. The half-intensity points occur at frequency ω above and below resonance ω_0, in other words when

$$1 + T_2^2(\omega_0 - \omega)^2 = 2$$

i.e. when $T_2^2(\omega_0 - \omega)^2 = 1$

or

$$\omega_0 - \omega = \pm \frac{1}{T_2}$$

The separation of the half-intensity points is then twice this in radians. It is, however, more usual to express linewidths in hertz (frequency $= \omega/2\pi$), giving

$$\nu_{1/2} = \frac{1}{\pi T_2} \tag{5.6}$$

where $\nu_{1/2}$ is the frequency separation of the half-height points. The linewidth thus normally gives us $1/\pi T_2^*$. The intensity of the absorption curve depends upon linewidth and, in fact, the product of linewidth and peak height is proportional to the area under the curve, which is proportional to the initial FID intensity and so to the nuclear concentration. These relationships are summarized in Fig. 5.18.

In general, any function of time will have an equivalent representation in the frequency domain, and it is of interest to consider some examples, which are illustrated in Figs 5.19, 5.20 and 5.21. We can see that:

1. An infinitely long wave of constant intensity is represented by a single monochromatic frequency. The precision with which we may measure this frequency depends upon how long we are willing to spend in its measurement.
2. Any distortion of this wave in the time domain results in the production of extra frequency components in the frequency domain. If the wave persists only for a limited time, then a packet of frequencies is produced, which gives the line definite width in the frequency domain. The line shape depends upon how the wave decays in the time domain. If the decay follows the law

$$I_t = I_0 \exp\{-(nt)^p\} \tag{5.7}$$

then if $p = 1$, we have an exponential decay as illustrated in Figs 5.19 and 5.20(a). If $p = 2$, we have a Gaussian decay; and if $p = 4$, we have a super-Gaussian decay. These and their frequency-domain equivalents are shown in Fig. 5.19.
3. If several frequency components are present in the time-domain signal, then these interfere and produce a beat pattern such as shown in Fig. 5.21. Each component gives a line in the frequency domain with width determined by the rate of decay of the time-domain signal. Note that when there are many, regularly spaced components present, the time-domain signal starts to resemble a series of short pulses.

This introduction of new frequencies with distortion of the wave is a general phenomenon, and one way of stating the Fourier theorem

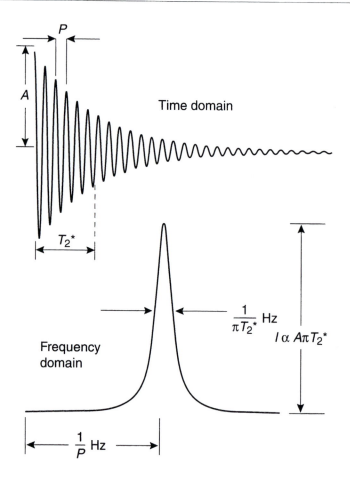

Figure 5.18 The relationships between time and frequency domains. The period P of the FID gives the position of the line. The rate of decay T_2^* gives the linewidth and the initial amplitude A gives the line its area and therefore its intensity proportional to $A\pi T_2^*$.

is to say that any periodic function of time, however complex, may be synthesized from a suitable combination of pure cosine waves.

One waveform that we will use consistently in producing spectra is the rectangular pulse generating $\boldsymbol{B}_1$. Usually a chain of pulses is used to produce a series of responses, which are added together. A single rectangular pulse (Fig. 5.21) has a very rapid decay, which can be represented as following the law of equation (5.7) with the exponent p being very large. The flanking waves in the frequency domain are now much more pronounced, and the intensity distribution follows a sinc curve, where $\mathrm{sinc}(x) = \sin(x)/x$. The width of the central part of the response is the inverse of the length of the pulse and, since NMR spectrometers are designed to have 90° pulses of length generally less than 100 μs, the width of the frequency coverage of the $\boldsymbol{B}_1$ field is at

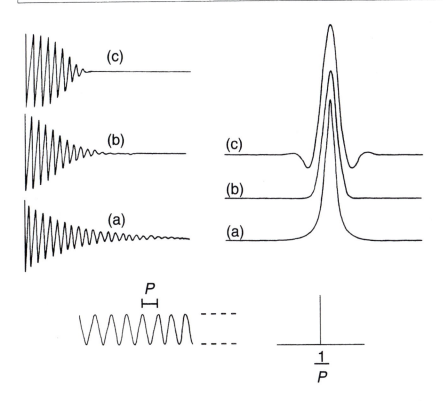

Figure 5.19 Equivalent time- and frequency-domain plots, the latter being shown on the right of the figure. The lowest trace is a cosine wave of period P, which gives a single infinitely narrow line at frequency $1/P$. (a) An exponential decay $I_t = I_0 e^{-nt}$, which gives a Lorentzian line with sharp top but wide skirts. (b) A Gaussian decay $I_t = I_0 e^{-(nt)^2}$, giving a line with narrow skirts but a thickened top. (c) A super-Gaussian decay $I_t = I_0 e^{-(nt)^4}$, which introduces flanking waves at the base of the peak. (From Akitt (1978) *J. Magn. Reson.*, **32**, 311, with permission.)

least 10 kHz. Thus the use of short $\boldsymbol{B}_1$ pulses ensures that all the nuclei in a sample, whatever their chemical shift, are swung around $\boldsymbol{B}_1$ by the same angle. Intensity distortion only occurs if the chemical shift range is very large. The same comments apply to a train of pulses, except that the frequency coverage is now discontinuous. This has little effect since the time between pulses will normally be longer than T_2^* and the spacing between the peaks of energy will be less than a linewidth. In all cases we will have present simultaneously in the output, signals due to all the chemically different nuclei in a sample, and the output will be complex and so require mathematical analysis to obtain a spectrum.

It is also possible to produce a frequency-selective pulse. The spectrometer output has to be much reduced so that a 90° pulse has to be very long. In such a case, the frequency spread of the pulse will be only a few hertz and one can choose nuclei of a particular chemical shift to precess around $\boldsymbol{B}_1$ without affecting any other nuclei in the

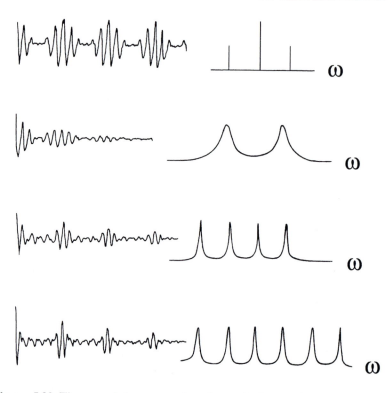

Figure 5.20 Time- and frequency-domain equivalents of multiple-frequency responses containing three, two, four and six components. The time-domain signals for the lower three are relatively simple because the components have been chosen to be of equal amplitude and regularly spaced. If the spacing and amplitudes are irregular, the time-domain signal becomes much more complex.

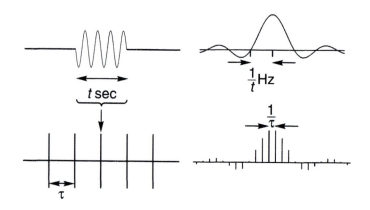

Figure 5.21 The frequency-domain equivalents of short, rectangular pulses. The upper pair of traces show a single pulse of length t seconds whose frequency is spread out over a range of $1/t$ Hz. The lower pair of traces show a train of such pulses separated by an interval τ seconds. In this case the frequency-domain signal is not continuous, but consists of peaks of energy separated by $1/\tau$ Hz with the same spectral envelope as the single pulse.

sample. Alternately, the pulse can be shaped to reduce its frequency spread.

It should be clear from all this that there is an inverse relationship between time intervals in the time domain and spread of frequency in the frequency domain. The degree of resolution in the spectrum (frequency domain) is thus determined by the time for which the response is collected following the B_1 pulse. The minimum linewidth is determined by T_2^*, but if the time of collection is shorter than T_2^* then this maximum degree of resolution cannot be attained.

5.6.3 Free induction decay, data size and zero filling

The time during which a single FID may be collected is limited by the finite size of the computer memory. If there are M locations, then they will all have been traversed after $M(DW)$ seconds. Since, strictly, we should wait until $5T_1$ seconds have elapsed and the system is again at equilibrium before we apply the next pulse, we may also have to introduce a waiting time. Thus a standard FT experiment can be summarized by Fig. 5.22. The size of the memory does not affect the spectral width. After transformation of the accumulated FID, the dispersion and absorption parts of the spectrum each occupy $M/2$ locations of the same memory (this gives maximum utilization of memory space), and the smallest frequency interval that can be detected is $2F/M$ Hz. Memory size therefore determines the resolution. If lines are closer than $2F/M$ they can never be separated. We note that $2F/M$ is equal to $1/M(DW)$, the time for which we observe an individual response, and that it is a fundamental fact that if we observe a response for only t seconds then our resolution cannot be better than $1/t$ Hz. Any line narrower than this limit will be represented by a single point in the transform and its absorption intensity may be zero if its frequency falls between those defined by two locations, though the broad wings of the dispersion part will still be visible. Adequate computer memory resolution is therefore essential for full definition of the spectrum. Each line must be represented by several points and so the memory size should be as large as necessary, which is not a limitation in view of the low cost of computer memory.

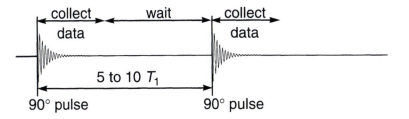

Figure 5.22 The ideal Fourier transform experiment, which allows the spins to relax to equilibrium before successive pulses are applied. The data are collected until the memory is completely traversed and all memory locations contain data.

In practice, the data size for the FID should be chosen so that the FID is undetectable between 40 and 80% of the way through the dataset being collected. If the FID finishes earlier, then the noise in the remainder of the dataset will be converted into extra noise in the final spectrum. Also if $T_1 = T_2$, time is being wasted during the noise collection. Alternatively, if the FID continues beyond the end of the collected data, resolution is being lost and care is necessary if zero filling is going to be used, as described in the next paragraph, since this will produce a step in the data.

The quality of the spectrum is affected by the number of data points used for both acquisition and Fourier transformation. These do not need to be the same as it is possible to increase the data size after the end of accumulation of the FID by adding a block of zeros. This is called zero filling. If an inadequate number of datapoints is used for the acquisition of the FID, then resolution is lost; compare Fig. 5.23(a) with 5.23(d). In the case shown in Fig. 5.23(a), the FID is non-zero at the end of the data set and attempts to improve the resolution by adding zero filled datapoints does result in a sharpening of the signal, but also in ringing on either side of the signal (Fig. 5.23(b)). The result of increasing the number of datapoints used for acquisition results in the FID being fully contained in the dataset, and the resolution gives what at first sight is an acceptable spectrum (Fig. 5.23(c)). However, after zero-filling, the peak shape is better mapped, and more importantly the relative heights of the two signals is now correct (Fig. 5.23(d)). A further increase in data size does not improve resolution, but can improve the appearance of spectra as peaks which were previously defined by a few data points will now be defined by more and will have a smoother appearance, the technique being equivalent to interpolating points in the frequency domain.

5.7 RAPID MULTIPLE PULSING

The waiting time needed to allow the spin system to come to equilibrium is a waste of our equipment since it is quiescent while waiting. It was quickly found that the signal-to-noise ratio can be maximized in a given time if we adopt the procedure: pulse – acquire data – pulse – acquire data, and omit the waiting time. Thus we allow the computer acquisition time to determine the pulse repetition rate. This time will usually be much shorter than the ideal of Fig. 5.22, so that some transverse magnetization will still exist at the time of each pulse after the first. In fact, the later 90° pulses will tip the magnetization through the xy plane and the intensity will be reduced. The signal strength is then optimized by reducing the length of the pulse, and so the pulse angle, until a maximum steady-state output is achieved. The pulse lengths commonly found correct for this type of data collection are in the range of 40° to 30°, depending on nuclear isotope

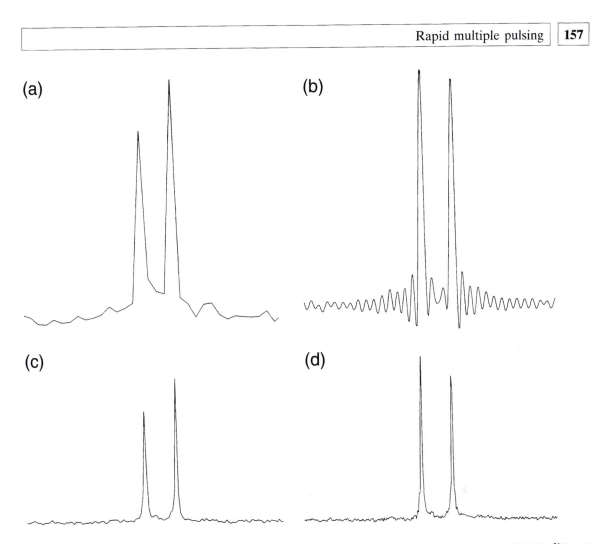

Figure 5.23 Examples of the effect of zero-filling. In each case only one doublet from a 162 MHz ^{31}P AX NMR spectrum is shown. A spectral width of 10 204 Hz was used. (a) A 16K FID transformed into a 8K real spectrum. The angular appearance of the spectrum is due to the digitization interval being 1.3 Hz, while the separation of the doublet is 5 Hz. (b) A 16K FID, increased to 128K by adding zeros, transformed into a 64K real spectrum. The extra digitization has resulted in a reduction of the linewidth, but the FID had not gone to zero by the end of the dataset and the result is that there is ringing on either side of the signals. (c) A 64K FID transformed into a 32K real spectrum. This is much better, but due to the lack of digitization, the relative heights of the two signals are wrong. (d) A 64K FID, increased to 128K by adding zeros, transformed into a 64K real spectrum. This spectrum finally gives true heights and linewidths to the signal.

observed, chemical shift range, relaxation time T_1 and memory size. These parameters are related by the Ernst equation

$$\cos(\alpha_0) = \exp\left(-\frac{T_p}{T_1}\right)$$

where α_0 is the pulse angle (determined by its length) and T_p is the time between pulses. Now, the relaxation times of nuclei situated in different parts of the molecule will usually be different, so that the optimum pulse length will be different for each. This means that there will be a distortion of the intensity of the signal from each type of nucleus, which fortunately is not too grave a disadvantage in many cases since we know a molecule is made up of integral numbers of each type of atom. If precise quantitative data are essential, then the pulse sequence of Fig. 5.22 must be used or a means found of reducing T_1 in our sample.

5.8 MANIPULATION OF COLLECTED DATA

FT NMR involves the collection of data in a computer memory. Now a computer is an infinitely variable instrument; it will do anything that you wish to the data that it contains, provided that you possess a suitable program. The computers used for NMR spectroscopy have a section of memory containing programs as well as memory used simply for data accumulation, together with a backing store in the form of a disc or discs. Thus a set of data that may have taken several hours to acquire can be stored in permanent form and recalled to allow a variety of mathematical processes to be carried out with the object of improving the final spectrum.

5.8.1 Line-shape manipulation

5.8.1.1 Exponential multiplication

The technique most commonly used in the time domain is to multiply the data progressively by an exponentially decaying function,

$$k = e^{-t/T_2^\dagger}$$

where t is the time corresponding to each point in the FID dimension, and $T_2^\dagger$ is a constant which behaves the same as the spin–spin relaxation time. Hence if a signal has a T_2 of 1 s, then this is a good value to use for $T_2^\dagger$ in exponential multiplication. In practice, it is usual to express $T_2^\dagger$ as a line broadening in Hz, which is related to $T_2^\dagger$ as in equation (4.1). This treatment increases the rate of decay of the FID and so broadens the lines obtained after Fourier transformation. This loss of resolution is normally acceptable because the signal-to-noise ratio is significantly improved. This comes about because the signal and noise have different shapes in the time domain; the signal is intense at first but decreases with time and may be undetectable towards the end of the FID, whereas the noise has no decay in intensity with time and may be the only signal present at the end of the decay. The exponential multiplication thus decreases the noise most where there is no signal to be attenuated. The effect of the process when applied to a noisy spectrum is shown in Fig. 5.24. It allows the signal-to-noise ratio to be increased in a time that is very much shorter than that needed to accumulate sufficient extra data. The amount of noise that

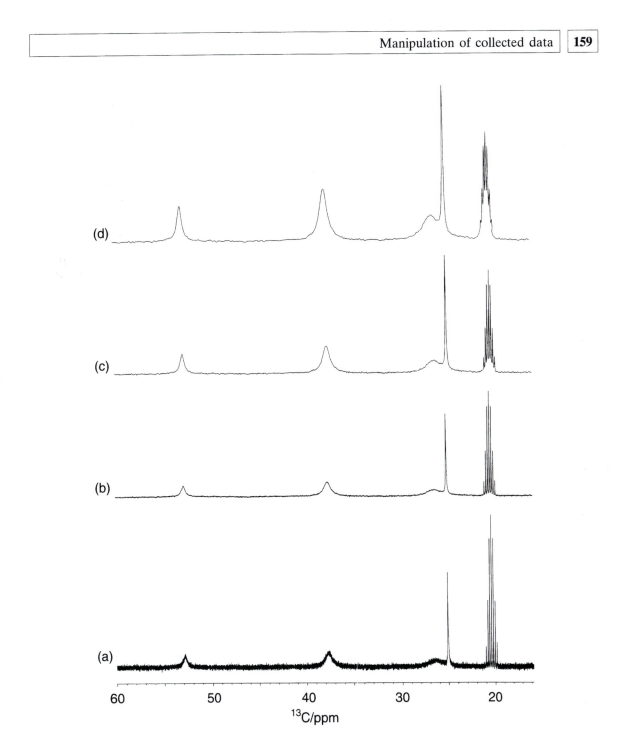

Figure 5.24 The effect of the exponential multiplication process on the 100.62 MHz ^{13}C NMR signal of bromocyclohexane in $(CD_3)_2CO$ at $-16°C$. (a) A transformed spectrum where no window function has been applied. (b). A transformed spectrum which has been broadened by multiplying by a decaying exponential with a time constant corresponding to a broadening of 1 Hz. (c) As b, but 4 Hz broadening. (d) As b, but 10 Hz broadening. Note that as the broadening is increased, the broad signals at δ 26.5, 38, and 53 become more clearly seen, but the coupling on the CD_3 signal of the solvent at δ 20.4 becomes more and more obscured.

is accumulated with a spectrum is determined by the rate at which data are accumulated, and there is relatively little flexibility in spectrometer setting possible. Thus, after a given time of accumulation, one has a certain signal-to-noise ratio, and this can be improved either by continuing with accumulation of data or by exponential multiplication if the loss in resolution is acceptable. If it is decided to continue accumulation of data, it is necessary to continue for at least as long as has already been done to obtain a significant improvement. This is a simple rule of thumb and is reasonable in view of the square-root law for signal-to-noise improvement, which predicts a 40% improvement if the number of accumulations is doubled. In contrast, exponential multiplication is almost instantaneous.

5.8.1.2 Lorentzian–Gaussian transformation

It would often be useful to be able to sharpen the lines and artificially improve the resolution. The use of multiplication by the rising exponential $k = e^{-t/T^*_2}$ with a negative value for T^*_2 does work, but the signal-to-noise ratio is badly degraded. The term becomes very large at large t values, i.e. at the right hand side of the FID where there is predominantly noise. The generally preferred function is the Lorentzian–Gaussian transformation equation

$$k = e^{-t/T^*_2}e^{-bt^2} \quad , \tag{5.7}$$

The function initially increases if a negative value is chosen for T_2, but if a positive value is chosen for b, the function then decreases to give a bell shape. FIDs before and after multiplication are shown in Figs 5.25(a)–(d). The multiplication transforms the lineshape from that of the ideal Lorentzian NMR signal to a Gaussian lineshape and has the effect of sharpening the signal and reducing the linewidth. This can be seen by comparing the multiplets in Figs 5.25(a) and (c) or (d).

A more extreme example is the determination of the long-range coupling between the protons in 2-chlorobenzaldehyde, a study of which permits the study of the barriers to rotation of the formyl group. Figure 5.26 shows two of the lines due to the proton H^4 obtained with an instrumental resolution better than 0.05 Hz. Structure is visible in one of the lines, and after resolution enhancement the small coupling $^6J(F–H^4)$ is seen with a value of about 0.024 Hz. The variation of this very small coupling between the formyl proton and ring proton H^4 with temperature enables the formyl group rotation to be studied. The wiggles in the baseline in both these spectra arise because of the rapid truncation of the data that occurs in the tail of the FID with the application of the noise-reducing function and which introduces a super-Gaussian element into the spectrum (Fig. 5.7(c)). For this reason, resolution enhancement is also known as a Lorentzian–Gaussian transformation. The technique thus has to be used with care, since these wiggles can in extreme cases be mistaken for hidden peaks.

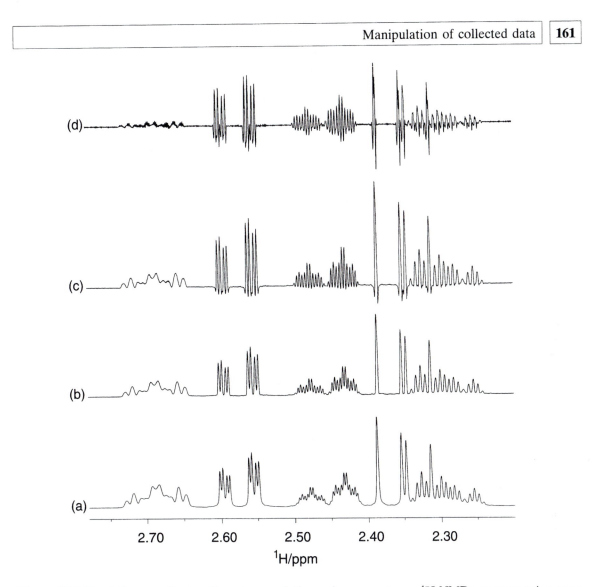

Figure 5.25 The influence of using Gaussian resolution enhancement on a ^{1}H NMR spectrum using equation (5.6). (a) No enhancement. (b) A T_2 of –0.637 s, and b = 0.157. (c) A T_2 of –0.318 s, and b = 0.076. (d) A T_2 of –0.59 s, and b = 0.039.

5.8.1.3 The sine-bell weighting functions

Although the exponential and Lorentzian–Gaussian transformation weighting functions are commonly used in one-dimensional NMR spectroscopy, in two-dimensional NMR spectroscopy, see Chapter 9, the sine-bell function is frequently used. It is common in two-dimensional NMR spectroscopy to have truncated FIDs. With the simple sine-bell function, the FID is multiplied by a sine wave going from $\sin(0°)$ to $\sin(180°)$. There are variations, where the sine bell is shifted, and, for instance, a shift of 30° simply means that the multiplying sine-bell starts at $\sin(30°)$ and runs to $\sin(180°)$. Also a sine-bell squared function is used where the multiplying sine function is squared.

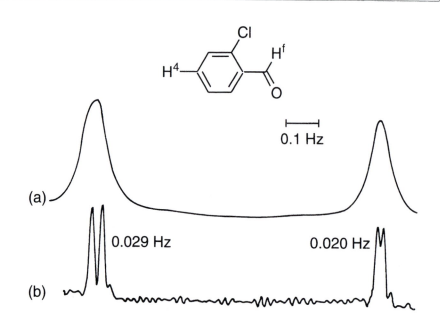

Figure 5.26 The ^{1}H spectrum of a fragment of the spectrum of 2-chloroben-zaldehyde showing how resolution enhancement can reveal the very small coupling between the formyl proton and ring proton H4. (From Laatikainen *et al.* (1990) *Magn. Reson. Chem.*, **28**, 939–46; copyright (1990) John Wiley and Sons Ltd, reprinted with permission.)

5.8.2 The Fourier transform

The production of the frequency-domain spectrum from the time domain FID is done by Fourier transformation. Equation (5.5) for the relationship between time and frequency domains is

$$F(\omega) = \int_{-\infty}^{\infty} f(t)[\cos(\omega t) - \mathrm{i}\,\sin(\omega t)]\mathrm{d}t$$

It can be shown that it is valid to replace the continuous functions by discrete ones such as we collect in a computer, i.e. we can replace the integral symbols by summations and can carry out valid transforms in a computer. The process of Fourier transformation can be thought of as carrying out a series of multiplications of the time-domain data by sine and cosine functions that are stepped along the data. The results of this series of multiplications are summed and, if there is present in the data a wave of the same period, then there is an output; otherwise there is none. The use of both sine and cosine functions means that intensity and phase information are retained. The process is repeated for all possible frequencies up to the Nyquist frequency, so that all the signals present in the data may be detected. This would obviously be a very time-consuming process, even in a modern computer, and would require a vast amount of extra computer memory in order to store all the intermediate data needed. Fortunately, Cooley and Tukey have shown that, provided all the data are contained in a memory with number of locations equal to a power of 2 (e.g. 1024, 2048, 4096, . . ., etc., or 1K, 2K, 4K, . . ., etc., locations), then the process can be factorized in such a way as to remove many redundant multiplications and can be carried out in the memory containing the data, with only a few extra locations being needed to store numbers temporarily. The process is thus fast and economical of memory space. The existence of the Cooley–Tukey algorithm and of minicomputers were both necessary before FT NMR could become a commercial possibility. In fact, the current generation of dedicated computers are so advanced that they can perform a transform on accumulated data in another part of memory in order to observe how the accumulation is proceeding while at the same time more data are being added to the FID. The sine and cosine multiplications produce, in effect, the two components of the nuclear magnetization in phase with and normal to B_1, and these are kept separate in the memory. The final result of the transform is thus two spectra, which should in principle be the absorption and dispersion spectra. This assumes that the B_1 signal fed to the phase-sensitive detector is exactly in phase with the nuclear signal, a condition that it is very hard to realize in practice. Thus the two sets of spectra contain a mixture of dispersion and absorption. In addition, we cannot be sure that we have started to collect data immediately after the B_1 pulse has ended; indeed, to avoid breakthrough of the strong pulse signal, we may have purposely delayed the collection of data, commonly by one dwell time. This means that the starting point for each frequency component differs, since each type of nuclear magnet will have precessed by a different amount by the time data collection starts, their phases are different and the degree of admixture of the two spectra varies across the spectral width. The two components thus have to be separated, a procedure known as phase correction, which is an essential part of FT spectroscopy. The two halves of the sets of spectra are weighted and mixed according to the formulae below.

new disp = (old disp) $\cos\theta$ – (old abs) $\sin\theta$

new abs = (old abs) $\cos\theta$ + (old disp) $\sin\theta$

where 'abs' means one half of the data and 'disp' the other. The multiplication is carried out point by point through the data and θ is adjusted until one line is correctly phased, i.e. purely absorption in one half of memory and purely dispersion in the other half. If not all the lines are then correctly phased, it is necessary to repeat the correction but allow θ to vary linearly as a function of position in memory. It is best to arrange that $\theta = 0$ for the correctly phased line and allow it to vary to either side. This second process is then continued until all the lines are correctly phased. Phase correction can be done automatically by the computer, though the operator may intervene if needed. The process is illustrated in Fig. 5.27.

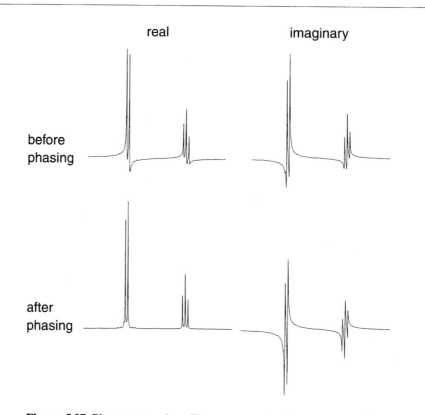

Figure 5.27 Phase correction. The upper pair of spectra are the two results of the transform operation with scrambled phase of the lines of the AB_2 spectrum. A two-parameter phase correction gives the purely absorption spectrum in one half and the purely dispersion spectrum in the other.

Sometimes it is not possible to make a perfect phase correction. This may arise because the behaviour of the spin system is affected by the way the spectrum has been obtained; we have already mentioned that too rapid pulsing if spin–spin coupling is present may result in distortion of line intensities, and in such a case phase anomalies are also inevitable.

5.8.3 Manipulations in the frequency domain

Once we have the fully phase-corrected spectrum in memory, we now have to extract from it the information that we require. We will also need to make some sort of permanent record. Again, the computer allows us to carry out a variety of processes.

5.8.3.1 Visualization and plotting data

The contents of the data memory, whether the data are time or frequency domain, can be displayed continuously on a monitor screen connected to the computer. The X axis is swept through the data memory and the numbers stored there are transferred sequentially to the Y axis. The result is a steady display of FID or spectrum, which allows the various manipulations to be monitored. A more permanent record is provided by plotting the data using some sort of printer which can print what is on the screen and much other data stored in the computer for each spectrum. Different sets of spectral data can be drawn in different colours, together with expansions or integrals (see below), and they can also reproduce text and so calibrate the paper, give spot chemical shifts for individual lines and numerical integral values.

5.8.3.2 Integration

The initial intensity of a FID is proportional to the number of nuclei contributing to that signal and this transforms as the area of the Lorentzian absorption. An integral of the spectrum (we always, of course, refer to the absorption spectrum) then will tell us how many nuclei contribute to a given line and can give us invaluable quantitive data about a molecule – in the absence of relaxation effects discussed in Chapter 4. The integration is carried out simply by adding the numbers in successive memory locations. Where there is no resonance, this sum will remain constant, but will increase in the region of any resonance. The integral then forms a series of steps, rising at intervals, with each rise corresponding to a resonance, see Fig. 5.28. For the plot to be vertical and horizontal as shown, it is necessary that the baseline is at zero intensity. Means are thus provided to correct the baseline to give the most acceptable integral.

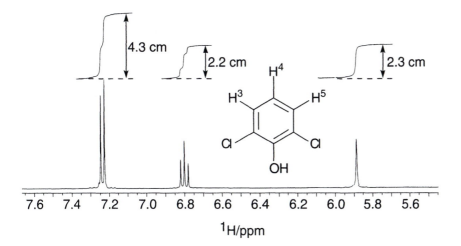

Figure 5.28 A 400 MHz ^{1}H NMR spectrum and its integral of 2,6-dichlorophenol in CDCl$_3$. The ratio of the heights of the steps is approximately 2:1:1 as required for H^3,H^5:H^4:OH.

5.8.3.3 Expansion

The typical monitor screen is rather small to be able to observe much detail, and means are provided to enable a part of the spectrum to be selected and expanded to fill the screen to assist in the various operations needed to improve a spectrum and, for instance, show up small splittings.

5.8.3.4 Storing data

Data can also be recorded on some form of magnetic storage system, usually a hard disc, at least for a limited period. It is usual to store the FID, which can then later be recalled and reprocessed to, say, show up some unexpected feature. Archiving is now possible using that contradiction in terms a writable CD ROM or magnetic tape.

5.9 QUESTIONS

5.1. It is required to run a series of spectral accumulations for a nucleus and a group of its compounds where the range of chemical shifts is likely to be 150 ppm. In order to accommodate all resonances without folding, a spectral width is chosen of 200 ppm. The spectrometer frequency is 90 MHz. What dwell time should be used? The longest relaxation time that is likely to be encountered is 20 s. A resolution of about 1 Hz is required. Calculate the appropriate memory size to be used to accumulate data (1K, 2K, 4K, 8K, 16K, 32K, 64K, ..., where 1K = 1024 are the permitted memory sizes) and the time required to traverse the memory and collect each FID following the read pulse. Will the nuclei be fully relaxed when the memory has been traversed and should the pulse length of the read pulse be 90° or less? What is the maximum permissible length of the stimulating read pulse if the nuclei are to be turned towards the xy plane by an equal amount whatever their chemical shifts?

5.2. At time 0, the z-magnetization is M_z of a nucleus with $T_1 = 1$ s. You apply a 90° pulse. What is now the value of M_{xy} in terms of M_z? You now wait 1 s. What is now the value of $\dot{M}_z$?

5.3. The lock operates on a very precise frequency to ensure exact calibration of the spectra. What happens if the phase is set incorrectly so that the dispersion signal is lop-sided?

5.4. What is the gain in signal-to-noise ratio for a quadrature detector operating with a 16 sample digital filter compared to a simple phase sensitive detector operating with analogue filters?

5.5. Use the Ernst equation to calculate the ideal pulse angle for maximum signal : noise ratio for a nucleus with $T_1 = 1$ s and a time of 1 s between pulses.

Making the spins dance

6

We are now going to discuss how we can predetermine the behaviour of the nuclear magnetization using pulses of various lengths and phase and carry out a variety of NMR experiments.

6.1 DECOUPLING

If we consider a nuclear system $A_n X_m$ in which all nuclei A are equivalent as are all nuclei X, and the two types of nuclei are spin-coupled, then the resonances of X and A will both be split into multiplets whose multiplicity depends upon the values of n and m and the spin quantum numbers of A and X. A and X may be the same isotope and so chemically shifted, preferably by a substantial amount, or they may be different isotopes and so with markedly different NMR frequencies. We will observe A while applying a second strong radiofrequency field B_2 to X, remembering that in the homonuclear case this may require some technical modifications of the experiment. B_2 has to have the same frequency as the centre of the X multiplet and so is stationary, or nearly so, in the rotating frame relative to these nuclei. The X nuclei then precess around B_2 at a frequency that depends upon the magnitude of B_2, and, if this precession is fast enough, rapidly and repeatedly reverse their direction so that their z magnetization effectively disappears at the coupled A nucleus whose multiplicity disappears and so becomes a singlet. A is said to be decoupled from X. For decoupling to be complete, it is necessary that

$$\frac{\gamma_X B_2}{2\pi} \gg J(AX)$$

An example of complete decoupling is shown in Fig. 6.1. In fact, there will always be a little residual coupling and a decoupled singlet is thus slightly broadened because of this. If B_2 is small, then the A spectra can become more complex and exhibit extra splittings.

Signals close to the B_2 irradiation frequency are displaced from their true frequencies. This is called the Bloch–Siegert shift. The Bloch–Siegert shift obeys the equation

$$\frac{\gamma B_2}{2\pi} = \sqrt{2(\nu_A - \nu_2)(\nu_{obs} - \nu_A)}$$

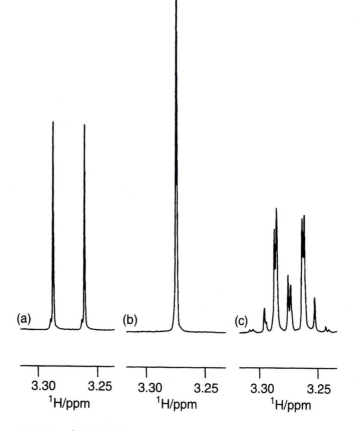

Figure 6.1 The ^{1}H NMR spectrum of $P(OCH_3)_3$. (a) Normal, with ^{31}P coupling. (b) With irradiation at the ^{31}P frequency to completely decouple the phosphorus. (c) With low-powered ^{31}P decoupling.

where B_2 is decoupler field strength, ν_A is the position of the signal in the absence of decoupling, ν_{obs} is the position of the signal in the presence of decoupling, and ν_2 is the decoupler frequency. This equation applies provided that $(\gamma B_2)^2 \ll (\nu_A - \nu_2)^2$. This equation can therefore be used to calibrate the strength of the homonuclear B_2 by carrying out irradiation near to a resonance.

6.1.1 Homonuclear decoupling

For homonuclear decoupling, A and X are the same nuclear isotope. We give an example of the decoupling of two proton resonances (Fig. 6.2). The chemical shift between the two coupled resonances of 2,6-dichlorophenol is some ten times larger than the coupling constant. It is thus easy to apply a B_2 strong enough to perturb the nuclei of one resonance without directly affecting the other. Thus by irradiating

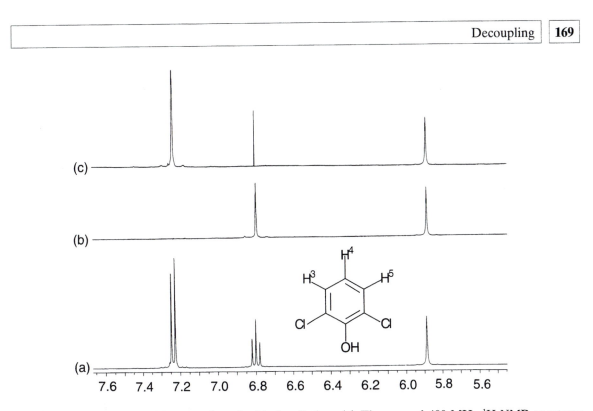

Figure 6.2 An example of homonuclear double irradiation. (a) The normal 400 MHz ^{1}H NMR spectrum of 2,6-dichlorophenol in CDCl$_3$. (b) As a, but with B_2 placed on the doublet due to H^3 and H^5 at δ 7.24. Note that the triplet at δ 6.80 due to H^4 has been reduced to a slightly broadened singlet. (c) As a, but with B_2 placed on the triplet due to H^4 at δ 6.80. Note that the doublet at δ 7.24 due to H^3 and H^5 has been reduced to a slightly broadened singlet. There is also a decoupling spike at δ 6.80 due to leakage of the decoupling frequency into the receiver. The OH proton at δ 5.89 is not affected.

one, the other becomes a singlet. Decoupling is established in such an experiment very quickly since precession around B_2 starts as soon as B_2 is applied.

At the same time as the precession of individual nuclei takes place under the influence of B_2, there is saturation of the X spin system. The word 'saturation' derives from usage in the old continuous wave NMR spectroscopy, where it is found that, if the irradiating power is too large, the intensity of the signal decreases. The spin system absorbs energy from the irradiating field so that the low-energy excess of spins becomes depleted, a process that is in competition with the T_1 relaxation processes. This also happens for B_2, and if this is large enough then the spin populations become equalized and the total magnetization of the irradiated resonances become zero. This is a non-equilibrium state and all the relaxation mechanisms present will work towards the re-establishment of equilibrium. If the predominant mechanism of relaxation is dipole–dipole, then the energy exchange required by this relaxation causes mutual spin flips of A and X, and the spin population of the A nuclei is disturbed, so that decoupling between $I = 1/2$ nuclei is usually accompanied by intensity modifications of the signal of the decoupled nucleus. This is the nuclear Overhauser effect (NOE). Because relaxation processes are involved in determining the NOE, this is established much more slowly than the decoupling.

In order to decouple, the decoupler must be on during the acquisition of the FID. The introduction of the B_2 homonuclear decoupling field is likely to cause interference with the detection of the FID. This can be avoided by using time-shared decoupling in which B_2 is only switched on while the receiver is resting between analogue-to-digital conversions. This mode of operation is made possible because the receiver is only required to operate for sufficient time to activate the analogue–digital converter at the end of each dwell time. The interruptions in B_2 need then be only quite a small proportion of the total time, see Fig. 6.3. This is the case if the sampling is made every dwell time though if over sampling is used and the digitizer is working at its maximum rate then the transmitter can be on for only about 10% of the time.

6.1.1.1 Single frequency heteronuclear decoupling

Heteronuclear decoupling occurs when A and X are different elements or isotopes (Fig. 6.4). The NH_2 group of formamide, H_2NCHO, consists of two signals due to restricted rotation about the N–C bond. This is just observable in the simple ^1H NMR spectrum but there is extensive broadening of the NH_2 signals by coupling with the quadrupolar ^{14}N. The CH proton is a doublet due to coupling to one of the NH protons. When the ^{14}N is decoupled, the two separate NH signals are well resolved at δ 6.85 and 7.1 and the coupling to the CH proton is also evident, as is the much smaller coupling to the other NH and the small coupling between the two NH protons.

In more complex molecules, there may be two or more types of X nuclei coupled to the A nuclei, and, provided the chemical shift

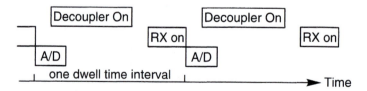

Figure 6.3 Time-shared homonuclear decoupling. Three parts of the spectrometer have to be switched on and off independently though in a particular order. The receiver alone is switched on for sufficient time to establish an output. This output is fed to the analogue-to-digital (A/D) converter, which starts to convert the voltage into a number at a precise time, set by the dwell time in use, and takes a finite time to make the conversion, 5 to 20 μs. The decoupler is switched on during the A/D conversion process and remains on until a little while before the receiver is switched on again. The effect of the decoupler may be modified by reducing the length of each transmitter pulse (decoupler on) as required.

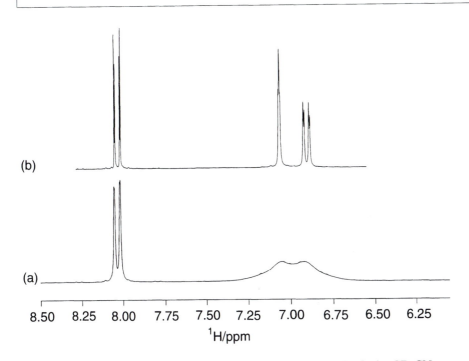

Figure 6.4 The 400 MHz ^{1}H spectrum of formamide, H$_2$NCHO, in CD$_3$CN. (a) The proton resonances of the NH$_2$ group are badly broadened by the ^{14}N nucleus and the CH resonance is slightly broadened. (b) The signals sharpen on irradiation of the ^{14}N nucleus.

between the types of X nuclei is larger than any coupling between them, then it is possible to carry out selective decoupling experiments in which the effects of each group of X nuclei can be examined in turn. A heteronuclear example is that of the ^{31}P spectrum of triethyl phosphite, (CH$_3$CH$_2$O)$_3$P, whose spectra are shown in Fig. 6.5. The phosphorus nucleus is coupled to both types of proton, that to the methylene protons giving a septet and that to the methyl protons giving a decet. The coupling to the methyl protons is, however, quite small, and the normal spectrum is a septet of broad lines that contain the decet structure due to the coupling to the methyl group. This broadening reduces the intensity of the resonances, to the extent that it is difficult to observe the two weak outer lines of the septet (Fig. 6.5(a)). Because the proton resonances of the ethyl and ethylene groups are separated by much more than the interproton coupling constant, it is possible to irradiate one with $\boldsymbol{B}_2$ sufficient to eliminate the coupling to the ^{31}P without affecting the coupling of the other type of protons. One effectively eliminates one of the coupling interactions, observes what the other produces on its own and then irradiates the other group to observe the effect of the first alone. The ^{1}H NMR spectrum of the P(OCCH$_3$)$_3$ fragment is in Fig. 6.5(c) and the well resolved methylene coupling in Fig. 6.5(b). It is, of course, possible in principle to apply sufficient $\boldsymbol{B}_2$ power that both types of proton are decoupled from the

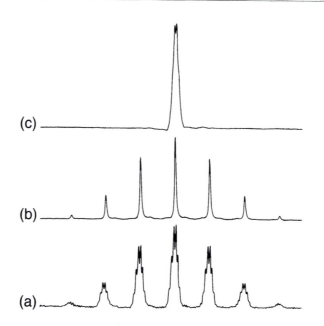

(c)

(b)

(a)

Figure 6.5 The 162 MHz ^{31}P spectrum of triethyl phosphite, $(CH_3CH_2O)_3P$. (a) Obtained without ^{1}H irradiation, showing coupling to the methylene protons broadened by the coupling to the methyl protons. (b) The methyl protons irradiated, which gives a well resolved septet due to coupling to the methylene protons. (c) The methylene protons irradiated, which selectively removes their influence from the spectrum.

phosphorus, whose resonance then becomes a singlet. In practice, it proves technically difficult to provide sufficient power to do this, and means have to be found to overcome this difficulty. The use of two separate B_2 frequencies at the same time is possible, but it has been found that the best method is to ensure a broad band of frequencies covering, in this case, the proton chemical shift range. This used to be achieved by frequency modulation of the B_2 frequency with a randomly varying waveform but now sequences of short B_2 pulses, such as WALTZ, see section 6.1.1.4, are used to swing the spins around. This gives such a spread of frequencies, that the power requirements are much less stringent. The technique is called broad-band decoupling and is used extensively, indeed almost invariably, when observing the nucleus ^{13}C in hydrogen-containing compounds. The hydrogen is broad-band irradiated and this removes all spin coupling to the carbon atoms, so simplifying their resonances to narrow singlets, increasing the intensity of the resonance of a given type of carbon atom and so reducing markedly the time necessary to obtain a given signal-to-noise ratio with this insensitive nucleus. In addition, there is also a valuable intensity enhancement due to the nuclear Overhauser effect, which combined with the simplification of the spectra gives a time reduction factor of the order of 250 times compared with a spectrum obtained without double irradiation.

6.1.1.2 Off-resonance decoupling

Single frequency decoupling works well when it can be placed on the frequency of the nucleus to be irradiated, but it does not produce satisfactory results when placed off resonance. For example, the 100 MHz ^{13}C NMR spectrum of carvone is shown in Fig. 6.6(c) and with the ^{1}H decoupling frequency placed at δ 10, Fig. 6.6(b). This has the effect of reducing the small $^2J(^{13}C^1H)$ and $^3J(^{13}C^1H)$ to a negligible size, while the multiplicity due to the larger $^1J(^{13}C^1H)$ can still be resolved. This technique used to be used to decide how many protons are attached directly to a given carbon atom and worked well on low field instruments, where the ^{1}H spectral width was less than 1000 Hz, but on high field instruments with spectral widths of up to 8000 Hz, the ^{1}H decoupling frequency is too far from many of the protons to produce a substantial reduction in $^{2,3}J(^{13}C^1H)$. The consequence is that this technique is now rarely used. It has been replaced by other techniques such as J-modulation, APT, PENDANT, INEPT and DEPT, see Chapter 8.

The shrinkage of the apparent $J(X^1H)$ can be used to calibrate the B_2 for ^{1}H decoupling of an X nucleus. Provided that $\gamma B_2/2\pi \gg |v_A - v_2|$ the equation (overleaf)

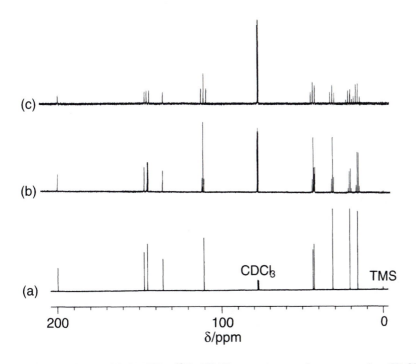

Figure 6.6 The 100.62 MHz ^{13}C NMR spectrum of carvone in CDCl$_3$. (a) Obtained with WALTZ decoupling. (b) Obtained with the ^{1}H decoupling frequency placed at δ 10. The ^{13}C nuclei attached to high frequency protons near δ 10 show big reductions in $^1J(^{13}C^1H)$. (c) Obtained without ^{1}H decoupling.

$$\frac{\gamma \boldsymbol{B}_2}{2\pi} = \frac{J(X^1H) \mid v_A - v_2 \mid}{J(X^1H)_R}$$

applies, where $J(X^1H)$ is the coupling in the absence of ^1H decoupling, $J(X^1H)_R$ is the coupling in the presence of ^1H decoupling, v_A is the position of the ^1H NMR signal in the absence of decoupling, and v_2 is the decoupler frequency.

6.1.1.3 Broadband heteronuclear decoupling

In order to be able to decouple all the protons from an X-nucleus, the single ^1H frequency was replaced by a band of frequencies. This technique works, but suffers from problems. The power required is relatively high and typically 10 watts were required to produce moderate decoupling. This produced considerable warming of the sample. As chemical shifts are temperature dependent, the signals were broadened by the temperature gradients and the signal-to-noise ratio was reduced by the poorer Boltzmann population difference at the higher temperatures. Secondly, as magnetic fields increased, and the ^1H frequency range increased, it became more difficult to decouple uniformly across the whole ^1H spectral width. This technique has now been abandoned in favour of the more efficient composite pulse decoupling techniques such as WALTZ and GARP, which we describe below.

6.1.1.4 WALTZ

In order to have some appreciation of how composite pulse decoupling works, let us examine the effect of applying a sequence of 180° pulses to ^1H while observing ^{13}C of a compound such as $CHCl_3$. In the absence of ^1H pulses or decoupling, the ^{13}C NMR signal will consist of a doublet, due to coupling to ^1H. The doublet arises because half the ^{13}C nuclei are attached to ^1H nuclei with $m_I = 1/2$ and the other half are attached to ^1H nuclei with $m_I = -1/2$. If we now apply a 180° pulse to ^1H, then we reverse the m_I values of the ^1H nuclei. The ^{13}C nuclei which were attached to ^1H nuclei with $m_I = 1/2$ are now attached to ^1H nuclei with $m_I = -1/2$ and vice versa. If we were to apply a sequence of 180° pulses, each ^{13}C nucleus would see a ^1H nucleus which was rapidly reversing its spin. The result is ^1H decoupling.

The experiment described above is inefficient and composite pulses are preferred, see section 6.2. Generally, the WALTZ-16 pulse sequence is used. It is built up of composite pulses based on the sequence $(90°)_x(180°)_{-x}(270°)_x$. This is given the shorthand notation $1\,\bar{2}\,3$, with 4 being used for a $(360)_x$ pulse. The numbers represent multiples of 90° and the bar over the 2 represents a reversal of phase to produce a 180° phase shift. The shorthand notation leads to the name WALTZ as the dance is based on a 1–2–3 step sequence. There are several variations depending on the length of the sequence used, but it is generally the WALTZ-16 sequence which is preferred.

WALTZ-16 = $\overline{3}$ 4 $\overline{2}$ 3 $\overline{1}$ 2 $\overline{4}$ 2 $\overline{3}$ 3 $\overline{4}$ 2 $\overline{3}$ 1 $\overline{2}$ 4 $\overline{2}$ 3 3 $\overline{4}$ 2 $\overline{3}$
1 $\overline{2}$ 4 $\overline{2}$ 3 $\overline{3}$ 4 $\overline{2}$ 3 $\overline{1}$ 2 $\overline{4}$ 3 $\overline{2}$

The WALTZ pulse sequence is repeated continuously to produce the decoupling. It decouples efficiently over a width of ± 4 kHz reducing a 150 Hz $J(^{13}C^{1}H)$ down to 0.15 Hz. This bandwidth is sufficient to produce a 10 ppm decoupling bandwidth even at 800 MHz. Typically the ^{1}H decoupler is attenuated so that the 90° pulse is around 100 µs. The power used is much lower than the older broadband decoupling, and consequently problems associated with sample heating have been considerably reduced.

There are circumstances where the bandwidth of the WALTZ pulse sequence is insufficient. The most common example is where ^{13}C is decoupled from ^{1}H in inverse ^{13}C detection, see section 9.7. Under such circumstances, the GARP composite pulse decoupling sequence is preferred.

6.1.1.5 GARP

In order to increase the bandwidth of the decoupling, it is necessary to use flip angles which are not multiples of 90°. This has the effect of broadening the bandwidth, but makes the decoupling efficiency less. The bandwidth doubles, but the residual splitting increases to around 0.3 Hz. The GARP-1 pulse sequence is based on the pulse sequence $RR\,\overline{R}\,\overline{R}$, where in terms of pulse angles

$R =$ 30.5 $\overline{55.2}$ 257.8 $\overline{268.3}$ 69.3 $\overline{62.2}$ 85.0 $\overline{91.8}$ 134.5 $\overline{256.1}$
66.4 $\overline{45.9}$ 25.5 $\overline{72.7}$ 119.5 $\overline{138.2}$ 258.4 $\overline{64.9}$ 70.9 $\overline{77.2}$
98.2 $\overline{133.6}$ 255.9 $\overline{65.5}$ 53.4

Typically the X decoupler is attenuated so that the 90° pulse is around 70 µs. The acronym GARP comes from **G**lobally optimized **A**lternating-phase **R**ectangular **P**ulses.

6.1.2 Measurement of T_1 and T_2

Different experiments are needed to measure T_1 and T_2 accurately, though, if they are likely to be equal, then measurement of T_1 alone usually suffices. However, in many systems $T_2 < T_1$ and both need to be measured. In addition, an understanding of the method used to measure T_2 will prove helpful when we come to discuss two-dimensional NMR.

6.1.2.1 Measurement of T_1 by population inversion

This is the most popular of several available methods. We have already seen how a maximum NMR signal is obtained after a 90° B_1 pulse. The same is obtained after a 270° pulse, except that the spins have been inverted relative to the 90° pulse and the output is 180° out of phase with that after the shorter pulse. If we arrange our computer to give us a positive-going absorption peak from the FID following a 90° pulse, then after a 270° pulse we will get a negative peak. If, instead, we use a 180° pulse, we create no M_{xy} but turn the excess low-energy spins into the high-energy state, $-M_z$. They will relax to their normal state with characteristic time T_1 and the magnetization will change from $-M_z$ through zero to $+M_z$ (Fig. 5.35). This, of course, produces no detectable effects. However, if, at some time τ (do not confuse this with τ_c) after the 180° pulse, we apply a second pulse of 90°, we will create magnetization M_{xy} equal in magnitude to M_z at that instant. The spectrum that results from processing the FID will be negative-going if M_z is still negative (as if part of the magnetization had undergone a 270° pulse) and positive-going if M_z has passed through zero (Fig. 6.7). A series of spectra are obtained in this way for a number of different values of τ and the intensity of the resulting peaks plotted as a function of τ, so allowing us to extract a value for T_1. The equations governing the behaviour of the transverse and longitudinal magnetization M_{xy} and M_z and their return to equilibrium following a 180° B_1 pulse are

$$(M_z)_t = (M_z)_\infty \left[1 - 2 \exp\left(\frac{t}{T_1}\right) \right]$$

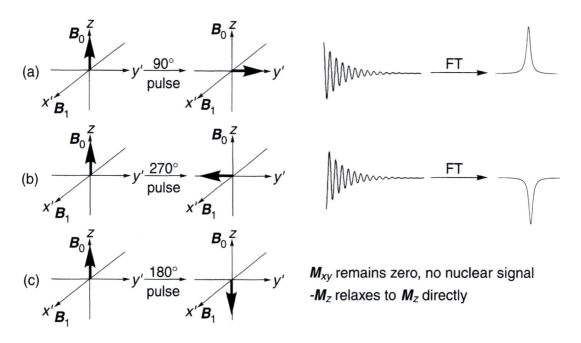

Figure 6.7 (a) Production of a nuclear response using a $(90°)_x$ pulse. The FID and its transform are represented on the right of the figure. (b) A similar response is obtained after a $(270°)_x$ pulse, but the FID is 180° out of phase, giving an inverted transform. (c) A 180° pulse gives no xy magnetization but places M_z in a non-equilibrium position opposing the field. This magnetization relaxes back to its equilibrium value and no transverse magnetization is produced at any time throughout this process.

Sufficient time must elapse between each 180°/90° pair of pulses to allow M_z to relax fully to its equilibrium value or incorrect results will be obtained. Usually a waiting period of five to ten times T_1 is used. If T_1 is very long, then the experiment can be very time-consuming, but other pulse sequences have been worked out that will allow the total time to be reduced (Figs 6.8, 6.9).

If there are several resonances in our spectrum, each relaxes at its own rate and it is possible with this method to measure the relaxation times of all the nuclei of the same isotope in a molecule and to obtain detailed information about molecular motion.

If the nuclei are spin-coupled then the apparent relaxation times may not be simply related to the real ones. Carbon-13 relaxation times are thus measured with the protons decoupled by double irradiation and this gives satisfactory values of T_1.

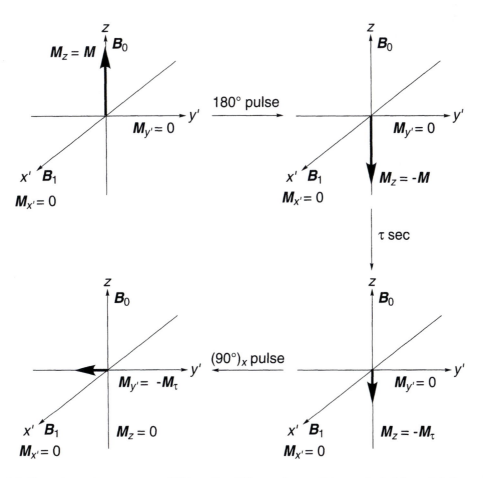

Figure 6.8 How the pulse sequence 180°–wait τ–90° can be used to perturb M_z and follow what then happens to M_z.

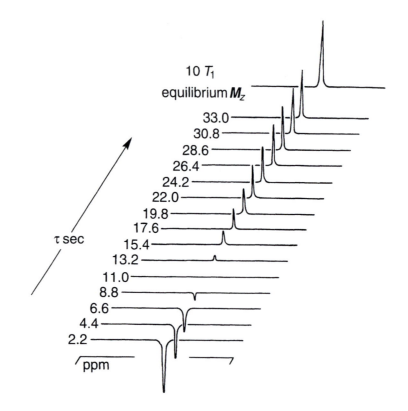

Figure 6.9 The full population inversion experiment. A series of spectra are obtained at different τ. A plot of their intensities as a function of time gives the rate of relaxation, from which T_1 can be derived. (After Martin *et al.* (1980) *Practical NMR Spectroscopy*, John Wiley and Sons Ltd, reprinted with permission.)

6.1.2.2 Measurement of T_2

The contribution of magnetic field inhomogeneity to T_2 will usually be of the order of 3.0 to 0.3 s for a high-resolution magnet. Thus, provided T_2 is less than about 0.003 s, it can be measured directly either from the linewidth or from the rate of decay of the FID. In the latter case we have to suppose that there is only a single resonance.

Accurate measurement of longer T_2 is made using a spin-echo experiment. This depends upon the rate at which the FID decays being faster than the rate of longitudinal relaxation $\boldsymbol{R}_1$. The decay is due to the T_2^* mechanism, which has two components. The loss of phase coherence between the spins in the xy plane is caused both by the random relaxation field and by the fact that the homogeneity of the magnetic field within the sample is not perfect, so that nuclei in different parts of the sample have different precession frequencies. There is an important difference between these two relaxation processes: the first is entirely random, and so unpredictable; whereas the second acts continuously and is constant at each part of the sample, so that we can in principle correct for this relaxation contribution. We assume that the random contribution to relaxation is negligible. If we apply a 90° pulse, all the magnetization M_{xy} is in the xy plane, but spins in different parts of the sample have different angular velocities due to the inhomogeneities in the magnetic field, and so some move ahead of the average and some lag behind. We wait a period of τ seconds for the spin distribution to evolve; the value of M_{xy} will decrease, and may even become zero if the field inhomogeneity is large. We then apply a 180° pulse, and all spins precess around B_1 to the other side of the xy plane. They do not take up their mirror positions. We have thus put the faster spins at the rear and the slower spins in front. They continue to precess, but after further time τ seconds they come into phase again and the magnetization M_{xy} is again a maximum. The 180° pulse is a refocusing pulse and the intensity of the spectrometer output rises following this pulse to a maximum τ seconds after and then decays again. Another 180° pulse will refocus the magnetization, which indeed can be refocused indefinitely, given perfect pulses. The refocusing does not work for the random relaxation process, and so the duration of the experiment is limited by the intrinsic transverse relaxation. Indeed, the train of echoes decays in intensity at a rate determined by the real T_2, which can be measured from a plot of intensity versus time. The experiment is summarized in Fig. 6.10. Each echo is effectively two back-to-back FIDs. It is possible to split them and Fourier transform each so that, if the sample contains several resonances, then T_2 can be measured separately for each from the resulting absorption spectra. If spin–spin coupling is present, however, then this does not work and modulation of the echoes is produced, which, as we shall see, proves useful in multidimensional spectroscopy.

6.1.2.3 Measurement of $T_{1\rho}$ and spin-locking

The Carr–Purcell pulse sequence can be turned into a spin-lock pulse sequence. All that is required is to make τ in Fig. 6.10 very short, so that the pulse sequence becomes $(90°)_x - \{(180°)_y\}_n$, where n is a large number, see Fig. 11.5. B_1 is now continuous and the magnetization is now locked in the y' direction, and the effective magnetic field is now no longer B_0 but B_1. The nuclei now precess about this field with the appropriate Larmor frequency. This is typically in the region of a few tens of kHz. Molecules which previously were tumbling slowly compared with an NMR frequency of $\gamma B_0/2\pi$ are now tumbling rapidly when compared with $\gamma B_1/2\pi$. This is exploited in a number of experiments. For example in the experiment ROESY, see section 9.6.3, it is possible to obtain NOE correlations between protons in large molecules where the conventional NOE measurements fail. The relaxation time, $T_{1\rho}$, the relaxation time in the rotating frame, is obtained from the diminution in signal strength with time of spin locking. In the

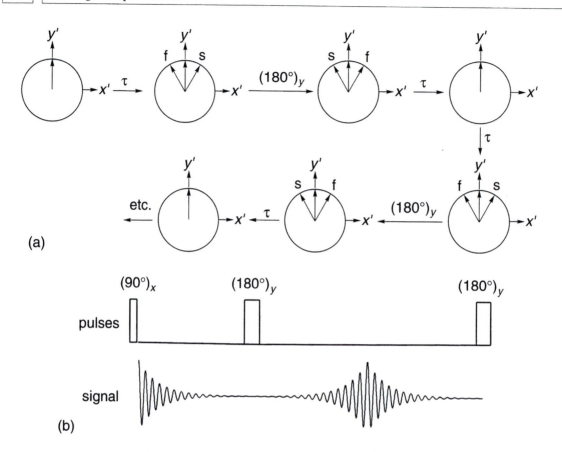

(a)

(b)

Figure 6.10 Illustrating the Carr–Purcell pulse sequence for measuring T_2. The behaviour of the spins is shown relative to the rotating frame, as if they were stationary. The $(90°)_x$ pulse produces magnetization M_{xy}, which then decreases as the spins move apart, s = slow, f = fast. The $(180°)_y$ pulse alters the relative positions of the slower and faster spins, which now close up again and M_{xy} increases, reaching a maximum. The spins are said to be refocused. The output then decreases again. Fig. 6.10(b) shows diagrammatically how the first and second (the echo) FIDs appear.

extreme narrowing region it is the same as T_2, but as the relaxation is occurring in a magnetic field strength of B_1, it is the frequency $\omega = \gamma B_1$ which applies, greatly extending the extreme narrowing region of relaxation.

6.2 COMPOSITE PULSES

Modern NMR spectroscopy relies on the use of a computer to apply accurately timed pulses to samples. We have earlier talked in terms of 90° and 180° pulses as if they were exactly that, but in practice it is difficult to apply an accurately set 90° or 180° pulse over the observation volume of the sample. At the edges of the coil, the effectiveness of the pulse can deviate markedly from that set. This is easily observed if the length of the 180° pulse is measured by gradually increasing the pulse width. The FID intensity increases up to the 90° point and then decreases, theoretically to zero at 180°. However, a perfect null is never achieved (Fig. 6.11) (unless a short sample is used and the loss in resolution can be tolerated).

The exact length of a 90° or 180° pulse depends on the sample and the exact tuning of a probe. It is therefore possible that the pulse length set differs significantly from the ideal value. This can lead to a loss in signal to noise ratio and to artefacts in spectra. A second problem arises from spectral width. A typical 180° pulse has a length of 20 μs. Such a pulse only produces acceptable excitation over ± 12.5 kHz. When it is remembered that the full spectral width of ^{19}F is 300 kHz and that of ^{31}P is less than 100 kHz at 9.4 T, it can be appreciated that such a spectral width is inadequate. The problem can be considerably reduced by using composite pulses. Only a few of the simplest ones are illustrated here, but much more complex and more efficient ones can be constructed.

6.2.1 Composite 180° pulse

The composite pulse $(90°)_x(180°)_y(90°)_x$ represents one of the simplest composite pulses. The 180° pulse is broken into two 90° pulses and an additional pulse is applied in the middle. Figure 6.12 illustrates the path of the magnetization vector if an error is made and the actual

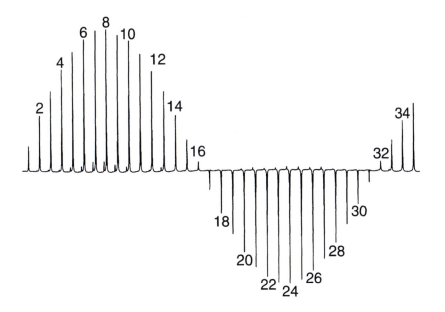

Figure 6.11 The determination of the length of a 180° pulse by gradually increasing the pulse length. The numbers given above or below each signal are the pulse lengths in microseconds used to obtain the particular signal. The 180° pulse is at 16 μs.

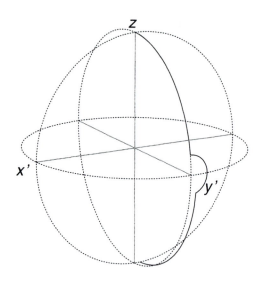

Figure 6.12 The path followed by the tip of the magnetization vector during a $(80°)_x(160°)_y(80°)_x$ pulse sequence is shown by the solid line. The pulse sequence has converted the z component of M_z to $-0.998M_z$ compared with $-0.940M_z$ produced by a single $160°$ pulse.

$180°$ pulse corresponds to a $160°$ pulse. The first pulse rotates the magnetization vector $80°$ around the x axis towards the xy plane. The second pulse rotates the magnetization vector $160°$ around the y axis. The third pulse rotates the magnetization vector $80°$ around the x axis and places it close to where it would have been if a true $180°$ pulse had been applied.

In practice, it has been found that a slightly different sequence, the $(90°)_x(240°)_y(90°)_x$ composite pulse sequence works best.

6.2.1.1 Composite 90° pulse

The pulse sequence $(90°)_x(90°)_y$ provides a simple way to minimize the effects of being off-resonance. If instead of placing the $90°$ pulse at the frequency of the resonance, it is placed $5\,kHz$ off-resonance, and if the $90°$ pulse lasts $10\,\mu s$, then during this time the nuclear magnetization will have moved $0.000\,01 \times 5000 \times 360° = 18°$ relative to the rotating frame set at the spectrometer frequency. This could produce significant errors if a pulse sequence were being used with a series of pulses at the spectrometer frequency. A subsequent $90°$ pulse along the y axis brings the magnetization close to the y axis, see Fig. 6.13.

6.3 REFOCUSING PULSE

Pulse sequences are frequently used with delays between pulses. This can cause very bad phasing problems. Consider what happens if a

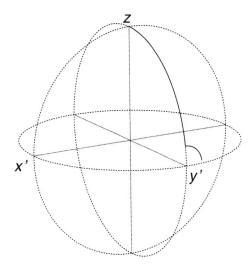

Figure 6.13 The path followed by the tip of the magnetization vector during a $(80°)_x(80°)_y$ pulse sequence is shown by the solid line.

$(90°)_x$ pulse is applied to a collection of nuclei in different chemical environments, say, the ^{1}H decoupled ^{13}C NMR spectrum of a typical organic compound, where the chemical shift range is large. Once the magnetization is in the rotating frame, the separate magnetization vectors for each type of ^{13}C fan out and there are phase shifts with respect to the instrument frequency. This is illustrated in Fig. 6.14.

The problem is easily avoided by placing a $(180°)_y$ pulse halfway between the $(90°)_x$ pulse and the beginning of the acquisition. In later chapters, you will find many pulse sequences where this technique is used. There is an additional advantage in doing this. If there is significant line broadening due to magnetic field inhomogeneity, then effects

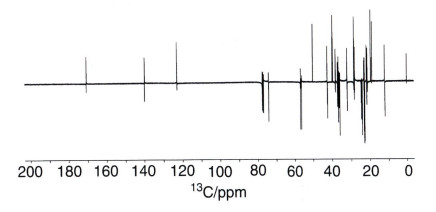

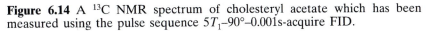

Figure 6.14 A ^{13}C NMR spectrum of cholesteryl acetate which has been measured using the pulse sequence $5T_1$–90°–0.001s-acquire FID.

due to this will also be refocused (Fig. 6.15). Comparison of this pulse sequence with that used for T_2 determination shows that it is the Carr–Purcell pulse sequence.

6.3.1 Selective pulses

It is often necessary to apply a pulse to a single resonance or even a line of a multiplet. When it is remembered that the bandwidth of a pulse depends on its length (Fig. 1.6), it can be seen that a selective pulse which has a length of 0.1 s, will uniformly excite nuclei within ± 2.5 Hz but will have effects several 10s of Hz further away from it. Hence, in order to apply a selective pulse, there are two problems to overcome. Firstly, the length of a 180° pulse must be increased from typically 20 μs to say 0.1 s. Secondly, the pulse needs to be made more selective.

6.3.1.1 DANTE pulses

Many of the older NMR spectrometers do not have the facility which enables the operator to change the power of a pulse. We are therefore forced to use short powerful pulses. The selectivity is then achieved using the DANTE pulse sequence. DANTE stands for **D**elays **A**lternating with **N**utation for **T**ailored **E**xcitation. Instead of using a single 180° pulse, the pulse is broken up into n short 180°/n pulses separated by a time τ. This is illustrated in Fig. 6.16. By a suitable choice of n and τ, the pulse can be made to last whatever time is required. Hence 11 short 16.4° pulses separated by 0.01 s produces a 180° pulse with a selectivity of ± 2.5 Hz. There are problems associated with the technique. One is that not only is a single excitation frequency produced at the spectrometer frequency, but additional ones (sidebands) are produced at ± m/τ, where m is any integer, on either side of the spectrometer frequency. Great care is necessary in the choice of τ to make sure that these sidebands do not fall on any other signal in the spectrum. Secondly, the pulse amplifier does not generate

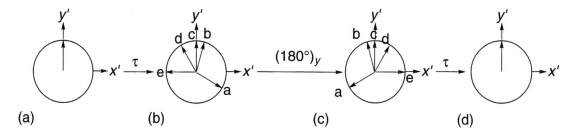

(a)   (b)   (c)   (d)

Figure 6.15 The effect of the $(180°)_y$ refocusing pulse on a collection of nuclei, a, b, c, d, and e, with different resonance frequencies. (a) The alignment of the nuclear spins immediately following a $(90°)_x$ pulse. (b) After a time τ, the nuclei have fanned out with nuclei a and b rotating faster than the rotating frame, and nuclei d and e rotating slower. Nuclei c are on resonance. (c) The effect of the $(180°)_y$ refocusing pulse. (d) After a time τ, the nuclei have refocused.

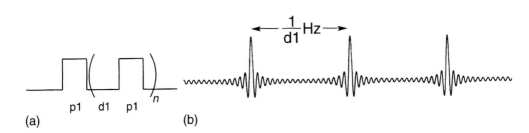

(a) p1 d1 p1 n

(b)

Figure 6.16 (a) The DANTE pulse sequence. (b) The frequency distribution generated by the DANTE pulse sequence.

a perfect square pulse. There is a rise and fall time. The 180° DANTE pulse has to be calibrated, and each component can be significantly longer than $180°/n$ when the required length is less than 1 μs.

6.3.1.2 Shaped pulses

On modern spectrometers it is possible to attenuate the pulse and hence lengthen the 180° pulse to that required by selectivity. Examination of Fig. 6.17(a) shows that a rectangular pulse is not particularly selective with significant perturbation of signals extending well outside the range of uniform excitation. This problem is considerably reduced by using shaped pulses. The relationship between the square pulse and the $\sin x/x$ shape in Fig. 6.17(b) is a Fourier transform. It is therefore of no surprise that if the power of the pulse is modulated following the $\sin x/x$ relationship, the result is highly selective. In practice, this can cause problems as the pulse becomes very long. Frequently a compromise is used, such as a pulse with a Gaussian shape, which produces a Gaussian frequency distribution (Fig. 6.17(c)).

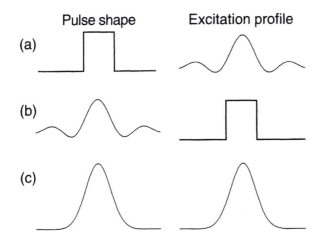

Figure 6.17 (a) A rectangular pulse and its Fourier transform. (b) A $\sin x/x$ pulse and its Fourier transform. (c) A Gaussian pulse and its Fourier transform.

6.3.2 The BIRD pulse sequence

The BIRD pulse sequence permits the selection of nuclei attached to a second coupled nucleus in preference to those not so coupled. For example, ^{1}H attached to ^{13}C or ^{15}N can be observed while the signals from the ^{1}H attached to ^{12}C or ^{14}N in the same sample are suppressed. The acronym, BIRD, stands for **BI**linear **R**otation **D**ecoupling. The experiment is essentially a 180°–τ–90° pulse sequence for the protons attached to the NMR inactive nuclei, e.g. ^{12}C, with τ chosen to give zero signal. The pulse sequence is given in Fig. 6.18. The first ^{1}H (90°)$_x$ pulse, p1, brings all the ^{1}H magnetization into the xy plane, pointing along the y' axis in the rotating frame. The two components of the ^{1}H doublet due to coupling to ^{13}C rotate with respect to the $x'y'$ rotating frame. One component, coupled to ^{13}C with an α spin, rotates more slowly and the other, coupled to ^{13}C with a β spin, rotates more rapidly. Each arm of the doublet is separated from the middle of the doublet by $J/2$. Hence one arm will be rotating at $+J/2$ Hz with respect to the rotating frame and the other at $-J/2$ Hz. After a time, $1/2J$ s, the two arms of the doublet will have rotated 90° and now point in line along the $\pm x'$ axes of the rotating frame (Fig. 6.19). (180°)$_x$ pulses are applied to ^{1}H and ^{13}C. In this case, the (180°)$_x$ ^{1}H pulse will have no effect, but if the ^{1}H NMR signal was off-resonance, it would act as a refocusing pulse, see section 6.3. The 180° ^{13}C pulse swops the spins of the ^{13}C nuclei. The result is that the protons which were attached to ^{13}C with α-spin are now attached to ^{13}C nuclei with β-spin and vice versa. As the sign of the ^{13}C spin determines the direction of rotation of the ^{1}H magnetization of the coupled protons with respect to the

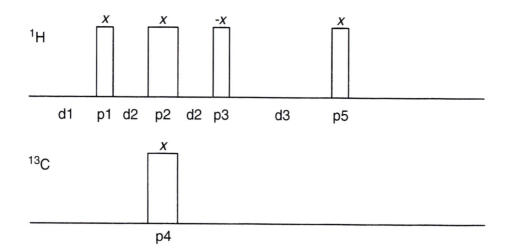

Figure 6.18 The BIRD pulse sequence. d1 is a relaxation delay, $5T_1$ being ideal. d2 = $1/2J$ s. d3 is a delay for the $-z$ magnetization generated in the ^{1}H nuclei attached to ^{12}C to decay to zero. p1, p3, and p5 are 90° pulses, while p2 and p4 are 180° pulses. The phases of the pulses are given above them. The actual BIRD pulse sequence comprises p1, d2, p2, p3, and p4. d3 and p5 have been added here to demonstrate the consequences of applying the pulse sequence.

rotating frame, the ^{1}H magnetizations now reverse their direction of rotation and refocus after $1/2J$ s. The final $(90°)_{-x}$ pulse returns the ^{1}H magnetization to the z direction. Compare this with what happens to the magnetization of the protons attached to ^{12}C (Fig. 6.19(b)). If the ^{1}H nuclei are on resonance, nothing happens during the $1/2J$ s waiting period. The $(180°)_x$ ^{1}H pulse, not only refocuses, but rotates the ^{1}H magnetization to point along the $-y$ direction. As the ^{1}H is attached to ^{12}C, not ^{13}C, the $180°$ ^{13}C pulse does nothing. The final $(90°)_{-x}$ pulse causes the ^{1}H magnetization to rotate into the $-z$ direction. We now have the situation where the magnetization of ^{1}H attached to ^{12}C is pointing in the $-z$ direction, while that of ^{1}H attached to ^{13}C is pointing in the $+z$ direction. During the following waiting period, d3, the magnetization of the ^{1}H attached to ^{12}C gradually recovers by the T_1 mechanism. d3 is chosen so that the magnetization of the ^{1}H attached to ^{12}C has recovered to approximately zero at the end of d3 (Fig. 6.18), though due to the spread of T_1 values in molecules, a compromise value of d3 has to be chosen. The final ^{1}H pulse gives an output for the ^{1}H–^{13}C protons but very little for the ^{1}H–^{12}C protons. Normally the BIRD pulse sequence is used as part of a longer pulse sequence combined with phase cycling to suppress more effectively the signals of ^{1}H attached to ^{12}C.

6.3.3 Magnetic field gradients

It may initially appear to be perverse to apply magnetic field gradients after having spent time shimming the magnetic field to obtain optimum resolution but this turns out to be beneficial. We will discuss simply in later chapters the application of pulses of field gradient during RF pulse sequences. They permit the selection of certain types of interaction with the elimination of others. This can be done

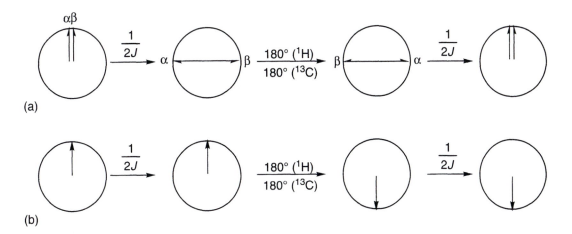

Figure 6.19 Part of the BIRD pulse sequence to show the behaviour of the nuclear spins in the xy plane. (a) The behaviour of the ^{1}H^{13}C doublet. (b) The behaviour of the ^{1}H^{12}C singlet.

using phase cycling of the type described above, with the difference that phase cycling requires many acquisitions to complete a cycle whereas field gradient pulses do the job in one acquisition. They are used particularly in biochemical spectroscopy.

6.4 QUESTIONS

6.1. Why do we use time shared decoupling for homonuclear decoupling but continuous decoupling for heteronuclear decoupling?

6.2. Where does the nuclear magnetization point after the three basic pulses used to make up the WALTZ decoupling sequence, i.e. $(90°)_x(180°)_{-x}(270°)_x$ assuming that the magnetization was in the z direction initially?

6.3. You have placed the ^1H decoupling frequency 1000 Hz from a proton which is coupled to a ^{13}C nucleus with $^1J = 125$ Hz. You measure the ^{13}C NMR spectrum and find an apparent ^1J of 100 Hz. What is the strength of the decoupling field? Note that $\gamma_C = 6.7263 \times 10^7$ rad T^{-1} s^{-1}.

6.4. In the inversion-recovery experiment used to measure T_1, the signal intensity is negative for short recovery times but positive for long recovery times. Calculate, in terms of T_1, at what recovery time the signal intensity is zero.

NMR spectra of exchanging and reacting systems

7

7.1 SYSTEMS AT EQUILIBRIUM

One of the most important contributions that NMR has made to chemistry is the insight it has given into the dynamic, time-dependent nature of many systems, particularly those which are at equilibrium or where simply intramolecular motion is involved. Spectroscopy based on higher-frequency radiation, such as classical infrared (IR) or ultraviolet (UV) spectroscopy, has given mostly a static picture because the timescale of many processes is slow relative to the frequency used. However, the lower frequencies used for NMR and the smaller line separations involved, coupled with the small natural linewidths obtained, means that many time-dependent processes affect the spectra profoundly. As an example, we consider the spectroscopic behaviour of ethanol in Fig. 7.1. The proton spectrum of 50% ethanol, $HOCH_2CH_3$, in $CDCl_3$ is a methyl triplet due to coupling to the CH_2, an OH triplet for the same reason and a doublet of quartets for the methylene protons. Any acidic impurity catalyses interchange of OH protons between molecules:

$$EtOH + EtOH^* \xrightleftharpoons{H^+} EtOH^* + EtOH$$

The exchange of the protons results in a short break in the CH_2OH coupling path and, since the total spin of the CH_2 protons in the two molecules between which the proton jumps may not be identical, then some of the OH protons will suffer an abrupt change in frequency. The result is to introduce uncertainty into their nuclear frequency and thus line broadening. In the spectrum of Fig. 7.1(b), which is of 50% ethanol, coupling of the OH and CH_2 proton signals is evident. Addition of acid causes acceleration of the rate of proton exchange to the extent that the OH–methylene coupling is completely lost and only an average frequency can be detected (Fig. 7.1(c)). The lines are now sharp. This is called the fast exchange region. When the lines are broadened (Fig. 7.1(a)) it is called the region of intermediate exchange rates and where coupling is fully developed is called the slow exchange region.

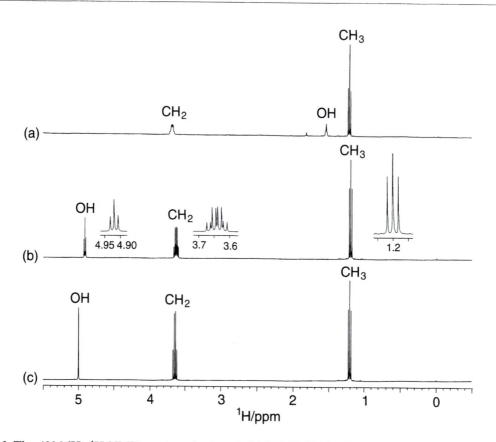

Figure 7.1 The 400 MHz ^{1}H NMR spectra of ethanol. (a) 2% EtOH in CDCl$_3$. (b) 50% EtOH in CDCl$_3$. (c) As (b), but with a drop of acid added.

Ethanol also allows us to demonstrate another aspect of fast exchange. In the concentrated solution, there is extensive hydrogen bonding between OH oxygen in one molecule and OH hydrogen in another. This interaction causes a high frequency shift of the OH proton resonance. If the alcohol is dilute (Fig. 7.1(a)) then the hydrogen bonds become dissociated to an extent depending upon the dilution, and there is a low-frequency shift of the OH resonance. The solution now can be regarded as containing two types of ethanol, hydrogen-bonded and non-hydrogen-bonded. These two species have different OH proton chemical shifts but are not observed separately because of the fast exchange between the two types of ethanol, which results in a signal of the average frequency being observed weighted by the concentrations of the two species. Thus chemical shift-dilution plots give information about the hydrogen bond dissociation. The lines in this solution are also broadened, presumably because of some acid impurity in the solvent.

Finally, we should note that in compounds such as ethanol there is very rapid rotation around the C–C bonds. At any instant, two of the methyl protons are close to the OH group and one is pointing away; see the Newman projection in Fig. 7.2. These differences are not observed in the spectrum, however, because each proton has an average chemical shift due to the rapid internal rotation. In certain molecules, such conformational changes can be quite slow and then the effects of the motion can be detected in the NMR spectra. Following the Karplus relationship (Fig. 3.1), the interproton coupling constants will also be different in an ethyl group; these are also averaged by the rotation to the value of the average of the Karplus curve.

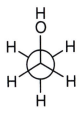

Figure 7.2 A Newman projection about the C–C bond of ethanol, showing the two environments for the methyl protons.

7.1.1 The effects of exchange on the lineshape of NMR spectra

We have described exchange by such words as 'fast' and 'slow'. It is now necessary to determine the timescale within which we can apply these terms correctly. A set of theoretical spectra for a two-site exchange with equal population in the two sites and no spin–spin coupling is shown in Fig. 7.3.

We start from the situation where the nuclei spend a long time in a given location in the molecule, Fig. 7.3(a), as shown in section 4.1, the linewidth is given by

$$W_{1/2} = \frac{1}{\pi T_2}$$

When chemical exchange occurs, the nuclei change their chemical shift as a consequence, and spend a time, τ_c, in a given location in the molecule. There is an extra contribution to the linewidth due to exchange and the linewidth of the exchange broadened line at half height, $(W_{1/2})_{ex}$, becomes

$$(W_{1/2})_{ex} = \frac{1}{\pi T_2} + \frac{1}{\pi \tau_{ex}}$$

This is a direct result of the Heisenberg uncertainty principle. As the lifetime in a given site gets shorter, the energy, and hence the linewidth becomes less well defined. τ_c is simply related to the rate of exchange, k, by

$$k = \frac{1}{\tau_{ex}}$$

Hence

$$(W_{1/2})_{ex} = \frac{1}{\pi T_2} + \frac{k}{\pi}$$

and k can be derived from the linewidth of the broadened line, $(W_{1/2})_{ex}$, by the simple calculation

$$k = \pi \left\{ (W_{1/2})_{ex} - \frac{1}{\pi T_2} \right\} = \pi \{ (W_{1/2})_{ex} - (W_{1/2})_0 \} \qquad (7.1)$$

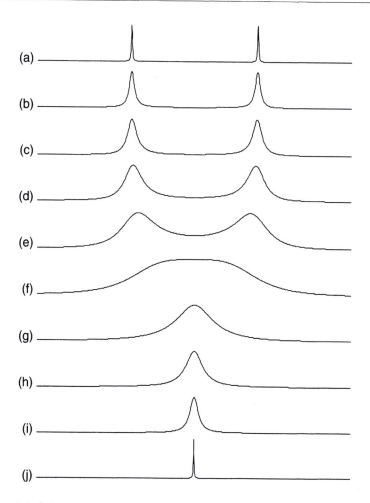

Figure 7.3 Calculated spectra for exchange between two equally populated sites separated by 40 Hz. T_2 values of both sites are 1 s. (a) $k = 0.1\,\text{s}^{-1}$. (b) $k = 5\,\text{s}^{-1}$. (c) $k = 10\,\text{s}^{-1}$. (d) $k = 20\,\text{s}^{-1}$. (e) $k = 40\,\text{s}^{-1}$. (f) $k = 88.8\,\text{s}^{-1}$. (g) $k = 200\,\text{s}^{-1}$. (h) $k = 400\,\text{s}^{-1}$. (i) $k = 800\,\text{s}^{-1}$. (j) $k = 10\,000\,\text{s}^{-1}$, where k is the rate of exchange between the two sites, and is normally varied by varying the temperature.

where $(W_{1/2})_0$ is the linewidth in the absence of exchange. Equation (7.1) is very simple and is general for any exchange problem, provided exchange is slow enough to result in separate signals for the exchanging sites. It does not matter if the populations are unequal, but caution is necessary when there is coupling, and the equation should only be applied if separate signals are observed for each line of the multiplet. Normally, $(W_{1/2})_0$ is estimated from a signal not involved in the exchange, and provided that $(W_{1/2})_{\text{ex}}$ is substantially larger than $(W_{1/2})_0$, an accurate value is obtained for k. The result of equation (7.1) is that for most compounds all signals involved in exchange initially broaden equally.

As the rate of exchange increases, the lines broaden, and the trough between the lines gradually fills. Once the lines are no longer well resolved, e.g. Fig. 7.3(d), the use of equation (7.1) becomes subject to error. The temperature where k is such that the trough between the two lines rises to be level with the peaks, Fig. 7.3(e), is known as the 'coalescence temperature'. At the coalescence temperature

$$k = \frac{\pi \left| \nu_A - \nu_B \right|}{\sqrt{2}} \qquad (7.2)$$

where ν_A and ν_B are the frequencies of the two sites in hertz at the coalescence temperature. This equation only applies to exchange between two equally populated uncoupled sites and $(W_{1/2})_0$ has to be negligible when compared with $\left| \nu_A - \nu_B \right|$. In principle, very accurate rates can be obtained at the coalescence temperature, provided that $(\nu_A - \nu_B)$ is known. As chemical shifts are temperature dependent, $(\nu_A - \nu_B)$ must be determined by extrapolation from lower temperatures, or, better, by complete lineshape fitting. Caution is necessary when extrapolating from low temperature, as when the signals begin to overlap and the trough between them fills, the maxima move together (Fig. 7.4).

It must be remembered that there is not a single coalescence temperature for a compound. $(\nu_A - \nu_B)$ depends on the magnetic field strength, the pair of nuclei exchanging, and the solvent. It is therefore meaningless to quote **the** coalescence temperature for a compound.

Above coalescence, the signals of the two lines are averaged, initially to a non-Lorentzian lineshape, but as the rate of exchange increases further, a Lorentzian lineshape is achieved (Fig. 7.3(g)–(j)). Then the equation

$$k = \frac{\pi(\nu_A - \nu_B)^2}{2\{(W_{1/2})_{ex} - (W_{1/2})_0\}} \qquad (7.3)$$

applies. This equation only applies to exchange between two equally populated uncoupled sites. The equation must be used with caution, as it is frequently difficult to obtain a reliable estimate of $(\nu_A - \nu_B)$ at the experimental temperature. The dependence of equation (7.3) on $(\nu_A - \nu_B)^2$ means that the line broadening of signals above the coalescence temperature can differ markedly for each pair of exchanging signals.

Frequently exchange processes affect several signals from a compound. This is illustrated here for *N*-methylaniline (Fig. 7.5). At –133°C, separate ^{13}C NMR signals are observed from all the six aromatic carbon atoms, showing that there is slow rotation about the Ph–N bond. At –126°C, the rotation produces a noticeable broadening of the signals due to C^2, C^3, C^5, and C^6 as can be observed by comparing the heights of these signals with the height of the signal due to C^4. On warming to –115°C, the signals due to C^3 and C^5 have averaged to give one signal, while one broadened signal is observed for C^2 and C^6.

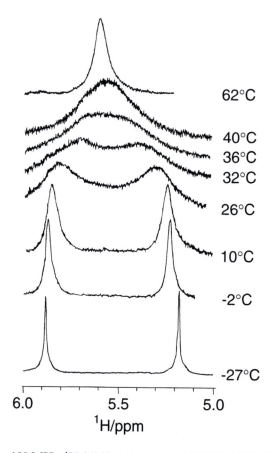

62°C

40°C
36°C
32°C

26°C

10°C

-2°C

-27°C

6.0   5.5   5.0

^{1}H/ppm

Figure 7.4 The 100 MHz ^{1}H NMR spectrum of $[Ti(\eta^1\text{-}C_5H_5)_2(\eta^5\text{-}C_5H_5)_2]$ in toluene at different temperatures. At –27°C the two types of cyclopentadienyl group are observed. For the η^1-group, there are rapid shifts of the titanium around the ring averaging all the protons of this ring. At higher temperatures, the two types of cyclopentadienyl ring exchange and an averaged signal is observed at 62°C. The spectrum at 36°C, where the doublet structure is just lost, is said to be at coalescence. (Reproduced with permission from Calderon *et al.* (1971) *J. Am. Chem. Soc.*, **93**, 3587 copyright, (1971) American Chemical Society.)

This is readily explained when it is remembered that the coalescence temperature depends on the signal separation. For C^3 and C^5, the signal separation is small, while for C^2 and C^6 the separation is large. Coalescence of the signals due to C^2 and C^6 in fact occurs at –115°C, while at this temperature, the averaged signal due to C^3 and C^5 is relatively sharp. The broadness of the signal due to C^2 and C^6 is substantial and such a signal can easily be missed. At –75°C, sharp averaged signals are observed for all the carbon atoms.

Exchange between unequally populated uncoupled sites results in differential broadening of the lines at exchange rates below the coalescence point (Fig. 7.6). The weaker signal from site A broadens more

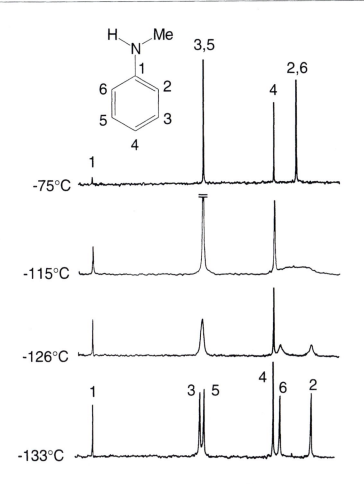

Figure 7.5 The variable temperature 25.16 MHz ^{13}C{^{1}H} NMR spectrum of N-methylaniline in Me$_2$O. (Reproduced from Lunazzi *et al.* (1979) *Tetrahedron Letts.*, 3031, copyright (1979), with permission from Elsevier Science.)

than the stronger one from site B. This is a direct consequence of the law of microscopic reversibility. This predicts that

$$\frac{k_A}{p_B} = \frac{k_B}{p_A} \quad \text{or} \quad k_A = k_B \frac{p_B}{p_A}$$

where k_A and k_B are the rates of leaving sites A and B, respectively, and p_A and p_B are the populations of sites A and B, respectively. Hence if $p_A = 3p_B$, then according to equation (7.1), the line broadening is proportional to k, the signal of B will broaden twice as much as the signal due to A. This is apparent in Fig. 7.6(c).

Above coalescence an average signal is observed at the weighted average chemical shift of the two signals, ν_{av}, where

$$\nu_{av} = \frac{\nu_A p_A + \nu_B p_B}{p_A + p_B}$$

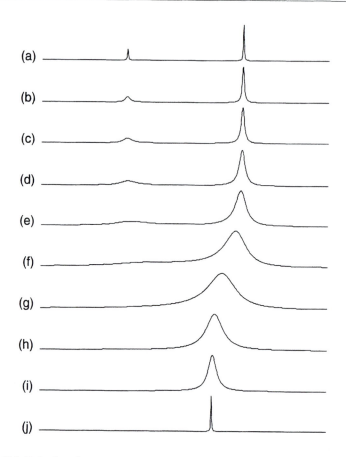

Figure 7.6 Calculated spectra for exchange between two sites with populations in the ratio 1 : 3 separated by 40 Hz. T_2 values of both sites are 1 s. (a) $k = 0.1 \, s^{-1}$. (b) $k = 5 \, s^{-1}$. (c) $k = 10 \, s^{-1}$. (d) $k = 20 \, s^{-1}$. (e) $k = 40 \, s^{-1}$. (f) $k = 88.8 \, s^{-1}$. (g) $k = 200 \, s^{-1}$. (h) $k = 400 \, s^{-1}$. (i) $k = 800 \, s^{-1}$. (j) $k = 10 \, 000 \, s^{-1}$.

Bromocyclohexane provides an example of a two-site exchange problem involving exchange between two molecular species of unequal populations, namely the conformers with axial and equatorial bromides.

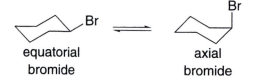

equatorial   axial
bromide   bromide

The variable temperature ^{13}C NMR spectra are shown in Fig. 7.7. The determination of the exchange rates and ratio of isomers where separate signals are observed is easy and accurate up to about −44°C. At higher temperatures, where only average signals are observed, the determination of accurate values of $(v_A - v_B)$ and ratio of concentration

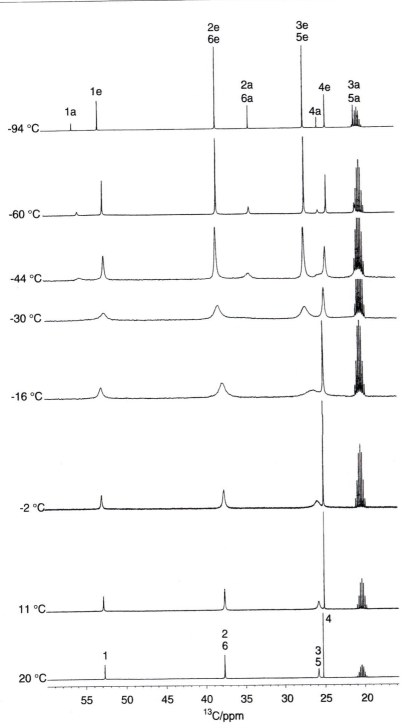

Figure 7.7 The variable temperature 100.62 MHz ^{13}C NMR spectrum of bromocyclohexane in $CD_3C_6D_5$. The 3a and 5a resonance is partially obscured at −44°C and below by the CD_3 solvent resonance. The a and e refer to the position of the bromine atom, axial or equatorial.

of isomers are subject to considerable error. Hence, exchange rates above coalescence have to be treated with considerable caution.

Examination of Fig 7.7 is instructive. Between 20 and –2°C, the signal due to C^4 is relatively sharp, but as the temperature is lowered through the range, there is differential broadening of the signals due to C^1, $C^{2,6}$, and $C^{3,5}$. Although for all four carbon nuclei the rate of exchange is the same, the broadening depends on the chemical shift separation between the axial and equatorial isomers. This is greatest for $C^{3,5}$ and least for C^4. At –44°C, separate signals are observed for the two isomers, and now the linewidths of all the signals for the major isomer are equal, as are those for the minor isomer. The relative linewidths of the major and minor isomers are 0.28 : 1, being inversely proportional to their relative concentrations. On further cooling, all the signals sharpen as the rate of exchange slows.

The analysis of spin-coupled systems usually requires computer fitting of the spectra. This can lead to accurate rate constants and valuable mechanistic information. For instance, two different mechanisms of exchange have been proposed for the trigonal bipyramidal structure shown in Fig. 7.8. Figure 7.9 shows the $^{31}P\{^1H\}$ NMR spectrum of $[Rh\{P(OMe)_3\}_5]^+$ at low temperature, along with spectra calculated on the basis of two mechanisms of exchange. Examination of the spectra shows that the better agreement is between the spectra calculated based on the Berry pseudo-rotation mechanism. The differences are only observed due to the second-order nature of the spectrum, which is A_2B_3X, where the X group is ^{103}Rh, 100% abundant, $I = 1/2$.

7.1.2 The use of magnetization transfer to study exchange

The difficulty of lineshape analysis is that the only information available is the rate of leaving each site. No information is directly available on the destination. In the case of a two-site exchange problem, the destination is normally obvious, but for multisite exchange problems, the destination is often far from obvious. There are a number of different magnetization transfer experiments used to examine such situations and they involve disturbing the Boltzmann population distribution in one or more sites.

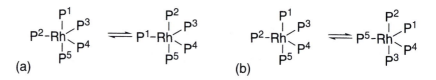

Figure 7.8 The two possible mechanisms of exchange in $[Rh\{P(OMe)_3\}_5]^+$, with the P(OMe)$_3$ ligands represented by individual phosphorus atoms. (a) The pairwise exchange mechanism which exchanges one axial phosphorus ligand with an equatorial one. (b) The Berry pseudo-rotation mechanism which exchanges the two axial phosphorus ligands with two equatorial ones.

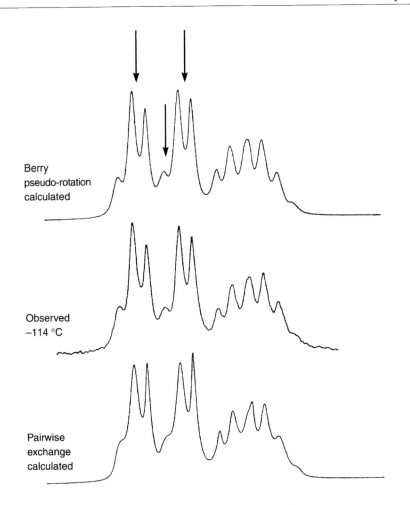

Figure 7.9 Observed and calculated 36.43 MHz ^{31}P NMR spectra of $[Rh\{P(OMe)_3\}_5]^+$ in $CHClF_2/CH_2Cl_2$, 9:1 at $-114°C$, with the effect of the protons removed by double irradiation. Arrows are used to draw attention to the regions of the spectrum where differences are most marked. (Reproduced with permission from Meakin and Jesson (1973) *J. Am. Chem. Soc.*, **95**, 7272, copyright (1973) American Chemical Society.)

The simplest experiment is saturation transfer. In this experiment, the decoupler is turned on to one site for several seconds. Then the decoupler is turned off and a general 90° pulse is applied to examine the effect. A second reference spectrum is taken where the decoupler is set on a region of the spectrum where there is no signal. A difference spectrum is taken to show what changes have been induced in the spectrum as a result of the irradiation. Let us analyse what is happening for exchange between two sites A and B. Before the decoupler was turned on, the spin system was at equilibrium with equilibrium magnetization, $M_z(0)$, at both sites. The decoupler, applied to site A, destroys the magnetization and $M_z^A = 0$. If there is exchange to the second site, nuclei from site A with no magnetization move to site B, while nuclei from site B, move to site A, where their magnetization is destroyed. The result is that the magnetization of site B is decreased by the exchange. The magnetization recovers by the T_1 processes. The experiment can be done dynamically, where the decoupler is turned on for various times, but here we will only analyse the case where the decoupler is turned on until equilibrium is achieved, typically $5T_1$. In a complicated molecule, it is easier to see the results of magnetization transfer in a difference spectrum, where the spectrum in the presence of magnetization transfer is subtracted from a reference spectrum so only changes are shown.

The time dependence of the magnetization at site B, $M_z^B(t)$, is given by

$$\frac{dM_z^B(t)}{dt} = \frac{M_z^B(0) - M_z^B(t)}{T_1^B} - k_B M_z^B(t) \tag{7.4}$$

The first term gives the T_1 recovery of the magnetization in site B and the second term is due to the loss of magnetization from site B by exchange. At equilibrium when $t = \infty$, the magnetization does not change, so

$$\frac{M_z^B(0) - M_z^B(\infty)}{T_1^B} - k_B M_z^B(\infty) = 0$$

or

$$k_B = \frac{1}{T_1^B} \frac{M_z^B(0) - M_z^B(\infty)}{M_z^B(\infty)} \tag{7.5}$$

If $k_B \gg 1/T_1^B$, then $M_z^B(\infty) \sim 0$. If $k_B \ll 1/T_1^B$, then $M_z^B(\infty) \sim M_z^B(0)$. Hence if $k_B \gg 1/T_1^B$, magnetization transfer causes the signal due to B to vanish as well as the signal due to A. If $k_B \ll 1/T_1^B$, the signal due to B is unaffected by magnetization transfer by exchange. If $1/5T_1^B \ll k_B \ll 5/T_1^B$, then k_B can be determined by measuring $M_z^B(\infty)$, $M_z^B(0)$, and T_1^B. This works well if $T_1^A \sim T_1^B$. Unfortunately if $T_1^A \neq T_1^B$, it is very difficult to measure T_1^A and T_1^B as exchange occurs during the waiting period of the $\pi - \tau - \pi/2$ pulse sequence, resulting in a partial averaging of T_1^A and T_1^B.

Considerable use is made of magnetization transfer measurements by decoupling in order to determine which signals are exchanging, and it can be seen accidentally during NOE measurements, see section 8.2. An example of its use is to determine how the osmium atom moves around the cyclooctatetraene ring in $[Os(\eta^6\text{-}C_8H_8)(\eta^4\text{-}C_8H_{12})]$ (Fig. 7.10).

Figure 7.10(a) shows the normal ^1H NMR spectrum of the cyclooctatetraene protons of $[Os(\eta^6\text{-}C_8H_8)(\eta^4\text{-}C_8H_{12})]$. The spectra in Fig. 7.10(b) and (c) are presented as difference spectra. A reference spectrum was also taken where the decoupling frequency was placed well away from any signal. This was then subtracted from the spectra where the decoupler was placed on H^2 in Fig. 7.10(b) and H^4 in Fig. 7.10(c).

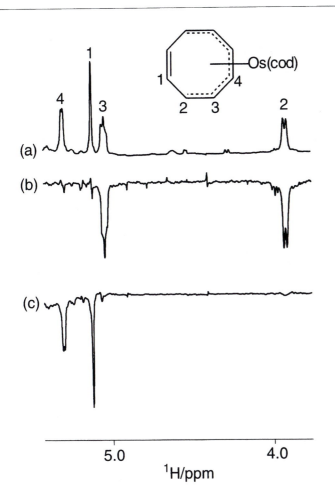

Figure 7.10 A partial 400.13 MHz ^{1}H NMR spectrum of $[Os(\eta^6\text{-}C_8H_8)$ $(\eta^4\text{-}C_8H_{12})]$ in $CD_3C_6D_5$ at 22°C. (a) The ^{1}H NMR spectrum of the $\eta^6\text{-}C_8H_8$ protons. The signals at δ 5.13 and 5.30 are due to H^1 and H^4, while those at δ 3.94 and 5.05 are due to H^2 and H^3 respectively. (b) A difference spectrum obtained by recording the spectrum with the decoupling frequency at δ 3.94 and then subtracting a second spectrum with the decoupling frequency set well away from any signal. (c) As (b), but with the signal at δ 5.30 irradiated. (Reproduced with permission from Mann (1988), *Adv. Organomet. Chem.*, **28**, 397.)

The resulting spectra show only the changes that occur as a result of the presaturation of H^2 or H^1. Examination of spectrum in Fig. 7.10(b) shows that when the magnetization at H^2 at δ 3.94 is destroyed a large negative signal results. There is magnetization transfer to H^3 at δ 5.05, which also gives a negative difference signal. The other signals have not changed so do not appear in the difference spectrum. Similarly, examination of spectrum in Fig. 7.10(c) shows that when the magnetization at H^4 at δ 5.30 is destroyed a large negative signal results. There is magnetization transfer to H^1 at δ 5.13. These results show that the dominant mechanism is a [1,5] metal shift (Fig. 7.11).

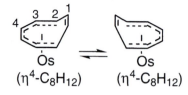

Figure 7.11 The [1,5] shift of the cyclooctatetraene ring in $[Os(\eta^6\text{-}C_8H_8)(\eta^4\text{-}C_8H_{12})]$.

The magnetization of one site can also be changed by a selective π-pulse. In this experiment, the pulse sequence, relaxation time – π(selective) – τ – $\pi/2$ is used. A series of measurements are performed with different values of τ, chosen to map the exchange and the subsequent spin–lattice relaxation. The resulting data are analysed using the family of equations for n magnetically nonequivalent sites, i,

$$\frac{\mathrm{d}M_z^i(t)}{\mathrm{d}t} = \sum_{j=1(i \neq j)}^{n} k_{ij} M_z^i(t) + \sum_{j=1(i \neq j)}^{n} k_{ji} M_z^j(t) + \left\{\frac{M_z^i(0) - M_z^i(t)}{T_{1i}}\right\} \quad (7.6)$$

The technique is particularly powerful, when applied to a multisite exchange problem. Typical experiments are shown in sections 7.1.2.1, 7.1.5.3 and 7.1.5.4.

7.1.2.1 Magnetization transfer in $[(\eta^5\text{-}C_5H_5)_2(H)Nb{=}CHOZr(H)(\eta^5\text{-}C_5Me_5)]$

If the rate of exchange is slow, so that linewidths are not significantly perturbed, then magnetization transfer is the ideal technique to use to examine exchange. This is applied here to $[(\eta^5\text{-}C_5H_5)_2(H)Nb{=}CHOZr(H)(\eta^5\text{-}C_5Me_5)]$ (Fig. 7.12). The formyl signal at δ 11.63 was inverted using a selective 180° pulse. Ideally, the inverted signal which is obtained after a short delay time of 0.002 s should be equal to that obtained after 3.0 s though of opposite sign. Due to imperfections, this was not achieved, but this is of no consequence as the magnetization has been substantially perturbed from equilibrium, and the return to equilibrium can be monitored. The formyl proton signal recovers to its normal intensity by two processes, spin–lattice relaxation and chemical exchange from the niobium hydride. The niobium hydride signal at δ –3.14 initially decreases in intensity because of exchange of the negative magnetization from the formyl proton. Both signals return to their normal sizes by spin–lattice relaxation. By fitting the evolution of the signal intensities with time to the differential equations, the two unknown quantities T_1 and the rate of exchange, k, can be determined. Values obtained at 32.5°C are $k = 15.7 \text{ s}^{-1}$ and $T_1 = 0.835$ s. Notice that only one T_1 value has been obtained although the two hydrogen environments will almost certainly have different T_1 values. This is because during one T_1 period of 0.835 s, the proton visits both sites a

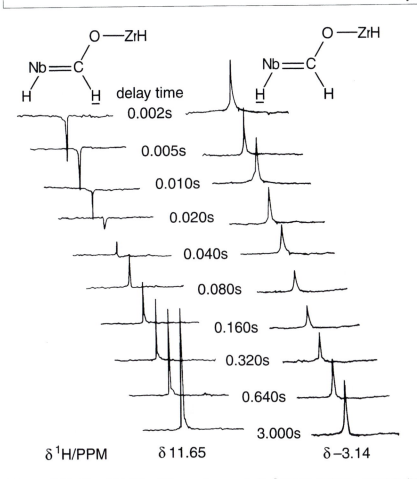

Figure 7.12 The 90 MHz ^{1}H spectrum of $[(\eta^5\text{-}C_5H_5)_2(H)Nb{=}CHOZr(H)(\eta^5\text{-}C_5Me_5)]$ in C_6D_6 at 32.5°C. A selective 180° pulse has been applied to the formyl proton (left set of spectra) and exchange transfers to the niobium hydride (right set of spectra). The delay time is that between the selective 180° and observing 90° pulse. (Reproduced with permission from Threlkel and Bercaw (1981) *J. Am. Chem. Soc.*, **103**, 2650, copyright (1981) American Chemical Society.)

number of times and T_1 is averaged over both sites. This makes the analysis of the data easier, but it also removes a possible extra complication. There is another mechanism by which one proton can perturb the intensity of another, namely the nuclear Overhauser effect, see section 8.2. This effect operates on the same time scale as T_1 and by choosing a temperature where k is fast compared with R_1, it can be neglected.

7.1.3 Temperature measurement

A factor that weighs heavily on the accuracy is, of course, the temperature of the sample. This is controlled by constructing a Dewar shield around the sample holder and passing over the sample cold nitrogen, boiled-off from liquid nitrogen, or air which have been adjusted to the required temperature by using a controlled heater. The heater is regulated by a thermocouple placed just below the sample tube. The method is prone to temperature gradients and so inaccuracies, which are minimized by calibrating the temperatures. This is done in two major ways. In the first method a capillary tube, containing MeOH for below room temperature and $HOCH_2CH_2OH$ for above room temperature, is placed inside the NMR tube. The separation between the OH and CH_3 or CH_2 groups, Δv ppm, is measured. The temperature is related to Δv by the equations

for methanol

$$T°C = 403.0 - 29.46\Delta v - 23.8\Delta v^2$$

for ethylene glycol

$$T°C = 466.0 - 101.64\Delta v$$

In the second method, the NMR tube is replaced by one containing a thermocouple in the same solvent.

7.1.4 The determination of activation energies from NMR data

We can relate τ to temperature very accurately at coalescence and can, for example, obtain a value for the free energy of activation at that temperature using the Eyring theory

$$\Delta G^{\ddagger} = RT\left\{23.759 + \ln\left(\frac{k}{T}\right)\right\} \tag{7.7}$$

where the rate constant k is obtained from $k = 1/\tau_{ex}$.

Equation (7.7) permits the calculation of $\Delta G^{\ddagger}$ values from NMR data and the values obtained are usually accurate to better than 1 kJ mol^{-1}, especially at and below coalescence. It is then very tempting to use the relationship

$$\Delta G^{\ddagger} = \Delta H^{\ddagger} - T\Delta S^{\ddagger}$$

Combining this equation with equation (7.7) permits the derivation of $\Delta H^{\ddagger}$ and $\Delta S^{\ddagger}$. A plot of $\ln(k/T)$ against $1/T$ gives a straight line with gradient $\Delta H^{\ddagger}/T$ and intercept $(\Delta S^{\ddagger}/R + 23.759)$. Unfortunately this approach is subject to hidden errors, substantially greater than given by regression analysis, and many of the values found in the literature are grossly in error. For example, although the published values of $\Delta G^{\ddagger}$ for rotation about the N–CHO in Me_2NCHO are in the range 93 ± 6 kJ mol^{-1}, the values of $\Delta H^{\ddagger}$ range from 25 to 115 kJ mol^{-1}. The errors in $\Delta H^{\ddagger}$ arise from many causes, some of which are temperature dependent chemical shifts, mis-estimates of the linewidth in the absence of exchange, poor temperature calibration, and signal broadening due to unresolved coupling.

There are several approaches to obtaining reliable values of $\Delta H^{\ddagger}$ and $\Delta S^{\ddagger}$, two of which are described here. Firstly, we can obtain data over a wide range of temperature if we can monitor coalescence points for a system at different spectrometer frequencies (or magnetic field strengths) and for different nuclei or groups with different chemical shifts that are affected equally by the exchange. This approach has been used to study hindered rotation of the *N*-ethyl groups of the iron complex $[(CH_3CH_2)_2N–C(S)SFe(CO)_2(\eta^5\text{-}C_5H_5)]$. Both the methyl and methylene protons of the ethyl groups are non-equivalent at low temperatures, but the chemical shift between methyl signals is less than that between methylene signals, so that we can monitor two coalescence points. In addition, we can obtain two more coalescence points from the ^{13}C NMR spectra of the two groups, and by using two different spectrometers we can observe a total of eight coalescences. The results are shown in Fig. 7.13, which gives the Arrhenius parameters $E_a = 66$ kJ mol^{-1} and log$A = 13.1$.

Alternatively, two independent methods can be used to determine the rate at substantially different temperatures. Lineshape analysis is reliable from the point when the rate of exchange is sufficient to produce line broadening which is at least ten times greater than the natural linewidth until coalescence. This can be complemented by magnetization transfer which is accurate from a rate of around $1/T_1$ until the linewidth due to exchange is around 10% of the signal

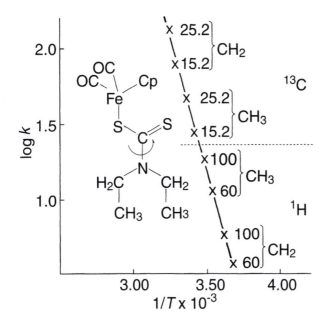

Figure 7.13 Determination of the activation parameters characterizing the hindered rotation around the C–N bond of $[(CH_3CH_2)_2NC(S)SFe(CO)_2(\eta^5\text{-}C_5H_5)]$. (From Martin *et al.* (1980) *Practical NMR Spectroscopy*, copyright (1980) John Wiley and Sons Ltd, reprinted with permission.)

separation. This procedure was applied to Me_2NCHO to give $\Delta H^{\ddagger} = 84.8 \pm 1.3$ kJ mol^{-1} and $\Delta S^{\ddagger} = -6.2 \pm 3.6$ J K^{-1} mol^{-1} (Fig. 7.14).

7.1.5 Some further examples of chemical exchange

7.1.5.1 Cis and trans isomers of a vinyl diamide

A more complex example of the hindered rotation in amides is shown in Fig. 7.15 for the *cis* and *trans* isomers of a vinyl diamide. Only the methylene proton resonances are shown in the figure, and they indicate that several different rotation processes take place. If we take the *trans* isomer first, we see that at the highest temperature recorded there is a well resolved methylene quartet overlying a broadened resonance. This latter broadens and splits at lower temperatures with a coalescence temperature of around 345K. Thus one of the amides is rotating more slowly than the other, which only shows coalescence between 273 and 295K. Exchange is no longer evident in the spectra at 228K. Only three quartets are observed because of overlap of the two high frequency quartets. In order to understand the spectra completely, it is necessary to identify which amide resonance is which, and this is done using nuclear Overhauser and two-dimensional proton carbon correlation spectroscopy, which we will describe later. It is sufficient to record here that it is the amide on the CMe carbon

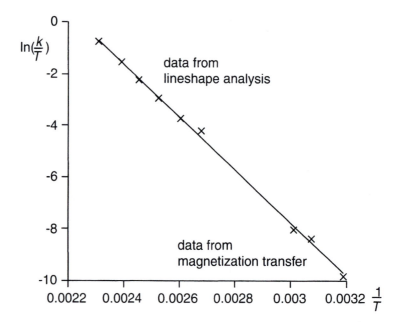

Figure 7.14 The Eyring plot as applied to exchange data for Me_2NCHO in DMSO/d_6-DMSO. (From Mann *et al.* (1977) *J. Magn. Reson.*, **25**, 91, reprinted with permission.)

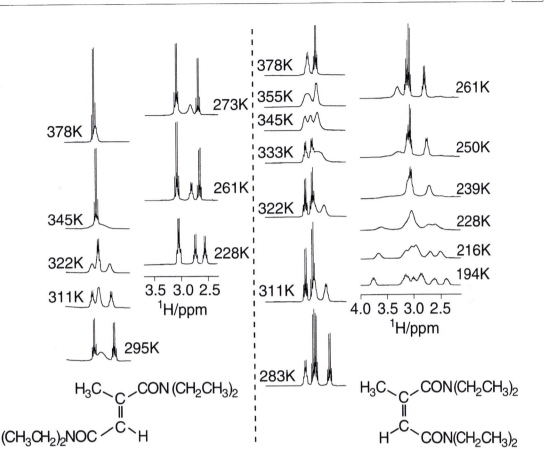

Figure 7.15 The 300 MHz ^{1}H NMR spectra of the vinyl diamides shown at a series of temperatures. The solvent is d_8-toluene. (From Szalontai *et al.* (1989) *Magn. Reson. Chem.*, **27**, 216–22, copyright (1989) John Wiley and Sons Ltd, reprinted with permission.)

that is rotating faster. The *cis* isomer shows much more complex behaviour, though at the higher temperatures there are two corresponding coalescence points at 355K and 333K. The quartets, however, split further at lower temperatures, and each resonance is transformed into a doublet of quartets due to the introduction of extra spin–spin coupling between geminal protons on the same carbon atom. The methylene protons have thus been rendered non-equivalent by some further restriction in the motion of the molecule. The activation parameters were obtained by calculating τ at the coalescence points. To do this accurately, it is of course necessary to know the chemical shift between the two coalescing signals. This may vary with temperature, and for the present example it was found necessary to measure the chemical shifts in the slow exchange limit over a wide range of temperatures to ensure that the correct frequency separation had been used.

7.1.5.2 [Co₄(CO)₁₂]

In the case of systems exhibiting fast exchange, it may simply be necessary to reduce the temperature sufficiently to slow down the exchange process and cause the slow exchange limit spectrum to appear, since this will be much more informative than the fast exchange spectrum, which may be just a singlet. A typical example is the ^{17}O spectrum of the cobalt carbonyl compound, $[Co_4(CO)_{12}]$. In this case, the ^{17}O NMR spectrum was recorded rather than the ^{13}C NMR spectrum due to extreme broadening of the ^{13}C NMR signals by scalar relaxation by ^{59}Co, see section 4.6. The carbonyl groups move around the cobalt cluster so that they all experience the full range of chemical environments and shifts in the molecule. Spectra at several temperatures are shown in Fig. 7.16, and it will be evident that the ambient-temperature trace is not very informative but that at –25°C all four types of carbonyl group can be distinguished. Note that the ^{59}Co spectrum of this compound contains two resonances in the intensity ratio $3:1$ so that the carbonyl exchange takes place on a cluster in which the positions and bonding of the cobalt atoms are invariant. The nucleus ^{59}Co is quadrupolar and the asymmetry of the environment around the metal atoms means that the relaxation time is very short. Linewidths are of the order of 7500 Hz but ^{59}Co chemical shifts are very large and so the resonances are resolved.

7.1.5.3 A multisite lineshape exchange problem, [(η⁴-C₈H₈)Ru(CO)₃]

A final example of fast exchange where the structure can be deduced only at low temperature is that of the intramolecular transfer of the $Ru(CO)_3$ group around the ring of the η^4-cyclooctatetraene complex, $[(\eta^4\text{-}C_8H_8)Ru(CO)_3]$. This compound, whose structure is shown in Fig. 7.17, contains four distinguishable types of hydrogen atom, and its proton spectrum should consist of four chemically shifted lines with intensity ratios $2:2:2:2$. In fact, only a singlet is observed at 25°C, as shown in Fig. 7.18. Cooling causes broadening, until, by –107°C, the expected structure starts to emerge as the exchange is slowed down. Interestingly, there is an obvious asymmetry in the spectrum of the outer triplets between –107 and –115°C, and this allows us to deduce the mechanism of the transfer of the $Ru(CO)_3$ group.

There are five possibilities for exchange in complexes of this type: (i) 1,2-shifts; (ii) 1,3-shifts; (iii) 1,4-shifts; (iv) 1,5-shifts; and (v) a random shift process, where all types of shift occur with equal probability. This is shown in Fig. 7.17, where the chemically shifted protons/carbons are numbered 1 to 4. Table 7.1 shows how the spin chemical shifts change for each process. Examination of Table 7.1 shows that when a 1,2-shift occurs, atom 1 and 1′ move to positions 1′ or 2′, i.e. for half the moves it remains in the same chemical shift position. Atom 4 and 4′ behave similarly, moving to positions 3 or 4. Contrast this with atoms 2, 2′, 3, and 3′ which move from their

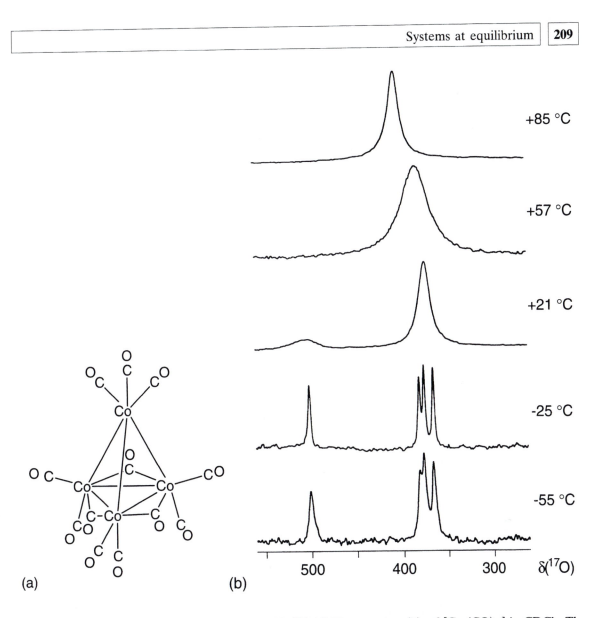

Figure 7.16 The structure, (a), and the 54.25 MHz ^{17}O NMR spectrum, (b), of $[Co_4(CO)_{12}]$ in $CDCl_3$. The carbonyl groups exchange positions, but this can be slowed down sufficiently by cooling to enable the four types of carbonyl group to be seen. The reference is water. (From Aime *et al.* (1981) *J. Am. Chem. Soc.*, **103**, 5920, copyright (1981) American Chemical Society, reprinted with permission.)

original position each time. Remembering that the line broadening below coalescence is proportional to the rate of leaving the site, clearly if the mechanism is a 1,2-shift then the signals associated with atoms 2 and 3 will be broadened twice as much as those associated with atoms 1 and 4. A similar analysis can be performed for the other mechanisms. For a 1,4-shift the reverse occurs with the signals associated with atoms 1 and 4 which will be broadened twice as much as those associated with atoms 2 and 3. For 1,3-shifts, 1,5-shifts and random shifts, all the

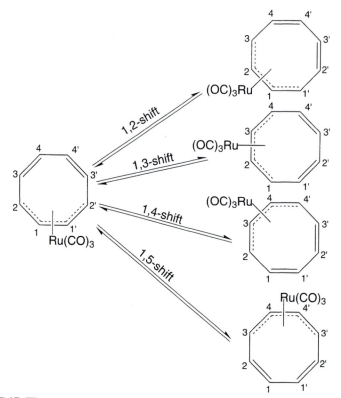

Figure 7.17 The possible mechanisms of fluxionality of the Ru(CO)$_3$ group in $[(\eta^4\text{-}C_8H_8)Ru(CO)_3]$.

signals broaden equally. In principle, the 1,5-shift can be distinguished as atom 1 exchanges exclusively with atom 4 and atom 2 with atom 3, so the high temperature limiting spectrum should consist of two signals in the ratio 4 : 4. However this presupposes that all the other possible mechanisms are of such high energy as not to be significant at 25°C.

Examination of Fig. 7.18 shows that between −115 and −107°C there is clearly differential line broadening. It is the triplets which broaden most. Strictly, this is an [ABCD]$_2$ spin system, but at the resolution achieved at low temperature, a simple analysis works. The doublets arise from H^1 and H^4, while the triplets arise from H^2 and H^3. It is therefore H^2 and H^3 that broaden most and the mechanism is a 1,2-shift. The derived $\Delta G^{\ddagger}$ is 32 kJ mol^{-1}.

When $[(\eta^6\text{-}C_8H_8)Cr(CO)_3]$ was examined, it was found that the four different ^{1}H NMR signals from the C$_8$H$_8$ ring broadened approximately equally. We have the same fluxional processes possible as for $[(\eta^4\text{-}C_8H_8)Ru(CO)_3]$, including random shifts. Examination of the table above shows that the rates of movement of a specific ^{1}H or ^{13}C from their original position to another is equal, hence equal broadening, for a 1,3-, a 1,5-, or a random shift. The 1,5-shift mechanism was dismissed

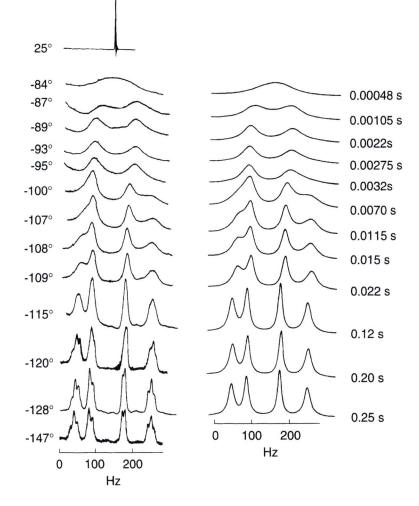

Figure 7.18 Variable temperature 100 MHz ^{1}H NMR spectrum of [(η^4-C$_8$H$_8$)Ru(CO)$_3$] in CFHCl$_2$/CF$_2$Cl$_2$. The left hand set of signals are the experimental, while the right hand set are calculated on the basis of a 1,2-shift mechanism. (From Cotton *et al.* (1969) *J. Am. Chem. Soc.*, **91**, 6598, copyright (1969) American Chemical Society, reprinted with permission.)

as it only exchanges 1 and 4 and 2 and 3 and will give rise to two signals at the high temperature limit, rather than the observed singlet. This left the 1,3- and random shift mechanisms, but line shape analysis cannot differentiate between these mechanisms as both mechanisms lead to equal broadening of the lines. This led to a disagreement between Cotton who favoured the 1,3-shift mechanism and Whitesides who favoured the random shift mechanism.

The argument was resolved by using saturation transfer (Fig. 7.19). The most informative spectrum is (b) where irradiation at site 4 produces a greater reduction in the intensity at sites 2 and 3 rather

Table 7.1 The predicted movements for the CH groups of a cycloocta-tetraene undergoing different shifts of the coordinated metal group

Starting position	Fluxional mechanism				
	1,2-shift	1,3-shift	1,4-shift	1,5-shift	Random shift
1	**1′**	2′	3′	4′	Any
2	1	1′	**2′**	3′	Any
3	2	1	1′	2′	Any
4	3	2	1	1′	Any
4′	**4**	3	2	1	Any
3′	4′	4	**3**	2	Any
2′	3′	4′	4	3	Any
1′	2′	3′	4′	4	Any

than 1. This is in agreement with a 1,3-shift which moves carbon 4 to sites 2 and 3. The intensity at site 1 also goes down, but this is from a double movement, e.g. 4 to 3 and then 3 to 1. A quantitative analysis of the spectra showed that the dominant mechanism is a 1,3-shift, but a 1,2-shift was also detected at approximately a quarter of the rate of the 1,3-shift. This clearly demonstrates the power of the magnetization transfer method to unravel complex exchange pathways.

Note that the intensity does not go to zero on irradiation of one site. Once the saturated nuclei move to another site, they are no longer being irradiated, and can recover their intensity, by re-equilibration losing the excess of energy to the lattice, i.e. through spin lattice relaxation. The size of the residual signals depends on a competition between exchange and relaxation.

7.1.5.4 A multisite magnetization transfer problem [Ir₄(CO)₁₁(PEt₃)]

One of the largest multisite problems ever to have been examined is that of carbonyl exchange in $[Ir_4(CO)_{11}(PEt_3)]$. This compound exists in solution as a mixture of two isomers, the major one with the PEt_3 ligand orthogonal to the carbonyl bridged Ir_3 face and the minor one with the PEt_3 ligand in the plane of the carbonyl bridged face (Fig. 7.20). There are 14 different carbonyl sites. The use of selective 180° pulses demonstrated the mechanism of exchange and the rates were determined. Fig. 7.21 shows one of many experiments. The carbonyl region of the ^{13}C NMR spectrum is shown in Fig. 7.21(a), obtained with a long τ so that the selectively inverted nuclei have regained their normal magnetization. This is then a normal ^{13}C NMR spectrum. A second spectrum is shown in Fig. 7.21(b) where a selective pulse was then applied to carbonyl a in the major isomer followed immediately, 3μs, by a general 90° pulse. The result is that the signal of carbonyl a is inverted. A third spectrum was recorded with a wait of 0.01 s between the 180° selective pulse and the 90° general pulse. The spectrum in Fig. 7.21(b) was subtracted from this third spectrum

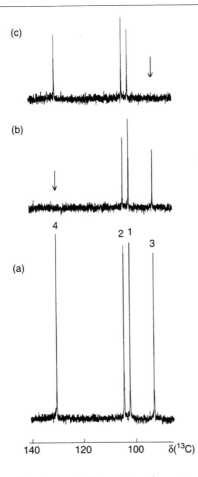

Figure 7.19 The application of the saturation transfer method to the 25.15 MHz ^{13}C NMR spectrum of $[(\eta^6\text{-}C_8H_8)Cr(CO)_3]$ in CD_2Cl_2/CH_2Cl_2 at $-15°C$. (a) Normal spectrum. (b) and (c) with irradiation as shown by arrows. (Reproduced with permission from B.E. Mann (1977) *J. Chem. Soc., Chem. Commun.*, 626.)

to give the spectrum in Fig. 7.21(c). Examination of Fig. 7.21(c) shows that the magnetization of C^aO has partially recovered. This is due to two factors, exchange bringing nuclei with equilibrium magnetization into this site, and T_1 relaxation. The signals due to C^AO, C^dO, and C^fO have decreased in intensity due to the exchange of C^aO into these sites, bringing negative magnetization with them. In the full study, a series of experiments were performed varying τ and the site of the selective 180° pulse, in order to map out the exchange network. It was concluded that two exchange mechanisms are operating, a mechanism involving a bridge-opened intermediate (Fig. 7.20(a)), and a mechanism involving a μ_3-CO intermediate (Fig. 7.20(b)) which interconverts the major and minor isomers and rates were determined for each mechanism.

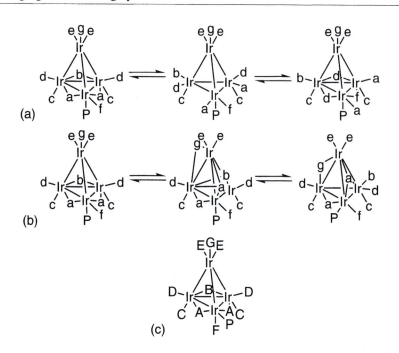

Figure 7.20 The two mechanisms, (a) and (b), of exchange identified for the fluxionality of the carbonyls and interchange of isomers of $[Ir_4(CO)_{11}(PEt_3)]$. Letters are used to denote the carbonyls and PEt_3 is abbreviated to P. (c). The lettering used to identify the carbonyls in the minor isomer in Fig. 7.21.

7.1.5.5 Solvent exchange on cations

One very wide field of interest is that of ligand exchange on complex cations. These processes have been studied in a variety of ways, and the rates measured for ligand exchange vary over many orders of magnitude and depend upon the nature of the cation chosen for study. In many instances, the rates fall, or can be made to fall, by changing the temperature, in the range accessible to NMR studies, so that this relatively new technique has been used to extend our knowledge of such systems. In addition, since NMR provides the means to study systems at equilibrium, our studies can be of the unperturbed system.

A particularly well studied system is that of solvent exchange on the cation $[Al(H_2O)_6]^{3+}$ in aqueous solution. This cation contains the NMR-active nuclei ^1H, ^{27}Al and, if isotopically enriched, ^{17}O, all of which have been used in its study. Such a highly charged ion is subject to hydrolysis and so solvent exchange can be quite a complex process:

$$[Al(H_2O)_6]^{3+} + H_2O^* \xrightleftharpoons{\quad k_{ex} \quad} [Al(H_2O)_5(H_2O^*)]^{3+} + H_2O$$

$$+ H^+ \Big\Updownarrow - H^+ \qquad\qquad\qquad + H^+ \Big\Updownarrow - H^+$$

$$[Al(H_2O)_5(OH)]^{3+} + H_2O^* \rightleftharpoons [Al(H_2O)_4(H_2O^*)(OH)]^{3+} + H_2O$$

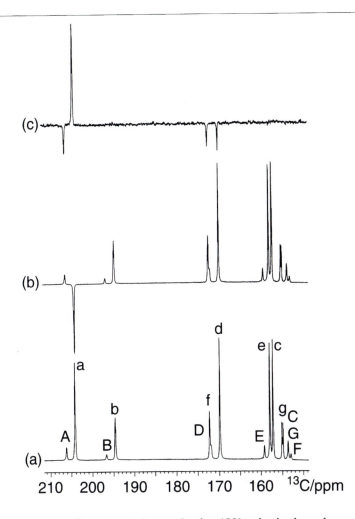

Figure 7.21 The effect of applying a selective 180° pulse in the pulse sequence, relaxation time–π(selective) – τ – $\pi/2$, to the bridging carbonyl in the major isomer of $[Ir_4(CO)_{11}(PEt_3)]$ in CD_2Cl_2. (a) The 100.62 MHz ^{13}C NMR spectrum of the carbonyl region of the spectrum acquired with $\tau > 5T_1$. (b) The 100.62 MHz ^{13}C NMR spectrum of the carbonyl region of the spectrum acquired with $\tau = 3$ μs. (c) The difference 100.62 MHz ^{13}C NMR spectrum of the carbonyl region of the spectrum after subtracting spectrum (b) from a spectrum with $\tau = 0.01$ s. (Note that this is at higher gain.) (Reproduced with permission from Mann *et al.* (1989) *J. Chem. Soc., Dalton.*, 889.)

The hydrolysis reaction causes exchange of hydrogen atoms, which in the present case is very fast, although the backward reaction is much faster than the forward reaction and the hydrolysis constant is only 10^{-5} (pK = 5). The oxygen exchange is much slower, though two pathways are possible via either the hydrolysed or the non-hydrolysed ions. We would like to know the rates of exchange of both hydrogen and oxygen, whether the latter is influenced by the acidity of the solution and so whether the hydrolysed species contributes to the exchange, and, if possible, the mechanism by which a water molecule replaces a complexed water molecule.

The rate of proton exchange can be studied using ^1H NMR spectroscopy. If a solution of an aluminium salt is cooled, this slows down the rates of exchange. If the temperature is low enough, and this can be achieved either by adding an antifreeze such as acetone or by using a very concentrated salt solution, then two proton resonances are observed due to solvent and complexed water, as shown in Fig. 7.22. Integration of the resonances together with knowledge of the aluminium concentration allows the hydration number (the number of water ligands attached to the cation) to be determined, and this is approximately equal to six, the method being less precise than the isotope fine-structure method depicted in Fig. 2.8. The chemical shift between the two resonances is 4.2 ppm, and we can calculate that coalescence will occur when the lifetime of the protons is about 1.8×10^{-3} s. We can therefore study the rate of exchange in dilute solution by measuring the proton relaxation times as a function of temperature and acidity, and this gives a protolysis rate constant of 0.79×10^5 s^{-1} at 298K. The ^{27}Al resonance is also influenced by this process since the hydrolysed cation has a much broader resonance than that of the non-hydrolysed species. The latter has a width of some 2 Hz, but the fast exchange mixes in the rapid quadrupolar relaxation of the hydrolysed cation and gives lines that are generally in the range 10–20 Hz wide. Note that, in this case, the exchange of one atom (H$^+$) results in apparent exchange of a second (^{27}Al).

The rate of oxygen exchange is best studied using the ^{17}O resonance. Here, the rate of exchange is known to be slower, so that resonances should be observed for bound and bulk water molecules. In fact, the chemical shift between the two is very small and, because of the quadrupolar broadening of the lines, they are not resolved. Separating the two resonances has been done in a variety of ways, but the one used for the most comprehensive studies was to add a paramagnetic cation, Mn^{2+}, for which the whole water molecule exchange rate is

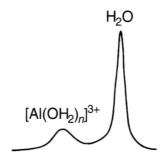

Figure 7.22 The 60 MHz ^1H NMR spectrum of 3 mol l^{-1} AlCl$_3$ at –47°C. The water complexed by the cation is seen 4.2 ppm to high frequency of free, bulk water. (After Schuster and Fratiello (1976) *J. Chem. Phys.*, **47**, 1554, with permission.)

very fast. The water not attached to aluminium thus is subjected rapidly to the very large magnetic field of the electron spins on the Mn^{2+}, and has its T_2 very much reduced and so its linewidth increased, to the extent that its signal disappears in the baseline of the spectrum. Only the oxygen bound to Al^{3+} is observed and, provided the exchange on Al^{3+} is slow enough, its relaxation is determined entirely by its own natural relaxation time and exchange lifetime (equation (7.1)). The problem in making such measurements with quadrupolar nuclei is that the relaxation time in the absence of exchange is not known and that it is very temperature-dependent. The relaxation time, or linewidth, thus has to be measured over a range of temperature wide enough that some data are acquired in the region where the exchange broadening is negligible. A set of measurements is shown in Fig. 7.23, where it will be seen that the temperature dependence of relaxation falls into two regions: lower temperatures, where the linewidth is dominated by the quadrupolar mechanism, and higher temperatures, where it is dominated by exchange. Extrapolation of the quadrupolar influence in principle allows the exchange rates to be determined as a function of temperature. In fact, because of the curvature, it is difficult to estimate the slope of the line giving k_{ex} with the required precision, and a separate experiment was needed in order to increase the range of values of k_{ex} observed. This was done using an injection or stopped-flow technique in which an aluminium salt solution in H_2O was injected rapidly into acidified H_2O enriched in ^{17}O, both solutions being at low temperature around 256K. Mn^{2+} was also present in the ^{17}O water. Thus, as the water exchanged with the aluminium solvation sphere, the ^{17}O signal of the mixture increased in intensity and the exchange rate could be obtained by plotting this as a function of time. This isotope exchange type of experiment is a common way of investigating systems that are at equilibrium. One is effectively perturbing the system but the perturbation is slight. The combined linewidth and stopped-flow results are shown in Fig. 7.23(b). In the case of aluminium it is possible to suppress hydrolysis completely by the addition of sufficient acid. Thus the data gave thermodynamic parameters for oxygen exchange on the non-hydrolysed cation: $k = 1.29$ s^{-1}, $\Delta H^{\ddagger} = 84.7$ kJ mol^{-1}, $\Delta S^{\ddagger} = +41.6$ J $K^{-1}mol^{-1}$. In addition, by measuring the exchange rates as a function of pressure at constant temperature, it was possible to obtain an activation volume for the oxygen exchange: $\Delta V^{\ddagger} = +5.7$ cm^3 mol^{-1}. We will discuss such experiments in a little more detail later, but the inference is that the transition state during the exchange involves a short-term increase in total volume. This occurs if the exchanging ligand leaves before the entering solvent molecule takes its place; in other words, it is a dissociative mechanism of exchange. In fact, the $\Delta V^{\ddagger}$ term is a little smaller than theory would demand if there were full dissociation, and the mechanism probably involves interchange during the dissociation, commonly called the I_d mechanism.

In the case of the gallium cation, $[Ga(H_2O)_6]^{3+}$, the hydrolysis of the ion is too strong to be suppressed completely by the addition of acid,

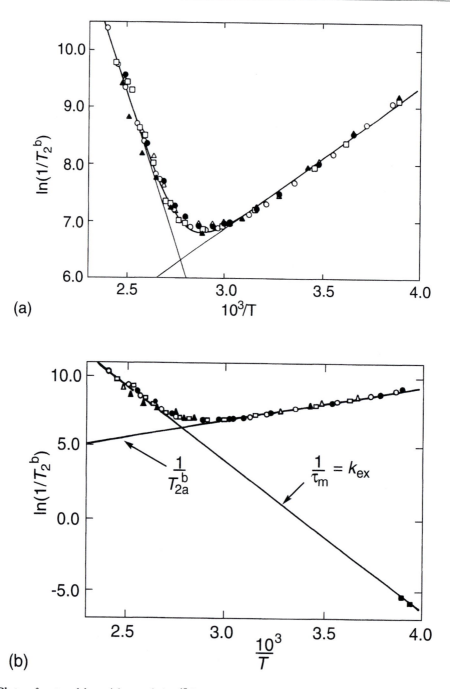

Figure 7.23 Plots of natural logarithms of the ^{17}O relaxation rates $1/T_2$ of $^{17}OH_2$ bound to the aluminium cation, as a function of the reciprocal temperature. The T_2 values were obtained from the ^{17}O linewidths. (a) How the linewidth decreases as the temperature is increased (from the right-hand side of the plot) and how these data can be extrapolated satisfactorily into the exchange perturbed region. The onset of detectable exchange causes the plot to curve upwards as the temperature is increased beyond 333 K. (b) This contains the same data but includes some fast injection data also, (■). (From Merbach *et al.* (1985) *Helv. Chim. Acta*, **68**, 545, with permission.)

and in this case oxygen exchange on both hydrolysed and non-hydrolysed species has to be measured. It is found that

$$k_{ex} = k_1 + \frac{k_2}{[H+]}$$

where $k_1 = 403$ s^{-1} and $k_2 = 14$ mol s^{-1}, both at 298K.

7.1.5.6 Effect of exchange on coupling patterns

The behaviour of t-butyllithium demonstrates the averaging effects of exchange upon coupling constants and coupling multiplicities. The compound was studied using ^{13}C spectroscopy and the spin–spin coupling of this nucleus to the lithium nuclei. The common isotope of lithium is ^{7}Li, which has a quadrupole moment sufficiently large that its quadrupole relaxation times are quite short, and so any coupling is not well resolved. On the other hand, the less abundant isotope ^{6}Li has a quadrupole moment that is some 50 times smaller, which is indeed the smallest among all the elements, and which thus behaves much more like a $I = 1/2$ nucleus than like a quadrupolar one. Further, ^{6}Li is easily available via the nuclear industry and so is proving to be a useful tool for the study of organolithium compounds by NMR methods. The structure of t-butyllithium is shown in Fig. 7.24, and the ^{13}C spectrum of the α-carbon atoms of a sample highly enriched in ^{6}Li is shown in Fig. 7.25 as a function of temperature. This compound forms tetramers in cyclopentane solution in which the lithium atoms are arranged in a tetrahedral cluster with the alkyl groups bonded to three lithium atoms by a multicentre bond and so situated over each face of the tetrahedron. The tetramer is, however, fluxional and the expected spectrum is only observed at low temperatures. The α-carbon atom of the t-butyl group is coupled equally to its three nearest-neighbour ^{6}Li atoms with a coupling constant of 5.4 Hz. For ^{6}Li, $I = 1$ so that a seven-line multiplet results, with intensities in the ratio $1:3:6:7:6:3:1$, and this is the spectrum observed at –88°C in Fig. 7.25. The coupling to the distant lithium nucleus is undetectably small. As the temperature is increased, fluxion of the structure occurs, and at 26°C this is fast enough for the α-carbon atoms to be influenced equally by all four ^{6}Li nuclei, which produces a nine-line multiplet with relative intensities of $1:4:10:16:19:16:10:4:1$. In addition, the close and long-distance coupling constants are mixed and the coupling constant of 4.1 Hz has three-quarters of its low-temperature value. Note that there can be no intermolecular exchange since this would involve breaking the coupling path and would collapse the multiplet to a singlet.

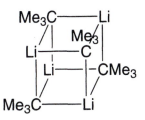

Figure 7.24 The structure of tetrameric t-butyllithium. The lithium atoms are arranged in a tetrahedron with the alkyl groups bonded equally to three lithium atoms and so positioned above a face of the tetrahedron.

7.1.5.7 High-pressure NMR spectroscopy

While it is normal to study the dynamics of a system as a function of temperature in order to obtain the thermodynamic properties of the

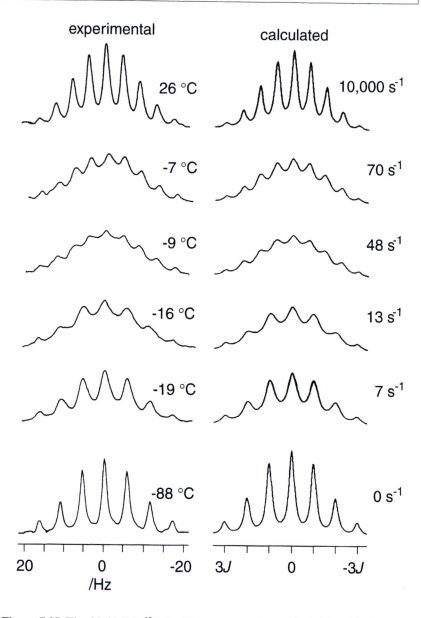

Figure 7.25 The 22.6 MHz ^{13}C NMR spectrum of $[(CH_3)_3C^6Li]_4$ with the effect of the protons removed by double irradiation at different temperatures in cyclopentane solvent. Only the resonances of the α-carbon atom are shown. (From Thomas *et al.* (1986) *Organometallics*, **5**, 1851, copyright (1986) American Chemical Society, reprinted with permission.)

changes observed, it is equally valid to use pressure as the variable, as this may give further insight into the mechanisms of the reactions observed. Often, if a full study is required, both temperature and pressure have to be varied.

In order to work at high pressure, a special probe has to be constructed. There are several variations on the structures of such probes, depending upon exactly what type of experiment it is required to undertake. A typical assembly consists of a pressure containment vessel of a size such that it can be placed in the probe space of an NMR spectrometer and with electrical leads through to the sample coil and with an inlet for the pressure fluid. This vessel contains a sample space surrounded by a detector coil and a temperature measuring device such as a platinum resistance thermometer. The sample is enclosed in a small glass tube with a close-fitting plastic piston to transmit pressure to the sample. Resolution of such probes can be good, even without sample spinning, and temperature control is excellent since the heat transfer medium is the pressurizing fluid.

Not surprisingly, the effect of pressure on the hindered rotation of amides has been quite deeply studied. ^{1}H spectra of the compound N,N-dimethyltrichloroacetamide are shown in Fig. 7.26 as a function of pressure at constant temperature. The expected doublet is coalesced near ambient pressure but splits as the pressure is increased. Thus the pressure decreases the rate of rotation, though there is relatively little change above pressures of 200 MPa. Now, the rate of hindered rotation in such compounds can be expressed in the usual transition-state form

$$k \propto e^{-E_0/RT}$$

where E_0 is some energy barrier. The pressure then has no direct effect on the amide rotation but operates via a change in the viscosity of the solvent, which increases with pressure. We thus have to write

$$k = F(\eta)e^{-E_0/RT}$$

where η is the shear viscosity and $F(\eta)$ is a function of η that does not vary linearly with either pressure or viscosity. It is possible to calculate $F(\eta)$ by making a variety of assumptions, and the results suggest that there is some form of coupling between the solvent motion and the rotation of the N,N-dimethyl groups.

If the exchange is very slow, so that the resonances of the various species present are well resolved, it may not be possible to obtain exchange rates, but the intensity of the resonances may vary, and this indicates a displacement of the equilibrium between the components, which can be caused by changes in either temperature or pressure. In the latter case, change occurs if there is a difference in volume between the species in equilibrium. As an example of such a study, we take the equilibrium:

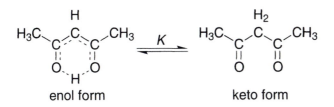

enol form   keto form

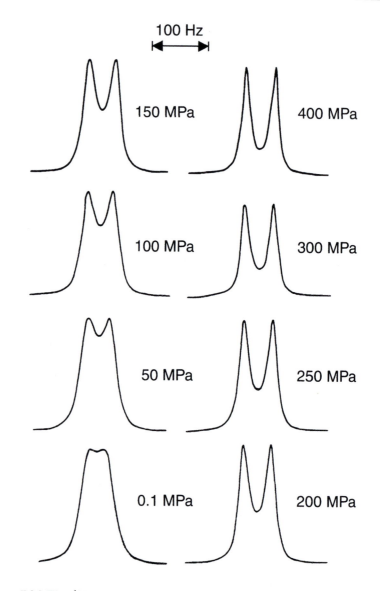

Figure 7.26 The ^1H spectra of the N-methyl resonances of $Me_2NC(O)CCl_3$ dissolved in pentane at 282.3 K as a function of applied pressure. (Jonas *et al.* (1990) *J. Chem. Phys.*, **92**, 3736; reprinted with permission.)

The value of the equilibrium constant K increases with pressure, so that the β-diketone is favoured at high pressures and its volume thus must be smaller than that of the enol. The difference in partial molar volumes is found to be -4.5 cm^3 mol^{-1} and is believed to arise because of the extra volume occupied by the hydrogen-bonded ring formed by the enol.

The exchange of ligands on cations is also studied by high-pressure NMR. Here we will consider the two complexes of beryllium,

$[Be(OSMe_2)_4]^{2+}$ and $[Be\{OC(NMe_2)_2\}_4]^{2+}$. Both ligands bond via the oxygen. Exchange is studied in the presence of excess ligand (c.f. the Al^{3+} water system described above), but with organic ligands it is possible to dilute the system with an inert solvent such as deuteriated acetonitrile, CD_3CN, and vary the ratio between the concentrations of free ligand and complex and so obtain data about the order of reaction. By 'inert' we mean here a solvent that will not compete with the ligand for sites on the cation but which may nevertheless form complexes in an even more inert solvent. Such systems are most easily studied using ^1H spectroscopy of the methyl groups. Two signals are observed due to bound and free ligand, and the exchange rates can be found from the shapes of the spectra, which for much of this work demands the calculation of full spectral envelopes. In the case of the $(Me_2N)_2CO$ complex, it was found that the rate of exchange was independent of the concentration of free ligand, so that the reaction is first order and this implies that the ligand must dissociate from the cation before exchange can take place. In the case of the $OSMe_2$ complex, the exchange rate did depend upon the concentration of ligand, so that the reaction is second order, which implies that a ligand must associate with the $[Be(OSMe_2)_4]^{2+}$ before another ligand will leave. Variable-temperature determinations then permit the thermodynamic parameters $\Delta H^{\ddagger}$ and $\Delta S^{\ddagger}$ to be obtained as well as the rate constants, and these are given in Table 7.2. The different signs of the entropy values are also in accord with the different reaction mechanisms in the present case, though there are many examples where such data are ambiguous. Determination of the exchange rates at different pressures, however, produces a very convincing demonstration of the reality of this difference. Some spectra are shown in Fig. 7.27, together with calculated envelopes, from which the exchange rates were obtained. All the spectra are on the slow exchange side of coalescence, but it will be evident that the spectra of $(Me_2N)_2CO$ are nearest coalescence at the lowest pressure, whereas the opposite is true for the Me_2SO. This difference occurs because a dissociation produces a temporary increase in volume, which is discouraged by an increase in pressure, whereas association produces a temporary decrease in total system volume. The variation of the rate constant with pressure allows a volume change of activation to be obtained from

$$\Delta V^{\ddagger} = - RT \left(\delta \ln \frac{k}{\delta P} \right)_T$$

Table 7.2 Thermodynamic data for solvent S exchange on $[BeS_4]^{2+}$

S	Rate at 298 K	$\Delta H^{\ddagger}$ (kJ mol^{-1})	$\Delta S^{\ddagger}$ (J K^{-1} mol^{-1})	$\Delta V^{\ddagger}$ (cm^3 mol^{-1})
Me_2SO	213 mol^{-1} s^{-1}	35.0	−83.0	−2.5
$(Me_2N)_2CO$	1.0 s^{-1}	79.6	+22.3	+10.5

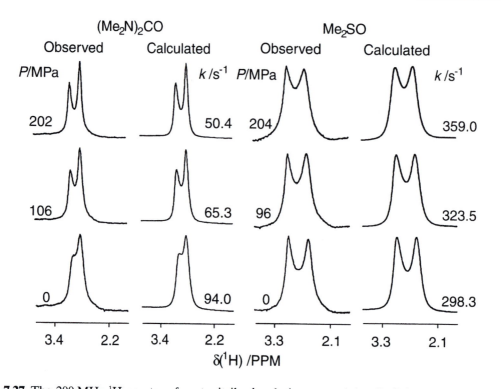

Figure 7.27 The 200 MHz ^{1}H spectra of acetonitrile-d_6 solutions containing Be(II) and $(NMe_2)_2CO$ (left) or Me_2SO (right), obtained as a function of pressure. The actual spectra are on the left and the calculated envelopes are on the right. (From Merbach (1987) *Pure Appl. Chem.*, **59**, 161, with permission.)

These values are also given in Table 7.2. $\Delta V^{\ddagger}$ is large and positive for $(Me_2N)_2CO$, and this is in accord with a fully dissociative mechanism for this ligand exchange. $\Delta V^{\ddagger}$ is negative for Me_2SO but its value is appreciably smaller, and this allows us to conclude that, while the mechanism is associative, there must be some tendency for the bound ligand to move away from the cation and that the mechanism is, more precisely, association and interchange of ligands near the cation, which limits the extent of the possible volume contraction were full association to occur. Note also the difference between the meaning of volume changes in this and the previous example. Here, the total volume before and after reaction is unchanged, and we observe only the adjustment necessary to allow reaction to proceed. There is thus no change in the position of equilibrium with pressure, in contrast to what happens when reaction produces a permanent volume change.

7.2 REACTION MONITORING OF SYSTEMS NOT AT EQUILIBRIUM

The previous example was just such an experiment, in which we were monitoring a physical change in a system, albeit under somewhat extreme conditions. In fact, NMR is used extensively to monitor what is happening during chemical preparations and is particularly useful because starting materials, intermediate, side products and main products may have resolved signals and the progress of a reaction may well be fruitfully examined without any previous purification.

7.2.1 Inter ion alkyl exchange

The slow equilibration of mixtures is easily followed by NMR. For instance, if the aluminate salts $NaAlMe_4$ and $NaAlEt_4$ are dissolved together in solution, it is found that the ligands scramble to form the mixed salts $NaAlMe_nEt_{4-n}$. If a coordinating solvent is used, the solutes ionize to a greater or lesser extent to give Na^+ and the regularly tetrahedral $[AlR_4]^-$, in which the ^{27}Al relaxation time can be sufficiently long to permit spin–spin coupling to the protons of the alkyl groups. In order to measure the rate of alkyl interchange, the two salts were dissolved in the non-coordinating solvent benzene and the coordinating solvent $(Me_2N)_3PO$, added in varying amounts. Immediately following dissolution, a sample of the solution was placed in an NMR spectrometer and the proton spectrum of the Al–Me groups measured as a function of time (Fig. 7.28). Any effect of the ^{27}Al was removed by double irradiation. The Al–Me signal then appears as a sharp singlet initially, but with time new signals are seen to emerge to low frequency corresponding to anions with one, two or three ethyl groups replacing the methyl substituents. By measuring the intensities of the resonances as a function of time, the rate of ligand scrambling is obtained. Interestingly, this is found to be a function of the concentration ratio of $(Me_2N)_3PO/Na^+$, and it is believed that this indicates that the scrambling takes place in aggregates containing two anions and sodium cations solvated by the $(Me_2N)_3PO$.

Reactions taking as little as 60 s can be studied in this way, and even faster processes can be followed using stopped-flow NMR spectroscopy. Two solutions containing the species it is proposed to react are held in reservoirs in the magnetic field so that the magnetization of the nuclei to be studied can reach its equilibrium value. Two power-driven syringes are then used to force measured amounts of the solutions into the NMR sample tube within a time of a few milliseconds. The collection of FID signals is initiated at the same time and each response stored separately in memory for later transformation into spectra. Reactions taking as little as 5 s can be studied in this way, one example having already been given for the exchange of water on Al^{3+}.

7.2.2 Oxidative addition to platinum(II)

Our next examples illustrate two reaction processes and how NMR can provide details about reactions not easily obtainable in other ways. The first example comes from the realm of organometallic chemistry.

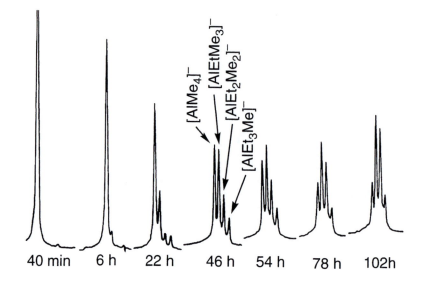

Figure 7.28 The 100 MHz ^{1}H–NMR spectra of the Al-Me protons with the effect of the ^{27}Al removed by double irradiation of a 1 : 1 mixture of NaAlMe$_4$ with NaAlEt$_4$ in benzene with added (Me$_2$N)$_3$PO, as a function of time. (From Ahmad *et al.* (1984) *Organometallics*, **3**, 389, copyright (1984) American Chemical Society, reprinted with permission.)

An oxidative addition was carried out on the platinum complex [PtMe$_2${Me$_2$PC$_6$H$_3$(OMe)$_2$}$_2$] using PhCH$_2$Br, with the objective of forming the octahedral complex [PtMe$_2${Me$_2$PC$_6$H$_3$(OMe)$_2$}$_2$(PhCH$_2$) Br], where Ph is a phenyl group. The proton-decoupled ^{31}P spectrum of the starting material is shown in Fig. 7.29 and is the typical 1 : 4 : 1 triplet. The initial product has a similar spectrum, though with different shift and coupling constant but, even after 1.5 h reaction, there is already a significant amount of another product present. After three days this becomes a predominant spectral feature and is seen to consist of a 1 : 4 : 1 set of AB sub-spectra. The initial product has isomerized to give non-equivalent phosphines, which have a small *cis* P–P coupling. The full structure can then be deduced from the methyl resonance pattern in the ^{1}H spectrum, where one large and one small coupling to ^{31}P are observed, so indicating the presence of a *trans* and a *cis* methyl group. The reaction that takes place is

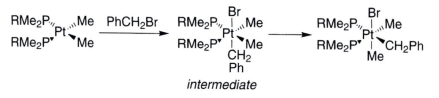

intermediate

where R = C$_6$H$_3$(OMe)$_2$ and Ph = phenyl.

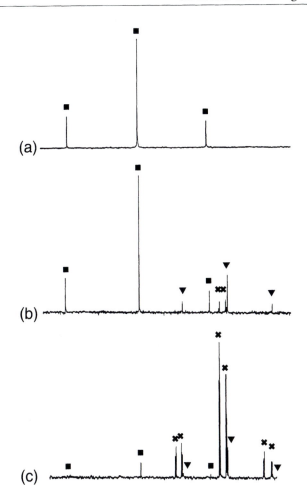

Figure 7.29 Three 40.5 MHz ^{31}P spectra obtained with double irradiation of the protons so as to simplify the spectra. The upper spectrum (a) is that of the starting material and is distinguished by squares in all three spectra. The central trace (b) was obtained 1.5 h later and shows the presence of another substance in which the two phosphorus nuclei are still equivalent and which is distinguished by triangles. After three days of reaction (c) both these spectral patterns are much diminished and a third, more complex, pattern (x) has taken their place. There is now a triplet of AB patterns, which shows that the two phosphorus nuclei in the complex are no longer equivalent. (Example supplied by Professor B.L. Shaw.)

7.2.3 The forced hydrolysis of aluminium salts

The second example examines the nature of the species contained in highly hydrolysed aluminium salt solutions. Aluminium salt solutions are highly acidic and will, for instance, dissolve many metals with the evolution of hydrogen. If we add sodium carbonate solution slowly so as to ensure that there is no precipitate formed, we find that we can add up to 1.25 moles per mole of Al^{3+} and form a solution whose stoichiometric composition is $[Al(OH)_{2.5}X_{0.5}]$, where X is the anion present. The rest of X forms the sodium salt and CO_2 is evolved. The problem has been to discover what exactly is present in such solutions, and ^{27}Al NMR has been able to provide useful new information about the nature of the molecular species formed. It was known that a cation $[AlO_4Al_{12}(OH)_{24}(H_2O)_{12}]^{7+}$ could be crystallized as the sulfate salt from the solutions to which the maximum amount of carbonate had been added. The structure of this cation has been obtained. It has a central aluminium atom tetrahedrally coordinated by four oxygen atoms and surrounded by 12 aluminium atoms, which are octahedrally coordinated and are linked by OH bridges and the oxygen atoms coordinated to the central Al. They also carry a terminal water ligand (Fig. 7.30). The four-coordinate aluminium atom is in a regular environment and so gives a narrow ^{27}Al resonance, which is diagnostic of the presence of this molecular species. The octahedral aluminium atoms are in a highly distorted environment, and so have a short relaxation time and a very broad resonance some 8000 Hz wide, which can only be observed under rather specialized conditions and so is normally not detectable. The system has been much investigated by pH titration methods but does not seem to be very amenable to these methods, although different workers had proposed the presence of some 24 total species in order to explain their results. Later workers, however, seemed to be agreed that only the tridecameric cation was present. Unfortunately, in partly neutralized solutions, the ^{27}Al NMR gives a

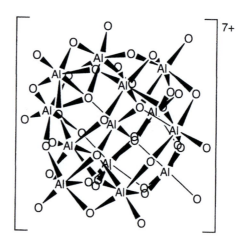

Figure 7.30 Structure of the tridecameric hydrolysed aluminium cation, $[AlO_4Al_{12}(OH)_{24}(H_2O)_{12}]^{7+}$, with the hydrogen atoms omitted. This consists of four groups of three AlO_6 octahedra, which share an apical oxygen atom and three edges to form a triangular structure. The apical oxygen is also coordinated to the central Al atom, which is thus in a tetrahedral environment since there are four Al_3 groups. Each Al_3 group is attached to the three others via double OH bridges. There are also three OH bridges in each triangular cluster. The remaining oxygen coordinated to the Al is fully protonated and is a water ligand.

spectrum containing three resonances, none very broad. These can be seen in the traces of Fig. 7.31, where the highest frequency line is assigned to the tridecamer, the lowest frequency line to unconverted Al^{3+} or $[Al(H_2O)_6]^{3+}$ and a rather broader line just to high frequency of the latter which originates from an oligomer. There is no doubt that this species exists but it is not detected in the pH titrations. A major difference between the two techniques is that pH titrations are carried out in dilute solutions whereas, because NMR is a rather insensitive technique, it requires much higher concentrations. (This difference is currently less important because of the existence of the very high-field spectrometers coupled with Fourier transform techniques.) The obvious question was asked: Was this difference in interpretation a concentration effect? It was found that, if a solution that had been hydrolysed so as to contain very little tridecamer and much oligomer was diluted, then the relative concentration of the tridecamer increased and that of the oligomer decreased, and that, with sufficient dilution, only the tridecamer could be detected. It was also possible to follow how the change occurred by simply diluting a solution with water and

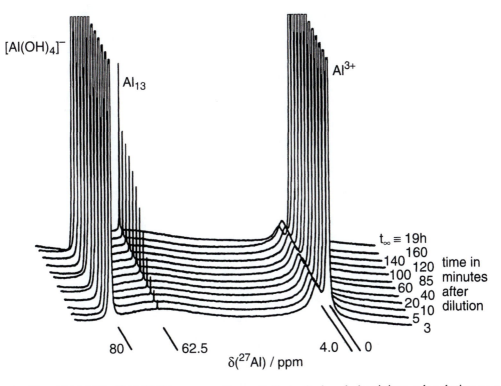

Figure 7.31 The 104.2 MHz ^{27}Al NMR spectra of a partially hydrolysed aluminium salt solution as a function of time immediately following dilution with water. The aluminium concentration before dilution was 1.0 M and this was reduced to 0.1 M by dilution. $[Al(OH_2)_6]^{3+}$ is the chemical shift reference. A capillary containing $[Al(OH)_4]^-$ was used to provide a concentration reference. (From Akitt and Elders (1988) *J. Chem. Soc., Dalton Trans.*, 1347, with permission.)

observing the ^{27}Al resonance as a function of time. A set of spectra are shown in Fig. 7.31. The oligomer resonance is reduced in intensity immediately upon dilution; a change that cannot be followed by this technique and which requires comparison of the spectra before and immediately following dilution. The tridecamer concentration then increases regularly over about 19 h. Simple dilution of these solutions, then, causes profound changes in composition and this explains some of the problems of the earlier workers. It is also remarkable that such a weak driving force as a quite moderate change in concentration should give rise to such a complex molecule as the tridecamer. It is suggested that the oligomer is, in fact, a mixture of species that have structures related to those of fragments of the tridecamer, one possibility being a fused $Al_3O_{13}H_{18}^+$ unit such as forms one of the triangular, flat faces of the tridecamer.

7.2.4 A technological application: crystallization of amorphous polyethylene

The proton spin relaxation time in amorphous, solid polyethylene, while shorter than that in liquids, is nevertheless long enough to obtain spectra by normal methods. When the material forms extended-chain crystals, the relaxation time becomes much shorter because the chain motion becomes more restricted, and this difference can be used to separate the signals of the two forms of the polymer. The technique used to do this was to apply the pulse sequence used to determine T_2, i.e. a $90° - \tau - 180° - \tau -$ echo sequence in which τ is made equal to 300 µs, too long to observe a response from the crystalline material but short enough to see an echo from the amorphous component. The echo intensity decreases with the time of crystallization as the amount of amorphous materials decreases. There is an induction period when no change in echo intensity is detected, and then the intensity falls regularly as crystallization proceeds. The rate increases as the temperature is decreased, and increased pressure allows crystallization to occur both at higher temperatures and at higher rates.

7.3 QUESTIONS

7.1. The ^{1}H decoupled ^{31}P NMR spectrum of $[Rh\{P(OMe)_3\}_5]^+$ is an A_3B_2X pattern at very low temperature and A_5X at room temperature. Is the mechanism of exchange intra- or inter-molecular?

7.2. You have two isomers in equilibrium. Their concentrations are in the ratio $10:1$. There is a contribution to the linewidth of a ^{1}H NMR signal due to the major isomer of 1 Hz from exchange. What is the exchange contribution to the linewidth of the minor isomer in the ^{1}H and in the ^{13}C NMR spectra, assuming that the signals are below coalescence? Accepting that the height of a signal is inversely proportional to its linewidth and ignoring non-exchange contributions to linewidth, what is the ratio of heights of the NMR signals due to the major and minor isomers?

7.3. Given the following rates as a function of temperature for the inversion on *cis*-decalin, calculate $\Delta H^{\ddagger}$ and $\Delta S^{\ddagger}$ for the dynamic process using the Eyring equation.

k (s^{-1})	T (K)	k (s^{-1})	T (K)	k (s^{-1})	T (K)
0.128	203	25.8	246	1000	282
0.57	214	60	253	1890	292
2.57	226	120	261.5	3450	300
11.1	237	350	271	7300	312

7.4 At –60°C, the ^{13}C signals of the σ-C_5H_5 group of $[Fe(\eta^5\text{-}C_5H_5)(\sigma\text{-}C_5H_5)(CO)_2]$ (see margin) are broadened by intramolecular chemical exchange. The linewidths of C^2 and C^5 are 11 Hz and of C^3 and C^4 are 6 Hz, while the linewidth in the absence of exchange is 1 Hz. Deduce the mechanism of exchange. Calculate $\Delta G^{\ddagger}$ for the carbon leaving the C^2 and C^5 positions.

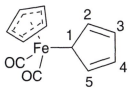

7.5. Bullvalene undergoes the degenerate Cope rearrangement. At –80°C, the ^{13}C NMR spectrum shows the expected four signals at δ 21.01, 31.01, 128.37, and 128.54, in intensity ratio 3 : 1 : 3 : 3. Above –50°C, the signals broaden due to the degenerate Cope rearrangement. The signals at δ 21.01, 31.01, and 128.54 broaden equally, but the one at δ 128.37 broadens by only one third of the amount. Use this information to assign all the ^{13}C NMR signals.

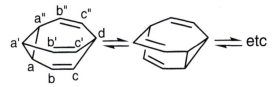

7.6. In the ^{13}C NMR spectrum of *cis*-decalin (see margin), at 226 K there is slow inversion of the two six-membered rings producing exchange $C^{1,5} \leftrightarrow C^{4,8}$ and $C^{2,6} \leftrightarrow C^{3,7}$. If the signal due to $C^{3,7}$ is presaturated for long enough so that the signal due to $C^{2,6}$ reaches an equilibrium value, its height is now 37% of its original value. Given that T_1 of $C^{2,6}$ and $C^{3,7}$ is 0.66 s, calculate the rate of inversion of *cis*-decalin and $\Delta G^{\ddagger}$ at 226K.

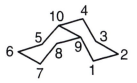

7.7. There are two ^{13}C NMR signals of equal intensity separated by 1 ppm, which exchange. The separation of the signals is temperature independent and the natural linewidth is negligible. At 25.2 MHz, the coalescence temperature was determined as 25°C, while at 100.62 MHz, the coalescence temperature was determined as 42°C. Explain why the coalescence temperature is magnetic field dependent and determine $\Delta G^{\ddagger}$. Assuming that $\Delta S^{\ddagger} = 0$, calculate the linewidth of the signal(s) at 0°C and 60°C at both magnetic fields.

Multiple resonance and one-dimensional pulse sequences $\boxed{8}$

Until the advent of computer controlled NMR spectrometers, there was a limited selection of NMR experiments available. However, with computer control, it is possible to carry out a much wider selection of experiments.

Many NMR experiments may be summarized by the sequence: Relaxation – Preparation – Evolution – Mixing – Acquisition. During the relaxation period, the nuclear spins are allowed to return towards their Boltzmann equilibrium. In principle, the relaxation period should be in excess of $5T_1$ for the slowest relaxing signal. In practice, we are impatient, and normally wait between T_1 and $2T_1$. It is during the preparation period that the spins of the nuclei are tuned to give us the information required. There is then an evolution period, which is often related to a coupling constant. It is usually necessary to apply a refocusing 180° pulse in the middle of the mixing period to make sure that the resulting spectrum can be phased. One or more other pulses are applied during the mixing period. At the end of the mixing period, the FID is acquired, frequently with decoupling. In the more complicated experiments, a string of pulses are applied, often to two or more different nuclei. A single such experiment is a one-dimensional NMR experiment. The data are collected in a single time dimension as the FID and are then transformed into the frequency dimension to give a conventional NMR spectrum. (Yes, strictly speaking there are two dimensions, intensity and time or frequency, but such spectra are described as one-dimensional, referring to the time dimension.) A second time dimension is introduced by varying a time during the mixing period. A series of FIDs are collected, each with a different value of time, a technique which gives rise to two-dimensional NMR spectroscopy. We will see in Chapter 9 that this provides a valuable tool for assigning signals.

In the present chapter we will be examining a range of one-dimensional NMR experiments. This is not intended to be an exhaustive coverage, but many of the important techniques will be examined. In all cases, the techniques give information on the connectivity between nuclei in compounds, and hence on the structure of compounds.

It is useful at this stage to introduce a commonly used nomenclature to indicate when decoupling is used. It is common to use irradiation to remove coupling or perturb a second nucleus. This is normally indicated by using curly brackets. Thus a $^{13}C\{^1H\}$ NMR spectrum is one where the ^{13}C NMR spectrum is observed with irradiation of protons. In this case, there will be broadband ^1H decoupling. In Fig. 8.1(b) we have a $^1H\{^1H\}$ NMR spectrum, where the ^1H NMR spectrum is recorded with irradiation at one of the hydrogen sites. Equally, Fig. 8.2(b) shows a $^1H\{^{31}P\}$ NMR spectrum, where an individual ^{31}P site is being irradiated.

8.1 DECOUPLING DIFFERENCE SPECTROSCOPY

Difference spectroscopy is a powerful technique used to show which signals change during a given experiment. A simple example is decoupling difference spectroscopy. In a typical experiment, the computer is instructed to take a series of NMR spectra with the decoupler set at different frequencies for each spectrum corresponding to different resonances. A reference is included with the decoupler set well away from any signal (Fig. 8.1(a)), which gives the single resonance spectrum. This reference spectrum is then subtracted from a spectrum obtained where some signal is decoupled (Fig. 8.1(b)). In the difference spectrum (Fig. 8.1(c)) the multiplets coupled to the irradiated signal show up as a mixture of fully coupled (negative) and decoupled (positive). The signals due to protons which are unchanged by the irradiation should subtract completely. In practice this does not happen perfectly. In part this is due to instrumental instability, but there are two other major contributions. Firstly, there will be nuclear Overhauser effects, see section 8.2. Secondly, there will be Bloch–Siegert shifts, see section 6.1. As a result of the Bloch–Siegert shifts, it is common to carry out the experiment with a relatively small value of $\boldsymbol{B}_2$ so that signals are not fully decoupled, and the difference spectrum consists of the coupled spectrum sitting on a broad hump.

Decoupling difference NMR spectroscopy is valuable as a means of dis-entangling overlapped signals, where it is important to derive coupling constants. If the required information is connectivity and assignment, then it is normal to use the two-dimensional technique, COSY, see section 9.2.

Decoupling difference NMR spectroscopy is not restricted to $^1H\{^1H\}$ NMR spectroscopy, but can be applied to any pair of nuclei. An example of $^1H\{^{31}P\}$ decoupling difference NMR spectroscopy is given in Fig. 8.2. The compound is $[Fe(\eta^5\text{-}C_5H_5)(PPh_3)\{1,2\text{-}(PMePh)_2C_6H_4\}]$ $[B\{3,5\text{-}(F_3C)_2C_6H_3\}_4]$. Due to the crowding of the molecule, rotation about the Fe–PPh_3 bond is slow on the NMR timescale. Due to the crowding in the molecule, the near neighbour anisotropy of the phenyl rings has spread these signals over a 1.8 ppm range, making the resolution of separate signals easier. The result is that all five PPh groups

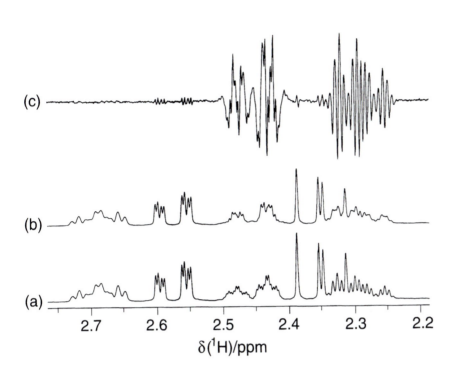

Figure 8.1 A partial 400 MHz ^{1}H NMR spectrum of carvone in CDCl$_3$. (a) The reference spectrum, where the decoupler was set at δ 10. (b) The spectrum where the decoupler was set on the signal at δ 6.77. (c) The difference spectrum, where spectrum (a) is subtracted from spectrum (b). Responses are observed only for the multiplets at δ 2.45 and 2.28 which have changed on decoupling. The unchanged coupled spectrum gives the negative signals, while the decoupled spectrum gives positive signals.

give separate ^{1}H NMR signals. Selective ^{31}P decoupling permits the identification of the PPh$_3$ phosphorus signal and the three phenyl groups attached to it. This is illustrated in Fig. 8.2. Fig. 8.2(a) shows the normal ^{1}H NMR spectrum of the aromatic hydrogen atoms. When the PPh$_3$ ^{31}P nucleus is decoupled (Fig. 8.2(b)), it is obvious that a large coupling is removed from the signals at δ 6.01 and 6.22 leaving these signals as doublets, and hence these are *ortho*-protons of two of the three phenyl groups, but the location of the third set of *ortho*-protons is difficult to discern. There is also the removal of a small coupling from the signals at δ 6.77 and 7.04, leaving triplets of intensity 2, making these *meta*-protons. The identification of the remaining signals is more readily made in the difference spectrum in Fig. 8.2(c). The missing *ortho*-protons are at δ 7.27 and can be recognized by comparison of the appearence of the difference signal with those assigned at δ 6.01 and 6.22. The other signals are more difficult to assign, but responses are detected at δ 7.54, 7.48, and 7.24.

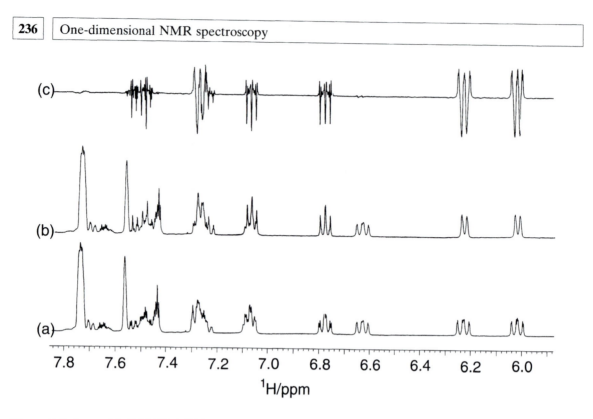

Figure 8.2 A partial 400 MHz ^{1}H NMR spectrum of [Fe(η^5-C$_5$H$_5$)(PPh$_3$){1,2-(PMePh)$_2$C$_6$H$_4$}][B{3,5-(F$_3$C)$_2$C$_6$H$_3$}$_4$] in CD$_2$Cl$_2$. (a) The reference spectrum. (b) The spectrum where the decoupler was set on the ^{31}P NMR signal of PPh$_3$. (c) The difference spectrum, where spectrum (b) is subtracted from spectrum (a). The coupled spectrum gives the positive signals, while the decoupled spectrum gives negative signals.

8.1.1 Selective population transfer

As artefacts occur in decoupling difference spectra due to Bloch–Siegert shifts, it would be advantageous to minimize these shifts by using very low powered decoupling. This can be achieved by selective population transfer. We illustrate the principle of this technique using an AX spin system (Fig. 8.3). The relative populations of the energy levels follow from the Boltzmann distribution. Relative to the $\beta\beta$ energy level, the $\alpha\beta$ and $\beta\alpha$ energy levels have an excess population of δN, while the $\alpha\alpha$ energy level has an excess population of $2\delta N$. When the decoupler is turned on at the frequency of the X, $\alpha\alpha$–$\alpha\beta$ transition, the populations of these two energy levels are made equal. Note what this does to the population differences associated with the A-transitions. For the $\alpha\alpha$–$\beta\alpha$ transition the population difference decreases from δN to $0.5\delta N$, i.e. the intensity of the signal halves. For the $\alpha\beta$–$\beta\beta$ transition, the population difference increases from δN to $1.5\delta N$, i.e. the intensity of the signal increases by 50%. The result is that a doublet changes from being a 1 : 1 doublet to a 0.5 : 1.5 doublet, and the changes are easily detected in a difference spectrum.

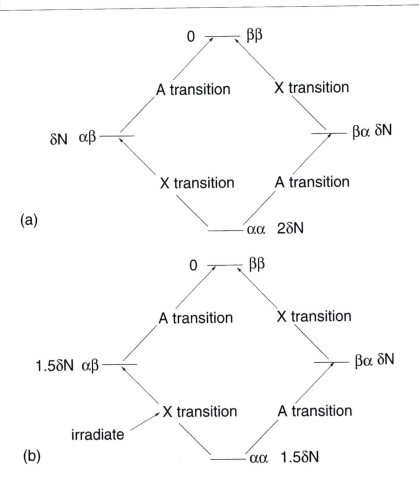

Figure 8.3 The energy level diagram for a homonuclear AX spin system. (a) In the absence of irradiation. (b) In the presence of irradiation at the $\alpha\alpha$–$\alpha\beta$ X transition.

8.2 THE NUCLEAR OVERHAUSER EFFECT

The nuclear Overhauser effect (NOE) provides information on connectivity through space. This contrasts with connectivity information from coupling constants, which normally only operate through a few bonds. The technique is valuable in determining the stereochemistry and conformation of molecules. For example, the stereochemistry of a highly substituted alkene can be found or the *cis–trans* relationship between ligands in square–planar metal complexes can be determined.

In outline the experiment is simple. When one nucleus is irradiated, the intensities of the signals due to other nuclei which are close in space change. In the case of ^1H with ^1H irradiation, the neighbouring protons can increase in intensity by 50%. In the case of ^{13}C with ^1H irradiation, the ^{13}C signals can increase in intensity by 199%.

The NOE arises directly from dipole–dipole relaxation, see section 4.2. The mechanism of operation can be explained for a nucleus I being relaxed by a nucleus S by dipole–dipole relaxation (Fig. 8.4). There are four energy levels for this spin system, $\alpha\alpha$, $\alpha\beta$, $\beta\alpha$, and $\beta\beta$. The transition probabilities between each energy level are given by

$$W_0 = \frac{\mu_0^2\gamma_I^2\gamma_S^2\hbar^2}{120\pi^2r^6}\frac{S(S+1)\tau_c}{1+(\omega_I-\omega_S)^2\tau_c^2} \tag{8.1}$$

$$W_1 = \frac{\mu_0^2\gamma_I^2\gamma_S^2\hbar^2}{80\pi^2r^6}\frac{S(S+1)\tau_c}{1+\omega_I^2\tau_c^2} \tag{8.2}$$

$$W_2 = \frac{\mu_0^2\gamma_I^2\gamma_S^2\hbar^2}{20\pi^2r^6}\frac{S(S+1)\tau_c}{1+(\omega_I+\omega_S)^2\tau_c^2} \tag{8.3}$$

where r is the distance between the two nuclei and ω_I and ω_S are the frequencies of the nuclei. The subscripts to W refer to the change in the total spin quantum number when the transition described by W occurs. The transition associated with W_0 is a zero quantum transition, that associated with W_1 is a single quantum transition and that associated with W_2 is a double quantum transition.

The theoretical maximum NOE, η_{max}, when nuclei, I, are observed and nuclei, I, are irradiated is given by

$$\eta_{max} = \frac{\gamma_S(W_2-W_0)}{\gamma_I(W_0+2W_1+W_2)} \tag{8.4}$$

The signal intensity is then $1+\eta_{max}$. In order for η_{max} to be obtained, the relaxation pathway has to be exclusively dipole–dipole. If any other relaxation mechanism operates, then the observed η is related to η_{max}, T_1 and T_{1DD} by

$$\eta = \frac{\eta_{max}T_1}{T_{1DD}}$$

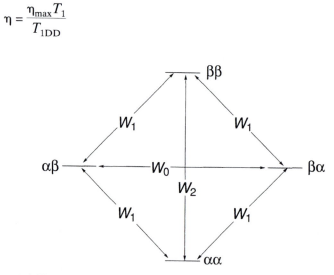

Figure 8.4 The relaxation pathways for two nuclei.

This equation is a very valuable tool to enable us to derive T_{1DD} for nuclei such as ^{13}C. The treatment becomes more complex for ^1H, as in most molecules there are many protons and there are many pathways for relaxation.

In the extreme narrowing limit, equation (8.4) simplifies to

$$\eta_{max} = \frac{\gamma_S}{2\gamma_I} \qquad (8.5)$$

When the extreme narrowing limit does not apply, η_{max} is dependent on both frequency and τ_c (Fig. 8.5).

A qualitative understanding of the origin of the NOE can be understood by examining the AX spin system with one of the protons undergoing double irradiation (Fig. 8.6). The relative populations of the energy levels follows from the Boltzmann distribution. Relative to the $\beta\beta$ energy level, the $\alpha\beta$ and $\beta\alpha$ energy levels have an excess population of δN, while the $\alpha\alpha$ energy level has an excess population of $2\delta N$. When the decoupler is turned on at the X frequency, irradiating both lines, transitions are induced between X spin states and the populations of the connected energy levels are equalized. Hence the populations of the $\beta\alpha$ and $\beta\beta$ energy levels become $\frac{1}{2}(0 + \delta N) = \frac{1}{2}\delta N$. Similarly, the populations of the $\alpha\beta$ and $\alpha\alpha$ energy levels become

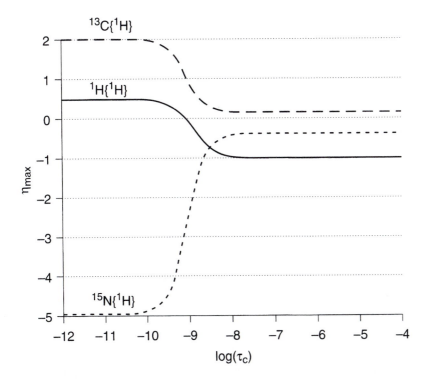

Figure 8.5 The variation of η_{max} with correlation time for $^1H\{^1H\}$, $^{13}C\{^1H\}$, and $^{15}N\{^1H\}$. The values have been calculated for $\boldsymbol{B}_0 = 9.4$ T.

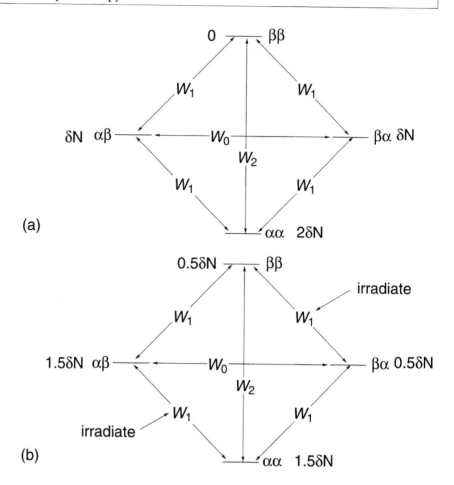

Figure 8.6 The energy levels for a homonuclear AX spin system. (a) The relative populations are given for each energy level at equilibrium. (b) The relative populations are given immediately after applying a radiofrequency decoupling field at the X nucleus.

$\frac{1}{2}(\delta N + 2\delta N) = \frac{3}{2}\delta N$. This is in the absence of relaxation. If we now take the case of the extreme narrowing limit, and examine equations (8.1) – (8.3), it can be seen that W_2 will be the largest term. The W_2 process will try to return the population difference of the energy levels $\alpha\alpha$ and $\beta\beta$ from δN towards $2\delta N$. In practice, at equilibrium, in the presence of irradiation of X, the population difference becomes $\frac{3}{2}\delta N$ as W_0 and W_1 do make a contribution. The effect of the radiation plus relaxation is to increase the population difference between the $\alpha\alpha$ and $\beta\alpha$ and between the $\alpha\beta$ and $\beta\beta$ energy levels from δN to $\frac{3}{2}\delta N$ with a corresponding increase of 50% in the intensity of the A transitions.

In contrast, when $\omega\tau_c$ is large, W_1 and W_2 become unimportant and W_0 is dominant. The result is to equalize the population difference between the $\alpha\beta$ and $\beta\alpha$ energy levels in the AX spin system with X irradiation. The consequence of the X irradiation is then to equalize the populations of all the energy levels, with $\eta_{max} = -1$ and all the signals vanish!

8.2.1 The experimental aspects of the $^1H\{^1H\}$ nuclear Overhauser effect

The NOE is very valuable in giving through-space connectivity information. Although theoretically we can expect enhancements of up to 50%, in practice the enhancements are frequently $< 5\%$. This makes them difficult to detect. The low value arises from a number of causes.

1. The NOE only arises through dipole–dipole relaxation. Any other relaxation pathway reduces the magnitude of the NOE. For ^1H, the most common competing pathway is paramagnetic relaxation due to the presence of paramagnetic transition metal ions or O_2. This is a particular problem with small molecules, < 200 D.
2. Examination of Fig. 8.5 shows that at 9.4 T, when the correlation time falls below 10^{-10} s, the value of η_{max} falls, becoming negative at 8.8×10^{-10} s. This can happen because of larger molecules, viscous solvents, or the presence of solids.
3. The NOE builds up by dipole–dipole relaxation. In a small molecule, T_{1DD} can be in excess of 10 s. In most molecules, T_{1DD} arises from a number of H–H dipole–dipole interactions, not a single one. In the NOE experiment, we are using individual H–H dipole–dipole interactions, and each one will have a T_{1DD} contribution with a time constant considerably larger than the measured T_{1DD}. As a consequence, the NOE can take many tens of seconds to build up to its maximum value. This is particularly a problem when the distance between the nuclei is substantial.

There are a number of procedures which are employed to facilitate the detection of these weak changes in intensity.

1. If the decoupler were to be left on during acquisition, it would be very difficult to distinguish between changes due to decoupling and those due to the NOE. The answer is simple. NOE builds up and decays over a period of many seconds. Decoupling is instant. The pulse sequence given in Fig. 8.7 is used. The signal is irradiated for the required period of time, the decoupler is switched off, an observing pulse is applied and the FID acquired.

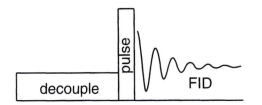

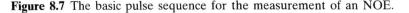

Figure 8.7 The basic pulse sequence for the measurement of an NOE.

2. The magnitude required for B_2 for irradiating the proton for an NOE experiment is very small. This raises a problem. How do we irradiate a multiplet, say a doublet? If the decoupling frequency is placed in the middle of the doublet, the power is so low, that neither line is perturbed. If the decoupling frequency is placed on one line of the doublet, selective population transfer occurs, see section 8.1.1. The result is that the signal which is coupled to the doublet undergoes substantial changes in the intensity of the lines of its multiplet, and this can obscure an NOE. The solution is very simple. Rather than irradiating at a single frequency, the irradiation frequency is cycled through the multiplet, pausing briefly at each line of the multiplet. This technique is now the normal method of irradiation in an NOE measurement.

3. By using the first two procedures, NOEs of 0.1% can be detected in favourable cases. The problem is that artefacts still occur. These can be considerably reduced by using field gradient pulses.

8.2.2 Some applications of the ^{1}H{^{1}H} nuclear Overhauser effect

So far we have only discussed the NOE in the presence of strong, broad-band irradiation or the generalized NOE that occurs if one transition is weakly irradiated. If the whole resonance of a proton is irradiated, whether it be a singlet or a multiplet due to spin–spin coupling, this will cause intensity changes to the resonances of other nuclei that are spatially close. The enhancements possible are quite small: $\gamma_H/2\gamma_H$ means that $\eta = 1/2$ only. It is usual then to obtain such spectra by using the difference mode, i.e. spectra are obtained with and without double irradiation, the FIDs subtracted and the difference Fourier transformed to give a spectrum that ideally contains responses only from those resonances perturbed in intensity by the double irradiation.

The NOE experiment is excellent for assigning protons. An example is given in Fig. 8.8. The ^{1}H NMR spectrum of tris-(2-carboxaldehyde)triphenylphosphine can readily be partially assigned as the signal of the aldehydic proton, H^7, at δ 10.56, H^3 and H^6 are at δ 8.02 and 6.90, though we do not know which is which, and H^4 and H^5 at δ 7.56 and 7.44, though again, we do not know which is which. The aromatic signals are assigned on the basis that the signals due to H^3 and H^6 show only one $^3J_{HH}$ of around 7 Hz, while H^4 and H^5 show two such couplings. The problem is to assign H^3 or H^6 and then H^4 and H^5 can be assigned by decoupling experiments. H^3 was assigned by use of the NOE from the formyl proton, H^7. On irradiation of H^7, the signal at δ 8.02 shows an 11% enhancement, and must be H^3.

An example is shown in Fig. 8.9 of a molybdenum complex that exists in solution as a mixture of two isomers, one with the phenyl group of the CHPh group pointing towards the cyclopentadienyl ring and the other with it pointing away. NOE difference spectroscopy enables not only the signal of the two isomers to be assigned but also those of the methyl signals between δ 2.5 and 3.0. In Fig. 8.9(b), the

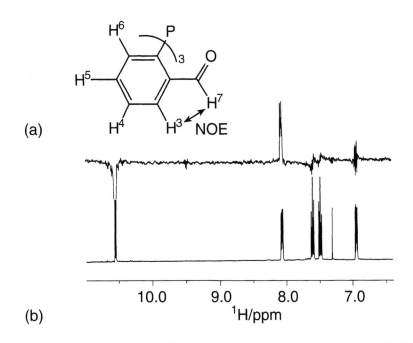

Figure 8.8 (a) The 400 MHz ^{1}H NMR spectrum of tris(2-carboxaldehyde) phenyl phosphine in CDCl$_3$. (b) A difference spectrum showing the enhancement of the signal at δ 8.02 after irradiation of the formyl proton at δ 10.56. (Reproduced from Whitnall *et al.* (1997) *J. Organomet. Chem.*, **529**, 35, copyright (1997), with permission from Elsevier Science.)

cyclopentadienyl protons of one molecule are pre-irradiated. Some of the phenyl proton resonances show a positive NOE. This shows that a phenyl group of this molecule is pointing towards the cyclopentadienyl ring. Contrast this with Fig. 8.9(e) where the cyclopentadienyl protons of the other molecule are pre-irradiated and no NOE is observed from the phenyl protons. At the same time an NOE is observed into H$^{8'}$ and CH$_3^{1'}$ enabling these signals to be assigned to the molecule with H$^{8'}$ pointing towards and the phenyl group pointing away from the cyclopentadienyl ring. The result in Fig. 8.9(e) does not permit a clear distinction between CH$_3^{1'}$ and CH$_3^{7'}$, but this comes from the spectrum in Fig. 8.9(d) where H$^{8'}$ is pre-irradiated and an NOE is observed from the phenyl *ortho*-protons on the same carbon atom as H$^{8'}$, C$_5$H$_5^{12'}$, and the adjacent CH$_3^{7'}$, showing that the methyl signal giving an NOE in Fig. 8.9(e) is by default CH$_3^{1'}$. It is also notable that the *ortho*-protons at δ 6.95 which show an NOE when H$^{8'}$ is pre-irradiated are different from those at δ 7.1 which show an NOE when C$_5$H$_5^{12}$ is pre-irradiated. The relative assignment of CH$_3^1$ and CH$_3^7$ comes from Fig. 8.9(c), where H^8 is pre-irradiated and both the phenyl groups on the same carbon atom and CH$_3^7$ show an NOE.

NOE difference spectroscopy is not restricted to proton–proton NOE measurements but can be used between different nuclei.

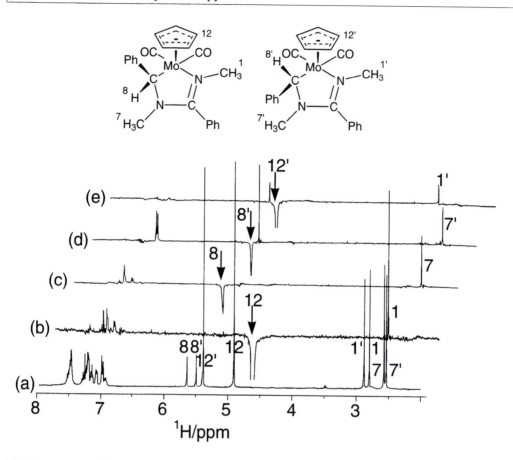

Figure 8.9 The 400 MHz ^{1}H NMR spectra of $[(\eta^5\text{-C}_5\text{H}_5)\text{Mo(CHPhNMeCPh=NMe)(CO)}_2]$ in $CDCl_3$. (a) The normal spectrum and (b) to (e), the NOE difference spectra with pre-irradiation at (b) δ 4.9, H^{12}, η^5-C_5H_5, (c) δ 5.6, H^8, (d) δ 5.5, $H^{8'}$, and (e) δ 5.4, $H^{12'}$, η^5-C_5H_5, with the shift scales progressively displaced to the right to give a clearer picture. (Reproduced from Brunner *et al.* (1983) *J. Organomet. Chem.*, **243**, 179, copyright (1983), with permission from Elsevier Science.)

8.2.3 Some applications of the X{^{1}H} nuclear Overhauser effect

The heteronuclear NOE can provide a very useful increase in signal strength. For the observation of ^{13}C, for instance, when protons in the molecule are double-irradiated, the ratio is 1.99 and $1 + \eta_{max}$ is effectively 3. This is a very useful gain in intensity for such low receptivity nuclei, and is one good reason why ^{13}C spectra are routinely obtained with proton broad-band irradiation. Some values of η_{max} are shown in Table 8.1 for various combinations of nuclei. We should note that some nuclei have magnetogyric ratios with negative sign and that η_{max} is then negative also. In such cases, provided η_{max} is more negative than -1 (^{15}N and ^{29}Si in the table), then the signal is inverted relative to the normal one and the total enhancement is less than η_{max}. If the relaxation mechanism contains other contributions than the direct dipolar one, then η_{max} will be reduced, and it is not uncommon to find

Table 8.1 Maximum nuclear Overhauser effects for several pairs of nuclei

Irradiate	^1H						^{19}F		
Observe	^1H	^{13}C	^{15}N	^{19}F	^{29}Si	^{31}P	^1H	^{13}C	^{19}F
η_{max}	0.5	1.99	−4.93	0.53	−2.52	1.24	0.47	1.87	0.5
$1 + \eta_{max}$	1.5	2.99	−3.93	1.53	−1.52	2.24	1.47	2.87	1.5

that $1 + \eta_{max} \sim 0$ and all signal is lost. In such cases, it is necessary to suppress the NOE in some way, and this can be achieved either instrumentally or by adding a paramagnetic salt (a relaxation agent), which completely dominates the relaxation process. Alternatively, INEPT or DEPT can be used, see section 8.3.

8.2.4 Application to ^{13}C NMR spectroscopy

Figure 8.10 illustrates the changes brought about in the ^{13}C spectrum of carvone by broad-band proton double irradiation. In the absence of irradiation, all the ^{13}C–^1H spin coupling interactions are observed. Thus the carbon atoms of the methyl groups, seen to low frequency, are split into quartets by the protons bonded directly to the carbon atom, and these lines are further split by longer-range couplings to the other ring protons. Only the quaternary carbon atoms give an apparent singlet, though there will be some long-range splitting. A triplet is seen in the centre of the trace at δ 77 due to the solvent $CDCl_3$ whose ^{13}C resonance is coupled to the ^2H nucleus. The CH and CH_2 resonances are doublets and triplets, respectively, due to the directly bonded hydrogen nuclei, and with longer-range two-bond coupling to the vicinal hydrogens, which produces fine structure in the resonances. Not all the resonances of these multiplets can be seen in the spectrum illustrated as the conditions were chosen to produce a noisy baseline. The difference in intensity between the phenyl quaternary carbon signal and the remaining phenyl carbon signals is due to the difference in relaxation time. That of the quaternary carbon atom is long and, at the pulse repetition rate used to obtain the spectrum, has insufficient time to recover its z magnetization after each pulse and so its signal is reduced in amplitude. The spectrum is much simplified by double irradiation, and each type of carbon nucleus appears as a singlet. In Fig. 8.10(b), the spectrum was recorded without NOE and Fig. 8.10(c) with NOE. The enormous improvement in the signal-to-noise ratio is immediately evident.

Double irradiation thus permits us to obtain ^{13}C spectra at natural abundance routinely with good signal-to-noise ratio but such spectra contain reliably only the chemical shift information. Often, this is sufficient, but there are instances where it may be useful to observe the coupling patterns, and if this could be done while retaining the NOE then much accumulation time could be saved. Alternatively, it might be useful to know the correct intensities, and for this we need to

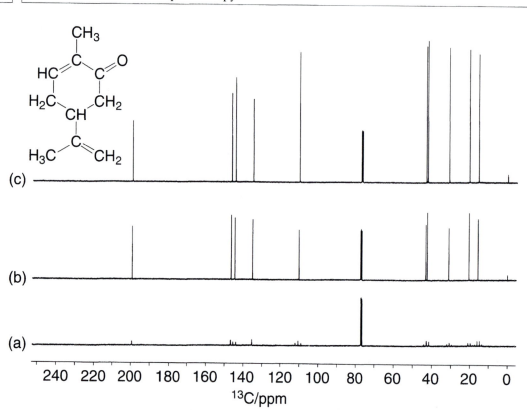

Figure 8.10 The 100.62 MHz ^{13}C spectra of carvone, dissolved in CDCl$_3$. (a) Obtained without proton irradiation. (b) With irradiation at 400.13 MHz with NOE suppression. (c) With irradiation at 400.13 MHz and full NOE, otherwise the spectrometer conditions were identical. A large improvement in signal-to-noise ratio is obtained with double irradiation, especially when the NOE is permitted.

decouple but suppress the NOE. Such techniques may be necessary if we are to observe quaternary carbon atoms with very long relaxation times. If we are happy to contaminate the sample, then the relaxation agents can be very useful since these effectively suppress the NOE and also shorten the ^{13}C relaxation times, so permitting a shorter delay between pulses. If contamination cannot be tolerated, or if the natural parameters of the molecule are required, then special pulse sequences are used, which depend upon the different timescales needed to set up decoupling and NOE. Thus we can have the following:

1. *Full decoupling without NOE.* This is achieved by decoupling only for the short time needed to collect the FID and then waiting sufficient time to allow any small NOE population changes to return to normal. The short pulse of B_2 does not allow appreciable build-up of the NOE (Fig. 8.11).

2. *No decoupling but full NOE.* Here the decoupling power is left on for sufficient time for the NOE to build up to its full value and is then switched off while the FID is collected (Fig. 8.12). This

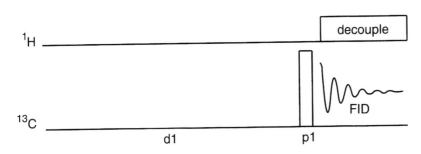

Figure 8.11 The NOE suppression pulse sequence. d1 is a relaxation delay of at least $5T_1$. p1 is a 90° pulse. Decoupler timing to allow a fully decoupled spectrum to be obtained without any distortion of intensity due to the NOE. There will be a small build-up of NOE during the B_2 pulse (typically 0.5 s) and this must be allowed to die away completely before the next pulse. The long delay time means also that the nuclear magnetization has decayed fully before the next 90° pulse and there are no intensity distortions due to relaxation effects.

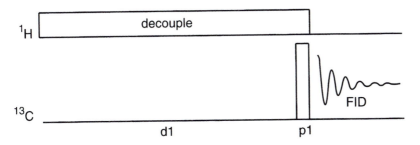

Figure 8.12 The pulse sequence used to record a ^1H coupled ^{13}C NMR spectrum with full NOE. d1 should be $5T_1$ or at least considerably longer than the acquisition time. Decoupler timing is arranged to allow a coupled spectrum to be obtained but with the benefit of the full NOE increase in intensities. There will be very little fall-off in NOE during the short time needed to collect the FID data. This technique allows up to nine times reduction in time over the basic method used to obtain a non-decoupled spectrum.

technique allows normal spectra to be obtained more quickly. If bad overlap of the multiplets occurs, a different approach may be needed, but normally it is possible to distinguish quartets, triplets, doublets and singlets by a comparison with the decoupled spectrum and so to assign resonances to CH_3, CH_2, CH and quaternary carbons. It is well worth taking the trouble to carry out this experiment because the NOE enhancement permits a time reduction in accumulation of as much as nine times.

NOE difference measurements can be very useful in determining the spatial relationship between ^1H and ^{13}C. This is particularly valuable where the ^{13}C does not have a directly attached proton so that its dipole–dipole relaxation arises from more remote protons. The assignment of connectivity between ^{13}C and directly attached protons is far more easily carried out using two-dimensional NMR spectroscopy, see

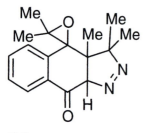

(8.1)

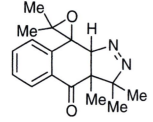

(8.2)

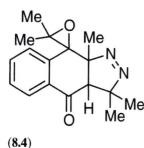

(8.3)

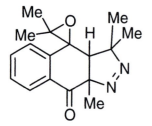

(8.4)

Chapter 9. The technique is illustrated by considering a problem that was solved using ^{13}C–^{1}H NOE measurements. When 2-methylnaphtho-1,4-quinone is treated with Me$_2$CN$_2$ a product was obtained which is an adduct incorporating a Me$_2$C and a Me$_2$CN$_2$ group.

There were four possible products. It was very easy to unambiguously identify the single hydrogen on the sp^3 carbon from it being a singlet of intensity 1 at δ 2.37. The ^{13}C NMR spectrum is shown in Fig. 8.13(a), while the NOE difference spectrum with pre-irradiation of this proton is shown in Fig. 8.13(b). Examination of the spectrum shows a clear NOE into the ^{13}C at δ 195.0, which must be due to the C=O carbon, showing that the structure must be (8.1) or (8.4). There are also NOEs to the carbon atoms at δ 90.0 and 90.5. Now, the CH group in (8.1) is attached to the C=O and one other carbon atom, whereas in (8.4) it is attached to C=O and two other carbon atoms. Hence the correct structure is (8.4).

A variety of other experiments is also possible. Off-resonance decoupling will give multiplet structure with apparently reduced coupling constants, which will reduce overlap in crowded spectra and permit assignment of the carbon spectrum and its correlation with the proton

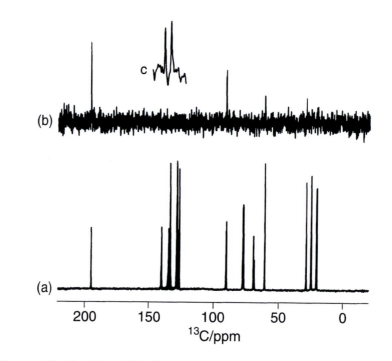

Figure 8.13 The 100.6 MHz ^{13}C NMR spectra of (8.4) in CDCl$_3$. (a) The ^{13}C NMR spectrum with no ^{1}H irradiation. (b) The difference spectrum obtained by pre-irradiating the proton at δ 2.37 at low power to develop an NOE in the ^{13}C NMR spectrum and subtracting spectrum (a). (c) An expansion of the signals at δ 90.0 and δ 90.5 in (b). (Reproduced with permission from Aldersley *et al.* (1983) *J. Chem. Soc., Chem. Commun.*, 107.)

spectrum. Such spectra obtained at several spot frequencies will indicate at which ^{1}H frequency a given ^{13}C multiplet becomes a singlet, and so permit an even more precise correlation between the two. Such experiments are, however, time-consuming and are now better carried out by two-dimensional techniques.

8.2.5 Detailed relaxation mechanisms by ^{13}C NOE and T_1 measurements

The relaxation time of ^{13}C nuclei can be obtained by the inversion-recovery method outlined in section 6.1.2.1, though this is not the only method available. The NOE factors η are obtained by comparing integrals with and without double irradiation, and if the relaxation mechanism is dipole–dipole then η should have a value near to 2. If η is less than this, then other mechanisms are present. An alternative method of measurement of T_1 and NOE in the same experiment is also possible and is done by a method that is a combination of the two previous ones. The NOE is allowed to build up for a time t prior to producing the FID and this is collected while irradiation continues. B_2 is then switched off, the system allowed to equilibrate and the process repeated. A series of experiments are run with different values of t (Fig. 8.14). Such experiments are known as dynamic NOE measurements and give both the Overhauser enhancement for each nucleus from the intensity of the signals when $t = 0$ and $t = \infty$, and the value of T_1 from the plot of the change in signal intensity as a function of t. The spectra obtained for such an experiment with biphenyl are shown in Fig. 8.15, and the results for biphenyl and some for toluene are illustrated in Table 8.2. The CH ring carbon atoms of biphenyl have an NOE only slightly less than 3 and so must be almost totally relaxed by the directly bonded hydrogen atoms. The carbon

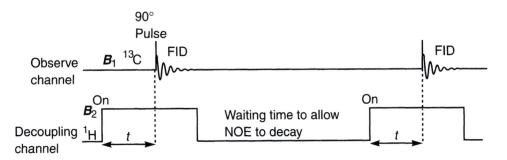

Figure 8.14 The dynamic NOE experiment. This is essentially a combination of the two experiments described by Figs 8.11 and 8.12. The decoupler is first gated ON to allow the NOE to build up. The amount of NOE increases if t is increased, reaching a maximum when $t > 9T_1$. At the end of time t, the FID is produced by the 90° pulse and collected with continuing irradiation. This is removed when the data collection is finished and the NOE allowed to decay to zero. Sufficient FIDs are collected to give the required signal-to-noise ratio, and the experiment is repeated with different values of t, but always the same number of FIDs are collected, so that spectral intensities can be compared directly.

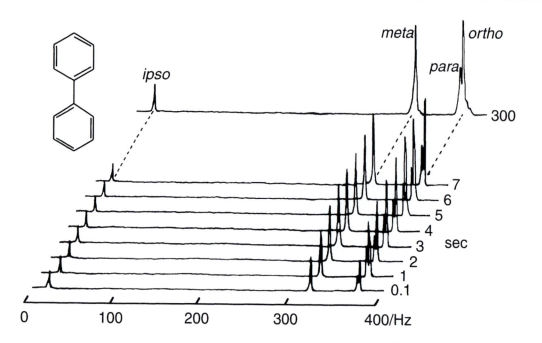

Figure 8.15 Fourier transform ^{13}C spectra of biphenyl in $CDCl_3$ solution excited by 90° read pulses separated by intervals of 300 s in order to ensure the re-attainment of equilibrium after each pulse. The spectra are stacked as a function of the time t of proton irradiation prior to the application of the pulse. (From Freeman *et al.* (1972) *J. Magn. Reson.*, **7**, 327, with permission.)

nuclei 2 and 3 have similar relaxation times but the T_1 of carbon 4 is appreciably shorter, and this is because the CH bond is not reoriented relative to B_0 by rotations around the axis of the molecule, whereas this motion reduces the correlation time of the carbon atoms 2 and 3, if only slightly. Much more marked is the long T_1 of the quaternary carbon nucleus, which relies on the long-distance effect of the protons. Its NOE, however, is reduced and indicates that only half the rate of relaxation is due to this dipole–dipole interaction and that another influence is operating, which we can suggest is that of chemical shift anisotropy. The T_1 values for the ring carbon atoms of toluene show similar behaviour, though the actual relaxation times are longer, as would be expected for a smaller, more rapidly rotating molecule. The relaxation time of the methyl carbon nucleus has, of course, to be multiplied by 3, the number of directly bonded protons, N_H, if it is to be meaningfully compared with the T_1 values of the CH carbon nuclei, and this gives a value of 48 s. This is really very long when compared with the ring CH carbons, and it is not surprising to observe that the NOE is also small, with only 30% of the relaxation rate arising from the dipolar mechanism. In this case, the dipolar effect must be reduced because of a very short correlation time, much shorter than for the rest of the molecule. This will be the free, almost unhindered, rotation of the methyl group, which in addition causes spin rotation relaxation

Table 8.2 Experimental spin–lattice relaxation times T_1, and nuclear Overhauser effect, η

Carbon atom	T_1 (s)	$I_\infty/I_0 = 1 + \eta$	η
(a) Biphenyl			
C4	3.4 ± 0.6	2.72 ± 0.16	1.72
C3	5.4 ± 0.7	2.80 ± 0.21	1.80
C2	5.2 ± 0.8	2.76 ± 0.19	1.76
C1	54 ± 0.4	2.00 ± 0.16	1.00
(b) Toluene			
CH$_3$	16 ($T_1 N_H = 48$)	1.61	0.61
C1	89		
C2	24		
C3	24		
C4	17		

and so accounts for the small NOE. One can calculate that the part of the relaxation which is purely dipolar, $3T_{1DD}$, has the value 160 s, which implies a reduction of τ_c for the methyl group of around eight times compared with the rest of the molecule.

A more complex and more informative example concerns the relaxation behaviour of the carbon nuclei in $[Cr(CO)_5(NC_5H_5)]$. This is a σ-bonded complex in which one CO ligand in $[Cr(CO)_6]$ has been replaced by an N-bonded pyridine molecule. The chemical interest in such molecules arises because it is uncertain what sort of energy barriers exist to the rotation of the $Cr(CO)_5$ moiety relative to the aromatic ligand. One way of probing the intramolecular motion is to measure the relaxation rates of the ^{13}C nuclei. Provided the mechanism of relaxation is known, then correlation times of individual atoms can be calculated. Two types of mechanism are expected to coexist in such molecules: dipole–dipole for the protonated carbon atoms and chemical shift anisotropy relaxation for the CO carbon atoms. The first mechanism can be confirmed if the NOE is high, and low NOE establishes its absence for the CO atoms. Another experiment is, however, needed if we are unequivocally to establish the presence of CSA relaxation, and this is achieved by measuring T_1 and η at different magnetic fields, though the rates of motion must be such that the extreme narrowing condition is met at the higher field and spectrometer frequency

The T_1 values were in this case obtained by a saturation-recovery technique in which the ^{13}C spin populations are equalized by the application of a series of closely spaced 90° pulses and the recovery of the signal intensity monitored as a function of time. The NOE was determined by comparing signal intensities obtained with continuous decoupling with those using gated decoupling of the type depicted in Fig. 8.11. CDCl$_3$ was used as solvent and solutions were degassed to ensure that there was no contribution to relaxation from solvent nuclei or dissolved paramagnetic oxygen. The two spectrometer frequencies used were 125.7 MHz (11.7 T, ^1H resonates at 500 MHz) and 50.29 MHz (4.7 T, ^{1}H at 200 MHz). Certain assumptions had to be made in order to calculate the chemical shift anisotropy of the carbonyl carbon atoms. Since the *trans* CO is on the molecular axis, its relaxation is not affected by rotation around this axis since the CSA will be axially symmetric. The same comment applies to the relaxation of the γ-carbon atom of the pyridine since axial rotation does not reorient the C–H bond. One can therefore assume that these two carbon atoms have the same correlation time of motion, that of the end-over-end tumbling of the complex. The correlation time for the γ carbon is then calculated from its T_{1DD} using equation (4.3) and this value is used to calculate $\sigma_\parallel - \sigma_\perp$ for CO using equation (4.6). The full results are shown in Table 8.3. First we see from the chemical shifts of the carbonyl carbons that the *cis* and *trans* carbonyl groups are very similar, so that we can assume similar chemical shift anisotropy in calculating the correlation times of the *cis* carbonyl carbons. The values of T_1, R_1 $(= 1/T_1)$ and η are given for each carbon at both magnetic fields. For the nuclei of the ring carbon atoms, the η values are quite substantial, and there is only a relatively small decrease in relaxation time with increase in magnetic field. The mechanism is predominantly dipolar, and we can extract R_{1DD} with reasonable precision, and so the correlation times. In the case of the carbonyl carbons, η is essentially zero, and there is little long-range dipolar interaction, T_1 is very field-dependent and T_{1CSA} can be calculated from the values obtained at each field and, thus, the correlation time of the *cis* carbonyls. It is evident that the rate of movement of the *cis* carbonyl groups is about half that of the reorientation of the plane of the pyridine ring, so that

Table 8.3 Relaxation and NOE data for the ^{13}C nuclei in two chromium carbonyl complexes

Atom	δ_C (ppm)	50.29 MHz			125.7 MHz			R_{1DD}	R_{1CSA}	τ_c (ps)
		T_1 (s)	R_1 (s^{-1})	η	T_1 (s)	R_1 (s^{-1})	η			
(a)	[Cr(CO)$_5$(NC$_5$H$_5$)]									
α	155.3	7.52	0.133	1.7	5.86	0.171	1.4	0.166	0.017[a]	5.5 ± 0.9
β	124.8	7.01	0.143	1.6	4.47	0.224	1.2	0.123	0.014[a]	6.0 ± 1.0
γ	137.1	2.44	0.410	1.6	2.16	0.463	1.3	0.319	0.102[a]	16.3 ± 2.8
cis-CO	220.7	10.3	0.097	0.0	2.63	0.38	0.0	0.000	0.338	Assumed same as above
trans-CO	214.3	25.6	0.039	0.1	3.96	0.253	0.0	0.000	0.143	12.8 ± 2.8
(b)	[(η^6-C$_6$H$_6$)Cr(CO)$_3$]									
CH	92.7	11.7	0.085	1.8	10.8	0.093	1.4	0.070	0.021[a]	3.6 ± 0.6
CO	232.8	31.3	0.032	0.1	6.61	0.151	0.1	0.005	0.142	6.3 ± 1.4

[a] These values probably embrace contributions from several mechanisms. R_{1DD} is calculated from the results at both frequencies and the average is taken. The calculated $R_{1DD} + R_{1CSA}$ should equal R_1 found by experiment at the higher frequency, but the averaging process destroys the equality. The values calculated for R_{1CSA} apply to the higher frequency only. The accuracy of the experimental T_1 values is approximately ± 7% and the error in η is ± 0.2.

Source: Data adapted from Gryff-Keller *et al.* (1990) *Magn. Reson. Chem.*, **28**, 25, copyright (1990) John Wiley and Sons Ltd, reprinted with permission.

the two halves of the complex are rotating independently around the N–Cr bond, with the rates of motion determined primarily by interactions of the groups with the solvent. The data for the (η^6-arene) complex, $[(\eta^6\text{-}C_6H_6)Cr(CO)_3]$, are also given in Table 8.3. They resemble closely the previous set of data, though the correlation times calculated are much shorter and the speed of rotation of the benzene is high indicating its very small interaction with the solvent.

8.3 ONE-DIMENSIONAL MULTIPULSE SEQUENCES

Several pulses can be used during the preparation period to extract information from NMR spectra. This opens up a wide range of experiments. Here we are only going to examine a selection of the more useful pulse sequences.

8.3.1 *J* Modulation, JMOD, and the Attached Proton Test, APT

This pulse sequence (Fig. 8.16) is used almost exclusively to edit ^{13}C NMR spectra. The resulting NMR spectrum has the ^{13}C signals from the C and CH_2 groups pointing up and CH and CH_3 ^{13}C signals pointing down (Fig. 8.17). This sorting of ^{13}C signals is a valuable aid to assignment. There is no convention as to whether the C and CH_2 signals are phase adjusted to be positive or negative, and it is usual to have to decide the phase used from the solvent signal. In Fig. 8.17, the $^{13}CDCl_3$ signal at $\delta\,77$ is positive, showing that the C and CH_2 signals have been phased positive.

How does this pulse sequence work? The spin dynamics are shown in Fig. 8.18. Figure 8.18 uses the rotating frame, and starts where a 90° pulse has been applied to bring the ^{13}C magnetization into the $x'y'$ rotating frame. The diagrams have been constructed on the basis that the ^{13}C NMR signal is on resonance, so that the magnetization vectors do not rotate with respect to the rotating frame. The simplest case is

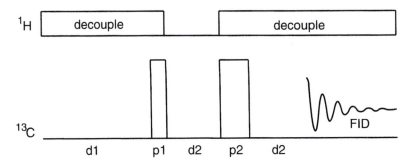

Figure 8.16 The pulse sequence for the *J* modulation pulse sequence. d1 is for relaxation, p1 is a 90° pulse, d2 is $1/J(^{13}C,^{1}H)$, and p2 is a 180° refocusing pulse.

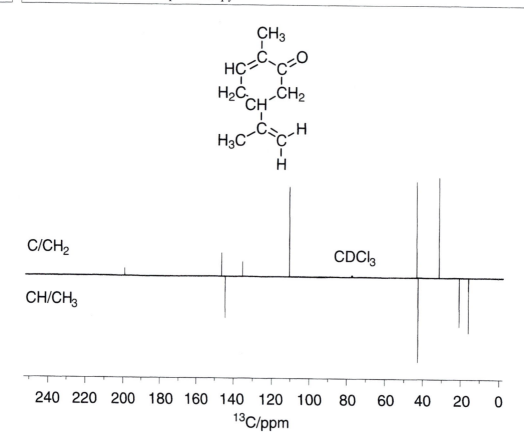

Figure 8.17 The application of the *J*-modulation pulse sequence to the 100.62 MHz ^{13}C NMR spectrum of carvone in CDCl$_3$. Note that the ^{13}C and ^{13}CH$_2$ ^{13}C NMR signals are displayed as positive signals, while the ^{13}CH and ^{13}CH$_3$ signals are displayed as negative signals.

a ^{13}C signal with no ^{1}H coupling (Fig. 8.18(a)). As the ^{13}C NMR signal is on resonance, it does not matter how long we wait after the initial 90° pulse as the signal will remain with its original alignment. However, in a normal ^{13}C NMR spectrum, many of the signals will be away from resonance, and will precess relative to the rotating frame at different frequencies. Thus, by the time we collect the FID, they will be very much out of phase and enormous phase corrections will be needed to rephase the spectrum. This is avoided by placing a 180° refocusing pulse in the middle of the sequence so that nuclei are all in phase again when the collection of the FID is initiated, see section 6.3. This is a general feature of these pulse sequences.

Let us now consider a ^{13}CH group, again on resonance (Fig. 8.18(b)). The rotating frame is at the frequency of the centre of the ^{13}CH doublet. One line of the doublet will be +*J*/2 Hz away and the other will be –*J*/2 Hz away. The result is that the ^{13}C nuclei attached to ^{1}H nuclei with α or β spins will be precessing at rates different from that of the rotating frame. After 1/2*J* s, one group will have precessed +90°

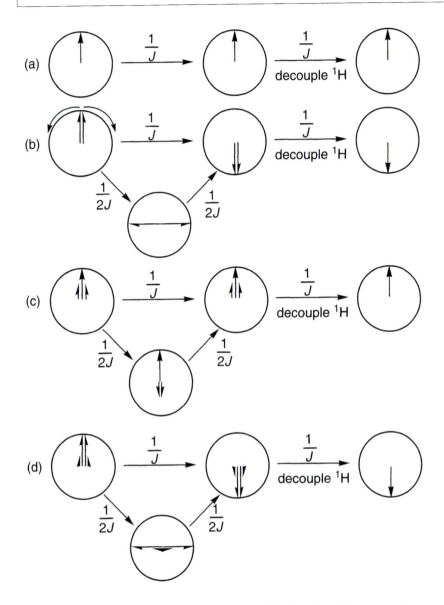

Figure 8.18 The behaviour of the nuclear spins in the $x'y'$ rotating frame during the J-modulation pulse sequence. In order to simplify the picture, it is assumed that the ^{13}C NMR signal is on resonance. (a) A ^{13}C not bearing ^{1}H. (b) A ^{13}CH group. (c) A $^{13}CH_2$ group. (d) A $^{13}CH_3$ group. The 180° pulse and second $1/J$ wait are omitted as they follow from the refocusing pulse description in section 6.3.

and the other −90°. After $1/J$ s, they will have precessed ±180° and refocused. The decoupler is now switched on and the doublet collapses to a singlet, which precesses at the same frequency as the rotating frame. The signal from the ^{13}CH group is 180° out of phase compared with the non-proton bearing ^{13}C.

A similar argument applies to the $^{13}CH_2$ and $^{13}CH_3$ groups. After $1/J$ s, the central line of the $1:2:1$ triplet of the $^{13}CH_2$ group has not moved, and the outer lines, separated by J Hz from the central line, have rotated 360° and are realigned with the central line. Decoupling then produces a singlet with the same phase as that of a non-proton bearing ^{13}C. After $1/J$ s, the inner pair of lines of the $1:3:3:1$ quartet of the $^{13}CH_3$ group have rotated 180°, like the ^{13}CH doublet, and the outer lines, separated by $3J/2$ Hz from the central line, have rotated 540° and are realigned with the inner two lines. Decoupling then produces a singlet, 180° out of phase compared with the non-proton bearing ^{13}C.

The choice of a value for J is normally easy. For most carbon atoms, $^1J(^{13}CH)$ is in the range 125–170 Hz. A compromise of 145 Hz or 142.86 Hz, where $1/J = 0.007$ s, is normally used. The technique does run into problems with sp ^{13}CH groups of alkynes, where $^1J(^{13}CH)$ ~ 250 Hz, and if d2 in Fig. 8.18 is set as 0.007 s, the arms of the ^{13}CH doublet rotate ±315° and give a signal with the same phase as a ^{13}C or $^{13}CH_2$ group!

Note that the decoupler is left on during the relaxation period as well as during acquisition. This is to allow the NOE to build up and to obtain the strongest possible signals with $^{13}C\{^1H\}$ NOE.

J modulation has also been used to obtain signals selectively from only the non-proton bearing ^{13}C nuclei, by using a d2 of $1/2J$ s. The way that this works is also apparent from Fig. 8.18. As mentioned earlier for non-proton bearing ^{13}C nuclei in Fig. 8.18(a), the length of the waiting time does not affect the alignment of the signal. In contrast for ^{13}CH, $^{13}CH_2$, and $^{13}CH_3$ groups, after $1/2J$ s, one half of the multiplet is pointing in the opposite direction to the other half. The result is after decoupling, they cancel each other, and there is no signal. In practice, this technique is difficult to use. Non-proton bearing ^{13}C nuclei give weak signals compared with proton bearing ^{13}C nuclei due to their longer T_1 values and reduced NOEs. There is normally a range of $^1J(^{13}C^1H)$ values, making the choice of d2 a compromise, and residual signals due to proton bearing carbon atoms are normally observed. Frequently it is preferred to obtain an INEPT or DEPT spectrum which does not contain signals due to non-proton bearing ^{13}C nuclei and identify them in the complete spectrum by comparison.

J modulation has been described here as it gives a relatively simple description of the use of multiple pulses and delays to edit NMR spectra. It is a technique which was introduced in the early days of multiple pulse NMR spectroscopy as it could be implemented on most spectrometers. To a large extent it has now been replaced by DEPT and PENDANT, *vide infra*.

A common variation on the sequence is the **A**ttached **P**roton **T**est or APT (Fig. 8.19). The advantage of APT over J modulation is that a shorter relaxation delay time, d1 can be used. This can easily be understood. In the absence of phase cycling, the J-modulation pulse program applies a total of a 270° pulse, leaving all the ^{13}C

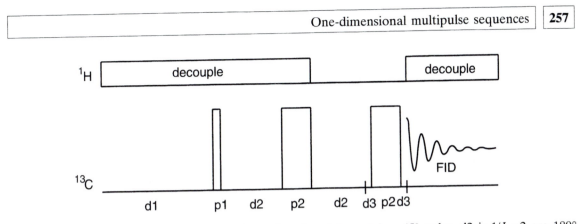

Figure 8.19 The APT pulse sequence. d1 is a relaxation delay, p1 is a 45° pulse, d2 is $1/J$, p2 are 180° pulses, and d3 is the pre-acquisition delay.

magnetization in the xy-plane which needs $5T_1$ to regain the z-direction. In contrast, the APT pulse program applies a total of a 405° pulse, equivalent to a 45° pulse, leaving $M_Z/\sqrt{2}$ magnetization in the z-direction, ready for the next pulse sequence to be applied.

8.3.2 INEPT

There is a family of pulse sequences which gain an enhancement in signal intensity from a higher frequency coupled nucleus, usually ^{1}H. Equation (1.3) gives the population difference between the upper and lower energy levels for an $I = 1/2$ nucleus. The population excess in the lower energy level is proportional to μ_I or γ_I. Hence for ^{1}H, with $\gamma_H = 26.7510 \times 10^7$ rad T^{-1} s^{-1}, the population excess is approximately four times greater than for ^{13}C, with $\gamma_C = 6.7263 \times 10^7$ rad T^{-1} s^{-1}. If it were possible to transfer this population excess from ^{1}H to ^{13}C, then the ^{13}C NMR signal strength would increase by a factor of approximately four times. This can be done, but the situation is not all win as the NOE is lost. Despite this, population transfer is generally preferred as the gains, γ_S/γ_I, are normally greater than those from NOE, $1 + \gamma_S/2\gamma_I$, see Tables 8.1 and 8.2, while the uncertainty associated with NOE, that dipole–dipole relaxation is required to be dominant, is removed. The gain is spectacular for very low frequency nuclei such as ^{103}Rh, especially as in this case NOEs are not normally observed.

The principle behind population transfer can be illustrated for an AX system consisting of a ^{13}C^{1}H pair of spins (Fig. 8.20). The relative populations of the energy levels are given. Before we apply a

Table 8.4 The intensity gains on polarization transfer from ^{1}H to several nuclei

Observe	^{13}C	^{15}N	^{19}F	^{29}Si	^{31}P	^{103}Rh
$\dfrac{\gamma_S}{\gamma_I}$	3.98	24.7	1.06	5.03	2.47	315

selective 180° pulse, the population differences for the two ^{1}H transitions are both 4δN, while for the two ^{13}C transitions they are both δN, in Fig. 8.20(a). In Fig. 8.20(b), after applying the selective 180° pulse to one ^{1}H transition, the population differences, and hence intensities of the ^{13}C transitions have changed to 5δN : (−3δN). Hence one transition has increased in intensity by 4δN units and the other decreased by −4δN units. If we were to now repeat the measurement, applying the selective 180° ^{1}H pulse to the other ^{1}H transition, we would get a ^{13}C doublet with intensities (−3δN) : 5δN. If these signals are now subtracted, the result is a 8δN : (−8δN) doublet compared with a 2δN : 2δN doublet if a simple ^{13}C NMR spectrum had been taken using two acquisitions. The gain has been a factor of four, or more accurately γ_H/γ_C.

This experiment is impractical to carry out routinely as it would be necessary to carry it out for each separate carbon environment in the molecule. However, exactly the same result can be achieved by the pulse sequence in Fig. 8.21 for all the ^{1}H coupled to X in a molecule, where X is an $I = 1/2$ nucleus. This is the **I**nsensitive **N**uclei **E**nhanced by **P**olarization **T**ransfer, or INEPT, pulse sequence.

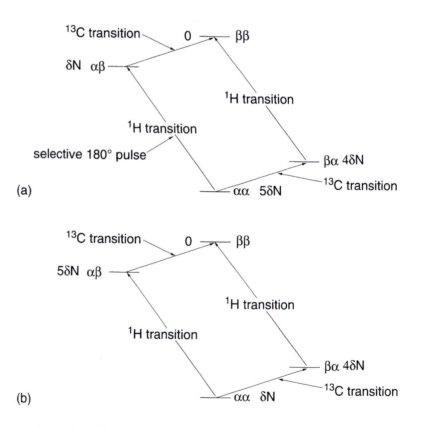

Figure 8.20 The effect of applying a 180° selective pulse to one line of the ^{1}H AX doublet of a ^{13}CH group. (a) Before. (b) After.

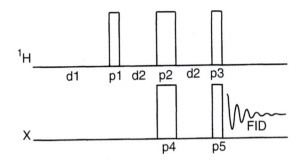

Figure 8.21 The INEPT pulse sequence. d1 is a relaxation delay, p1 is a $(90°)_x$ ^{1}H pulse, d2 is $1/4J_{XH}$, p2 is a 180° ^{1}H pulse, p3 is a $(90°)_y$ ^{1}H pulse, p4 is a 180° X pulse, and p5 is a 90° X pulse.

Let us examine the effect of these pulses on the ^{1}H nuclei (Fig. 8.22). We need to keep track of both the ^{1}H and X nuclei. In order to do this, ^{1}H nuclei attached to X nuclei with α-spin are labelled a, and those attached to X nuclei with β-spin are labelled b. The first $(90°)_x$ pulse brings the ^{1}H magnetization into the xy plane pointing in the y' direction. If the ^{1}H nuclei are on resonance, the two lines of the doublet rotate at $\pm J_{XH}/2$ Hz. After $1/4J_{XH}$ s, they will have rotated ±45°. A ^{1}H$(180°)_x$ refocusing pulse is applied. If nothing else were to be done, the two lines of the doublet would refocus in the $-y'$ direction after a further $1/4J_{XH}$ s. Instead a X(180°) pulse is also applied. This has the effect of exchanging the X spin states. The protons which were attached to X with α-spin are now attached to X with β-spin and *vice versa*. The result is that the ^{1}H nuclei precession direction in the rotating frame is reversed, and after the further $1/4J_{XH}$ s they end up aligned 180° with respect to each other. The final ^{1}H$(90°)_y$ pulse rotates

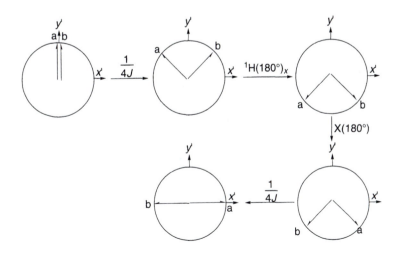

Figure 8.22 The effect of the INEPT pulse sequence on the ^{1}H magnetization in the xy plane. The diagrams start following the ^{1}H$(90°)_x$ pulse.

their magnetization back into the z-direction. One line of the doublet is pointing in the $+z$ direction, while the other is pointing in the $-z$ direction. This has achieved the same effect as the selective 180° pulse described above, but has produced the effect for all the ¹HX spin systems in the sample. The X(90°) pulse then produces an X FID with polarization transfer. Phase cycling is used to alternately invert the lines of the ¹HX doublet. For example, if instead of using a ¹H(90°)$_y$ pulse as the final ¹H pulse, a ¹H(90°)$_{-y}$ pulse was used, the phase of the ¹H doublet is reversed. Phase cycling of the receiver permits addition or subtraction of the spectra. The resulting ¹H coupled X spectrum does not have the usual 'Pascal triangle' intensities, but rather those given in Table 8.5. This is illustrated in Fig. 8.23 for some rhodium hydrides. Note that to distinguish between a 1 : (−1) doublet and a 1 : 0 : (−1) triplet, the separation of the lines has to be measured, and is J in the former case and $2J$ in the latter, with J being pre-determined from the ¹H NMR spectrum.

In order to be able to ¹H decouple, the pulse sequence has to be extended. This is because in a ¹H coupled INEPT spectrum there is as much positive intensity as negative intensity. The result of decoupling is to give zero intensity. Fortunately the answer is very straightforward. A delay is placed between the basic INEPT pulse sequence of Fig. 8.21 and acquisition (Fig. 8.24). Of course, it is neces-

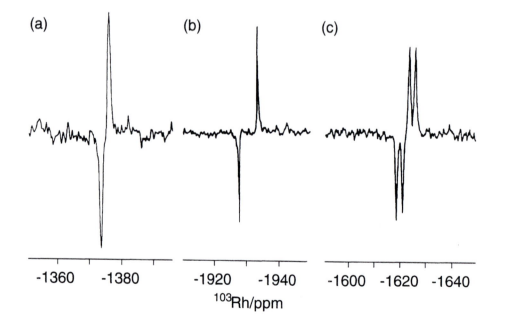

Figure 8.23 The 12.62 MHz ¹⁰³Rh NMR spectra obtained using the INEPT pulse sequence with $\tau = 0.0074$ s, corresponding to $^1J(^{103}\text{Rh}^1\text{H})$ of 34 Hz. The spectra are referenced to $\Xi = 3.16$ MHz. (a) $[(\eta^5\text{-C}_5\text{Me}_5)\text{RhH}(\text{SiEt}_3)(\eta^2\text{-C}_2\text{H}_4)]$ in $\text{CD}_3\text{C}_6\text{D}_5$, showing a −1 : 1 doublet. (b) $[(\eta^5\text{-C}_5\text{Me}_5)\text{RhH}_2(\text{SiEt}_3)_2]$ in C_6D_6, showing a −1 : 0 : 1 triplet. (c) $[(\eta^5\text{-C}_5\text{Me}_5)\text{RhH}_3(\text{SiEt}_3)]$ in $\text{CD}_3\text{C}_6\text{D}_5$, showing a −1 : −1 : 1 : 1 quartet. (Reproduced by permission of Academic Press from Mann (1988) *Adv. Organomet. Chem.*, **28**, 397.)

Table 8.5 The INEPT Pascal triangle. It is constructed in the same way as the normal Pascal triangle adding together the pair of numbers on the line above, except that it is extended by putting 1 and –1 at the ends of the next line.

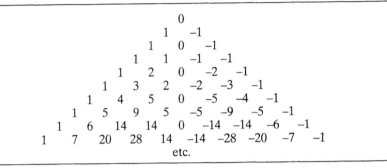

sary to include a refocusing 180° pulse for both ^{1}H and X, p4 and p7, halfway through the delay.

Figure 8.25 shows the response of the X-magnetization in the xy plane for an AX group. The first diagram shows the state of the magnetization following the X(90°)$_x$ pulse, p6. The two lines of the X doublet are 180° out-of-phase. It is necessary again to follow the spin states of the other nuclei. The X-nuclei attached to ^{1}H nuclei with α-spin are labelled a and those attached to ^{1}H nuclei with β-spin are labelled b. After $1/4J$ s, the magnetizations have rotated ±45° in the rotating frame, assuming again that the X signal is on resonance. As usual it is necessary to use an 180° X refocusing pulse. It is then necessary to use a 180° ^{1}H pulse to reverse the ^{1}H spin states and hence the labels on the magnetization vectors. The result is that they continue to precess in the same direction and refocus after a further $1/4J$ s. The ^{1}H decoupler is then turned on and the ^{1}H decoupled FID collected, enhanced by polarization transfer.

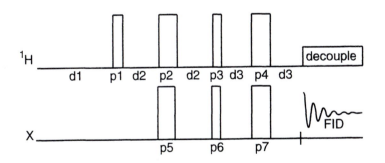

Figure 8.24 The refocused INEPT pulse sequence with ^{1}H decoupling. d1 is a relaxation delay, p1 is a (90°)$_x$ ^{1}H pulse, d2 is $1/4J_{XH}$, p2 and p4 are 180° ^{1}H pulses, p3 is a (90°)$_y$ ^{1}H pulse, d3 is a variable delay, see text, p5 and p7 are (180°)$_x$ X pulses, and p6 is a (90°)$_x$ X pulse.

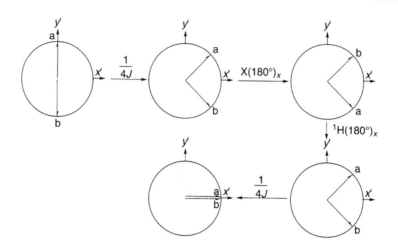

Figure 8.25 The effect of the INEPT pulse sequence on the X magnetization in the *xy* plane.

The introduction of the delay d3 into the refocused INEPT pulse sequence provides another method to distinguish between CH, CH_2 and CH_3 groups. The variation of their intensity with d3 is plotted in Fig. 8.26. It will be noted that when d3 is chosen to be $0.125/J$ all the signals are positive, when $d3 = 0.25/J$ only CH ^{13}C signals are detected, while when $d3 = 0.375/J$, CH and CH_3 groups are positive, while CH_2 groups are negative. In each case, only ^{1}H bearing ^{13}C atoms are detected. It is therefore possible to distinguish between each type of group.

In practice, INEPT is not used to sort CH, CH_2 and CH_3 ^{13}C NMR signals, but DEPT is normally used. The problem with INEPT is that it is far more sensitive to the value of J_{XH} than is DEPT. A typical ^{13}C NMR sample has a range of $^1J(^{13}C^1H)$ and the resulting refocused INEPT spectra show varying intensity and phase. Where INEPT wins over DEPT is when the coupling constant is small. During the delay(s) between the pulses, relaxation occurs if the delays are long. It is then advisable to minimize the delays, and INEPT has the shorter delays. INEPT is valuable for recording ^{1}H coupled ^{13}C NMR spectra and this is illustrated in Fig. 8.27.

8.3.3 DEPT

DEPT is the pulse sequence of choice to edit ^{13}C NMR spectra. The pulse sequence is given in Fig. 8.28. It is not possible to use a simple vector model to describe this pulse sequence. The ^{13}C editing is carried out by choosing a suitable p3. Values of 45°, 90°, and 135° are used. With a 45° pulse, the CH, CH_2, and CH_3 signals are all positive, a 90° pulse only gives CH signals, and the 135° pulse gives CH and CH_3 positive, while the CH_2 signal is negative. Suitable addition and

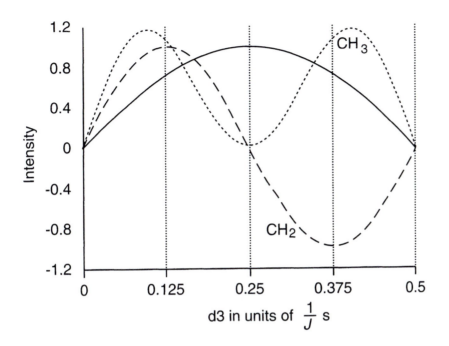

Figure 8.26 The response of CH, CH$_2$, and CH$_3$ groups as a function of d3 when the refocused INEPT pulse sequence is used.

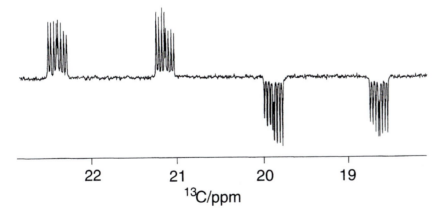

Figure 8.27 A partial 100.62 MHz ^{13}C NMR spectrum of one of the methyl groups of carvone in CDCl$_3$. The spectrum was recorded using the INEPT pulse sequence and shows ^{1}H coupling. The $-1:-1:1:1$ pattern arises from the INEPT pulse sequence which was set up for a $J(^{13}C^1H) = 145$ Hz corresponding to an average $^1J(^{13}C^1H)$. The doublet of doublet of doublets multiplet structure of each of the four lines arises from the much smaller $^3J(^{13}C^1H)$.

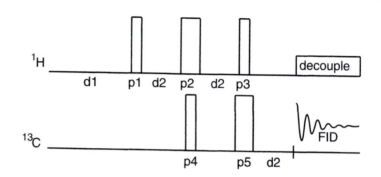

Figure 8.28 The DEPT pulse sequence with ^{1}H decoupling. d1 is a relaxation delay, p1 is a 90° ^{1}H pulse, d2 is $1/2J_{XH}$, p2 is a 180° ^{1}H pulse, p3 is a ^{1}H pulse of variable length, see text, p4 is a 90° X pulse and p5 is a 180° X pulse.

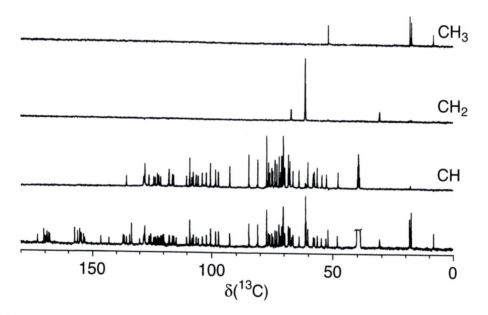

Figure 8.29 Edited ^{13}C DEPT spectra of ristocetin at 100.6 MHz. The lower spectrum is the normal broadband decoupled spectrum showing all the carbon resonances. Note that even the resonance of the CD$_2$H groups of the solvent dimethylsulfoxide-d_6 appears in its appropriate trace. (From Bruker, with permission.)

subtraction of DEPT-45, DEPT-90 and DEPT-135 spectra yield spectra only showing CH, CH$_2$ or CH$_3$ signals. An example is given in Fig. 8.29.

8.3.4 PENDANT

Although DEPT is an excellent pulse sequence to use for the observation of CH, CH$_2$, and CH$_3$ groups, it does not detect non-protonated ^{13}C atoms. J modulation or APT used to be the initial method of choice to measure a ^{13}C NMR spectrum, followed by DEPT as required to differentiate between C and CH_2 groups and between CH and CH_3 groups. Recently, J modulation and APT have been replaced by the pulse sequence, PENDANT, **P**olarization **EN**hancement **D**uring **A**ttached **N**ucleus **T**esting. PENDANT has nearly the same sensitivity as DEPT for CH, CH_2, and CH_3 groups, and also gives signals for quaternary carbon atoms. The pulse sequence is given in Fig. 8.30.

An example of a PENDANT ^{13}C NMR spectrum is given in Fig. 8.31. Comparison of this spectrum with the corresponding JMOD spectrum in Fig. 8.17 shows that the non-proton bearing carbon atoms give stronger signals. As these carbons generally give weaker signals than proton bearing carbon atoms due to their longer T_1 values, and possibly due to smaller NOE's, it is these carbons which set the number of acquisitions and hence time required for a given spectrum.

8.3.5 The use of spectral editing pulse sequences for ^{13}C assignments

For most compounds, it is now pointless to acquire a simple $^{13}C\{^1H\}$ NMR spectrum, when, in the same length of time a PENDANT spectrum can be acquired. The sorting of the ^{13}C NMR signals into C and CH_2 signals with one phase and CH and CH_3 signals with the opposite phase is often sufficient to assign the signals, when combined with the chemical shift information. Where there is ambiguity, the appropriate DEPT spectra can be measured.

Signals can be missing from spectra recorded using pulse sequences such as J modulation, APT, INEPT, DEPT, or PENDANT. In all these pulse sequences, there are delays between pulses and relaxation can occur during these delays. This is a severe problem where there are broad signals such as occur where there is substantial broadening due to exchange or scalar relaxation by a quadrupolar nucleus.

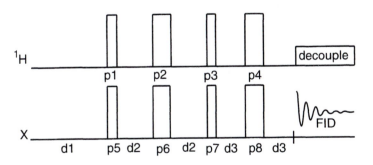

Figure 8.30 The PENDANT pulse sequence with ^1H decoupling. d1 is a relaxation delay, p1 is a $(90°)_x$ ^1H pulse, d2 is $1/4J_{XH}$, p2 and p4 are $(180°)_x$ ^1H pulses, p3 is a $(90°)_y$ ^1H pulse, d3 is $5/8J_{XH}$, p5 is a $(90°)_x$ X pulse, p6 and p8 are $(180°)_x$ X pulses, and p7 is a $(90°)_y$ X pulse.

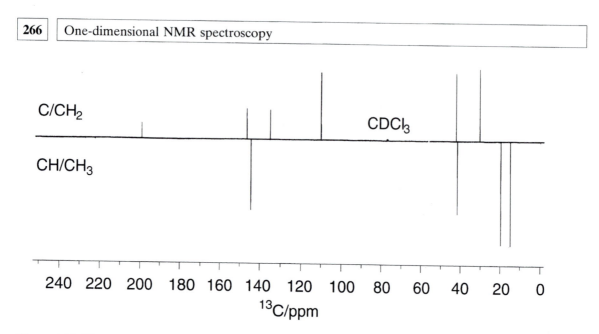

Figure 8.31 The 100.62 MHz ^{13}C NMR spectrum of carvone in CDCl₃ recorded using the PENDANT pulse sequence.

8.3.6 One-dimensional INADEQUATE

The observation of coupling between low abundance nuclei is diffi-cult. For example, ^{13}C is only 1.08% abundant, and the ^{13}C NMR spec-trum of an organic molecule consists of sharp singlets flanked by weak ^{13}C satellites. These satellites are often obscured by the strong central resonance. The one-dimensional INADEQUATE, **I**ncredible **N**atural **A**bundance **D**ouble **Qu**antum **T**ransfer **E**xperiment, pulse sequence is given in Fig. 8.32. This pulse sequence gives the ^{13}C coupling as an out-of-phase doublet. A longer pulse sequence is required to bring the doublet into phase (Fig. 8.33). It is this pulse sequence that was used to produce the spectra in Fig. 8.34. d2 can be chosen to select $^{1}J(^{13}C^{13}C)$ or a longer range coupling (Fig. 8.34). This is the experiment of choice to measure $J(^{13}C^{13}C)$, while two-dimensional INADEQUATE is the experiment to use where ^{13}C–^{13}C connectivity is required.

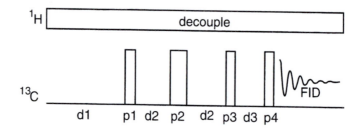

Figure 8.32 The one-dimensional INADEQUATE pulse sequence with ^1H decoupling. d1 is a relaxation delay, p1, p3 and p4 are 90° ^{13}C pulses, d2 is $1/4J_{CC}$, p2 is a 180° ^{13}C pulse, d3 is 3 μs.

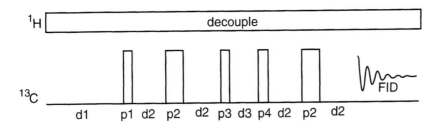

Figure 8.33 The one-dimensional INADEQUATE pulse sequence with ^{1}H decoupling and rephasing. d1 is a relaxation delay, p1, p3 and p4 are 90° ^{13}C pulses, d2 is $1/4J_{CC}$, p2 is a 180° ^{13}C pulse, d3 is 3 μs.

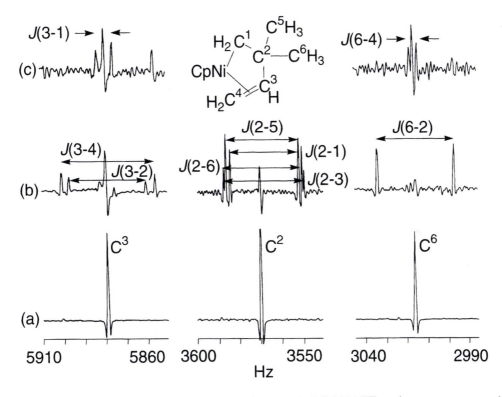

Figure 8.34 The application of the one-dimensional INADEQUATE pulse sequence to determine $J(^{13}C^{13}C)$ in $[(\eta^5\text{-}C_5H_5)Ni(1,3,4\text{-}\eta^3\text{-}2,2\text{-dimethylbutenyl})]$ at 100.62 MHz in d_8-THF. (a) The signals due to C^3, C^2, and C^6. (b) INADEQUATE NMR spectra with $d_2 = 0.0062$ s corresponding to $J(^{13}C^{13}C) = 40$ Hz. (c) INADEQUATE NMR spectra with $d_2 = 0.08$ s corresponding to $J(^{13}C^{13}C) = 3$ Hz. (Reproduced from Benn and Rufińska (1982) *J. Organomet. Chem.*, **238**, C27, copyright (1982), with permission from Elsevier Science.)

8.4 EXERCISES IN SPECTRAL INTERPRETATION

The determination of the structures of organic molecules using proton and carbon spectra together will now be considered in some exercises.

We have shown in the previous chapters that the carbon spectra are capable of yielding much information through a series of quite time-consuming experiments. These facilities are normally reserved for the more difficult samples, and often it will be sufficient to have simply the broad-band proton-decoupled spectrum. This gives the chemical shift information, details of coupling to nuclei other than hydrogen, and a carbon count, provided all the likely errors in line intensity are taken into account. It is also certain that the proton spectrum will be available, since this is so easy to obtain, and interpretation will be based on the two sets of data taken together. The carbon spectra, of course, give information about atoms not bonded to hydrogen, which is not available in the proton spectra.

The carbon chemical shifts are much more widely dispersed than are the proton shifts. The ranges within which different types of carbon atom resonate are shown on the chart in Fig. 8.35. The chemical shift of a given carbon atom in a family of compounds is sensitive to the influence of all four substituents, and for alkanes, for instance, can be predicted using the Grant and Paul rules

$$\delta_i = -2.6 + 9.1n_\alpha + 9.4n_\beta - 2.5n_\gamma + 0.3n_\delta$$

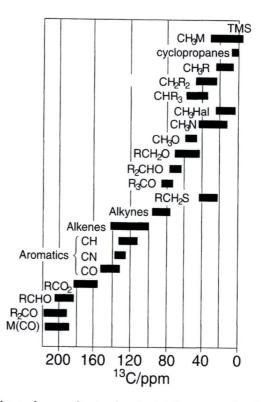

Figure 8.35 Chart of approximate chemical shift ranges of carbon atom nuclei in organic and organometallic compounds. M represents a metal and Hal a halogen.

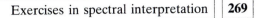

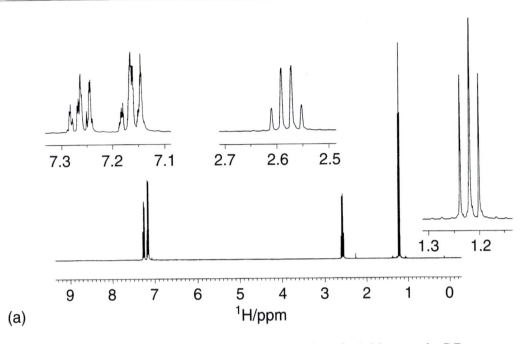

Exercise 4 (a) The 400.13 MHz ^1H NMR spectrum of a solution of ethyl benzene in C_6D_6.

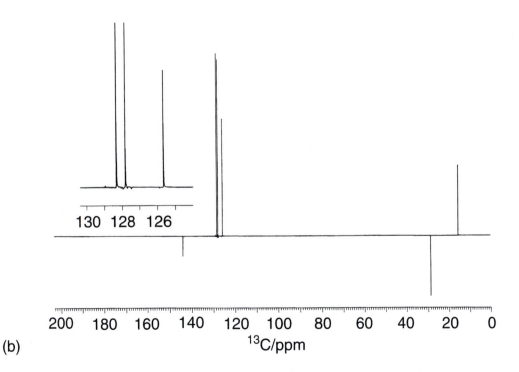

Exercise 4 (b) The 100.62 MHz ^{13}C JMOD NMR spectrum phased so that C and CH_2 groups are negative and CH and CH_3 signals are positive. In each case, expansions are given as insets.

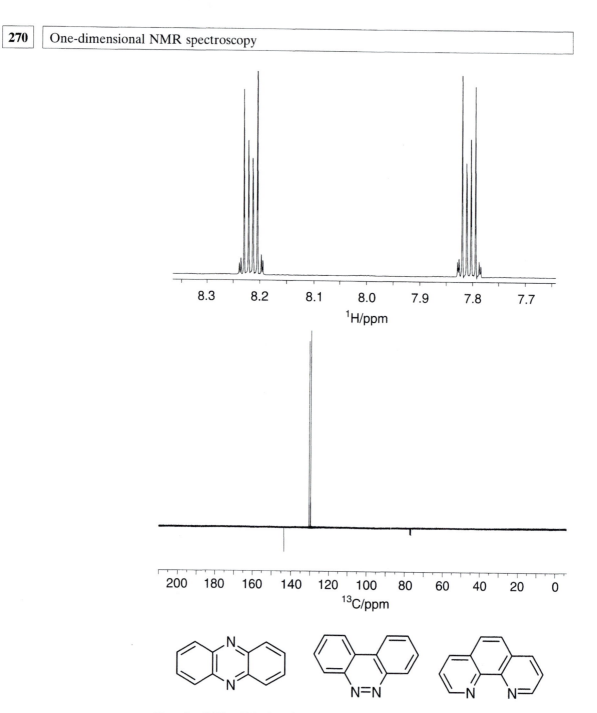

Exercise 5 The 400 MHz ^{1}H NMR spectrum (upper) and 100.62 MHz ^{13}C{^{1}H} JMOD NMR spectrum (lower) of the aromatic compound $C_{12}H_8N_2$. The ^{13}C NMR spectrum shows signals at δ 143.3, 130.4, and 129.6 due to the compound and 77.0 due to CDCl$_3$. The proton spectrum is diagnostic of an [AB]$_2$ four-spin system in which there are two pairs of protons with the same chemical shift δ_A, δ_B, but different coupling constants J(AB), J(AA'), J(AB') and J(BB'). Only three lines are observed in the carbon spectrum. This information is sufficient to identify the compound, some suggestions for which appear below the spectra. Example supplied by A. Römer.

The chemical shift of carbon i can be calculated from the number of directly bonded carbons (n_α) and the number of carbon atoms two (n_β), three (n_γ) and four (n_δ) bonds removed; –2.6 ppm is the chemical shift of methane. Similar rules exist for ethylenic hydrocarbons and substituent effects have been documented.

The use of the two sets of information together is illustrated in Exercise 4 for the simple case of ethyl benzene. The 400 MHz ^{1}H NMR spectrum contains the unmistakeable quartet–triplet pattern due to an ethyl group and five protons resonate in the aromatic region. The aromatic protons split into two sets of signals with an intensity ratio of 2 : 3. The signal of intensity 2 at δ 7.26 is a triplet with additional structure. This must be due to the *meta*-hydrogen atoms, which are coupled to the *ortho* and *para* ones. The *ortho*- and *para*-hydrogen signals overlap at δ 7.16. This is indeed sufficient to give the structure in this case, especially if the molecular weight were known. We nevertheless turn our attention to the carbon JMOD spectrum. We find the two carbons of the ethyl group and four aromatic resonances with appropriate polarities. The aromatic signal to high frequency is rather small, as would be expected for quaternary carbon and its polarity indicates a lone C. The remaining three are CH and two are significantly larger than the third so may arise from two carbon atoms each. The pattern has the form to be expected for a monosubstituted six-membered ring, and allows us to over determine the structure of this molecule. All its features are apparent, though we should note

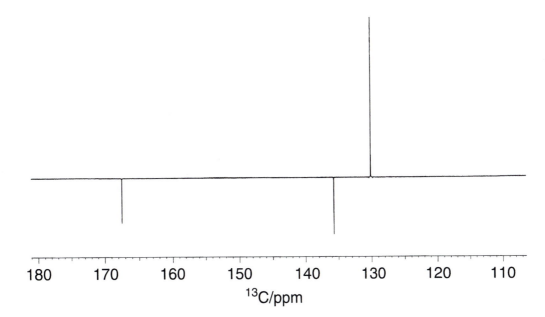

Exercise 6 The 100.62 MHz ^{13}C JMOD NMR spectrum of $C_{10}H_2O_6$ in $(CD_3)_2CO$. The spectrum is plotted so that signals due to CH and CH_3 groups are positive and those due to C and CH_2 groups are negative. Example supplied by A. Römer.

carefully the ambiguities of the intensity of the resonances. Which line should we take as representing one carbon? Some more complex examples are given in Exercises 5 and 6.

Exercise 6 shows the ^{13}C JMOD NMR spectrum of the compound $C_{10}H_2O_6$ with proton double irradiation. In a ^1H coupled spectrum, not shown, $J = 179$ Hz is observed to the resonance at 130.1 ppm and $J = 6$ Hz to the resonance at 135.6 ppm. The proton spectrum is a singlet in the aromatic region. What is the structure of the molecule, given that it is an anhydride?

Two-dimensional NMR spectroscopy $\boxed{9}$

Two-dimensional NMR spectroscopy has become an important tool for chemists. Several experiments are possible and new variants are being continually invented, such is the power of the method. The two dimensions are dimensions of time as previously discussed at the beginning of Chapter 8. One of these is already familiar, and is the time domain within which we collect the FID output from the spectrometer and which contains frequency and intensity information. The second dimension refers to the time elapsing between the application of some perturbation to the system and the onset of the collection of data in the first time domain. This second time period is varied in a regular way and a series of FID responses are collected corresponding to each period chosen.

In order to illustrate how two-dimensional NMR spectroscopy works, we will show how the one-dimensional J modulation described in section 8.3.1 can be converted into a two-dimensional experiment. The delay d2, which is related to J, is replaced by a variable delay $t_1/2$, where t_1 is made to assume many values. The pulse sequence is shown in Fig. 9.1 which is essentially that of Fig. 8.16 though we should note that in practice it is more common for the decoupler to be off during the first delay and on during the second, though this makes no difference to the outcome. Examination of the behaviour of the spin magnetization in Fig. 8.18(b) shows that the intensity of the ^{13}C signal from a CH entity is a function of the delay time and varies as $\cos(2\pi t_1/J)$. It will also decay by the T_2 process though the 180° pulse refocuses the magnetization and this is the true T_2; see section 6.3. Similarly, the ^{13}C signal of a CH$_2$ group will be seen from Fig. 8.18(c) to remain always positive and to follow a $[\cos(2(\pi t_1/J) + 1]$ law. In the two-dimensional form of the experiment, a series of FIDs are collected corresponding to different values of t_1. Each FID is then Fourier transformed individually to give a series of spectra, one for each value of t_1 chosen. So we have spectra varying as a function of f_2 (corresponding to the t_2 dimension) and covering a range of t_1 values as in Fig. 9.2. If we look at the intensity at δ 42.45, which arises from a CH, we will observe that it changes across the spectra as a decaying cosine wave. This is made clearer by the section through the spectra shown in Fig. 9.3. The frequency of this wave gives the C–H coupling constant. The signal is not however, a simple decaying cosine wave but is modulated by smaller, long distance couplings to give lower frequency waves. This is not very apparent for the signal at δ 42.45 as this unresolved coupling broadens the lines and gives the rapid decay. This modulation is much clearer for the signal at δ 43.10 as two and three bond couplings to single protons introduce negative components into the pattern which without these smaller couplings would remain always positive.

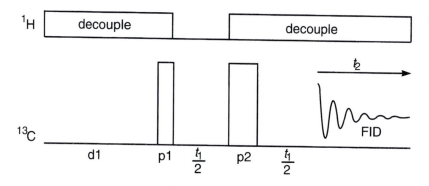

Figure 9.1 The *J* modulation pulse sequence. d1 is for relaxation, p1 is a 90° pulse, $t_1/2$ and t_2 are the two time dimensions for the two-dimensional experiment, and p2 is a 180° pulse.

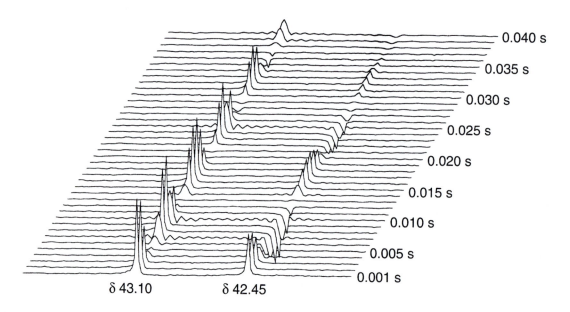

Figure 9.2 The individual spectra of a partial *J*-resolved 100.62 MHz ^{13}C NMR spectrum of carvone in CDCl$_3$ obtained using the pulse sequence in Fig. 9.1. The individual FIDs have been transformed in the t_2/f_2 direction and plotted for t_1 values from 0.001 to 0.04 s, incremented by 0.001 s, given on the right. Note the decaying cosine oscillations of each signal in the t_1/f_1 direction.

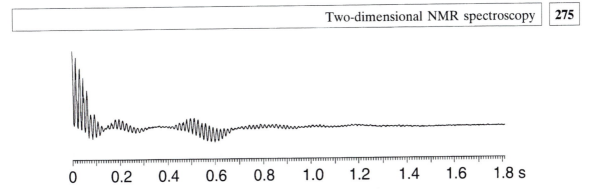

Figure 9.3 Cross-section through the signal at δ 43.10 after transformation in the f_2 direction of a *J*-resolved 100.62 MHz ^{13}C NMR spectrum of carvone obtained using the pulse sequence in Fig. 9.1. A partial stack plot is shown in Fig. 9.2. The *x*-axis corresponds to the delay, t_1. The high frequency components are due to $^1J(^{13}C^1H)$, while the low frequency components are due to $^2J(^{13}C^1H)$ and $^3J(^{13}C^1H)$.

It should be clear that chemical shifts are contained in the f_2 dimension and the coupling information is in the t_1 dimension. We can measure the coupling frequency from the plots of Fig. 9.3 but it is much more convenient to get the computer to do it for us by performing a Fourier transform across the spectra in the t_1 direction, especially as this will give us both the evident frequency and the smaller components. This is now the f_1 dimension. The result is shown in Fig. 9.4. The set of spectra shown in Fig. 9.4 is very difficult to use, especially when there are many signals. It is essentially the same problem that was solved many years ago by cartographers. In exactly the same way as mountains are represented on maps by contours, the intensity of the NMR signals of a two-dimensional NMR spectrum are represented by contours on a two-dimensional plot of f_2, normally horizontal, against f_1, normally vertical (Fig. 9.5). The result is that overlapping multiplets are separated (Fig. 9.6).

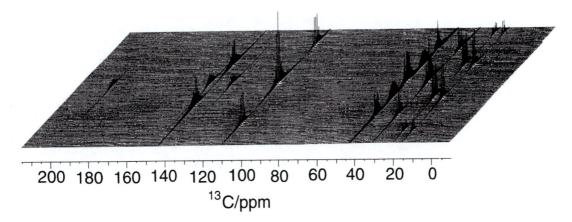

200 180 160 140 120 100 80 60 40 20 0

^{13}C/ppm

Figure 9.4 The individual spectra of a *J*-resolved 100.62 MHz ^{13}C NMR spectrum of carvone obtained using the pulse sequence in Fig. 9.1. The individual FIDs have been transformed in the t_2/f_2 and t_1/f_1 directions and plotted for a selection of t_1 values.

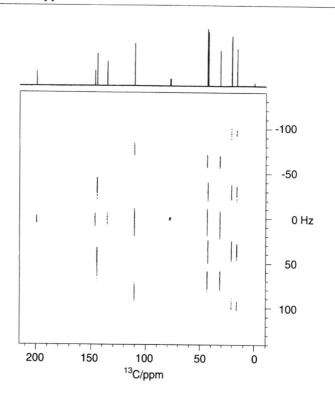

Figure 9.5 The contour plot of a *J*-resolved 100.62 MHz ^{13}C NMR spectrum of carvone in CDCl$_3$ obtained using the pulse sequence in Fig. 9.1.

There are some experimental considerations. Two-dimensional NMR spectra can produce very large data sets and can take very long times. The data set for the FIDs giving rise to the spectrum in Fig. 9.5 is 32 768 data points in the t_2 direction and 128 data points in the t_1 direction. The sweep width in both dimensions was kept to a minimum to maximize the digitization. As each word is 32 bits, this gives rise to a dataset of 16 megabytes. The major consideration is time. In order to complete the phase cycling to remove artefacts, a multiple of 4 spectra have to be acquired for each t_1 value. The acquisition plus d1 time needs to be long enough for recovery, 3.3 s in this case. The result is that the minimum time required for this experiment is $4 \times 3.3 \times 128$ s = 1690s = 28 m 10 s though on top of this there is the time associated with the pulse sequence. In practice, considerably longer times can be required. There is not normally a time penalty in increasing the f_2 resolution by increasing the acquisition time, as a corresponding decrease can be made in d1. However, to increase the f_1 resolution by a factor of two requires a doubling of time. The time for the experiment can be reduced by decreasing d1, but then artefacts become a problem. Real gains are made by using magnetic field gradient pulses. Then d1 can be shortened, and phase cycling is not required or is considerably reduced.

We will describe some of the experiments possible in more detail below. If the molecule to be studied is complex, it is necessary to carry out several different two-dimensional experiments in order to discover all the correlations necessary to provide an unequivocal structure. We will use the simple molecule carvone to introduce how several of the sequences work and use D-amygdalin, whose formula is in Fig. 9.7, to demonstrate the advantages of the two-dimensional technique.

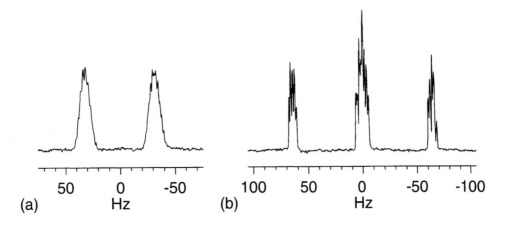

Figure 9.6 Cross-sections of the signals at (a) δ 42.45 and (b) 43.10 of a *J*-resolved 100.62 MHz ¹³C NMR spectrum of carvone in CDCl₃ obtained using the pulse sequence in Fig. 9.1, showing the multiplicity due to ¹H coupling. Note that due to the pulse sequence used the coupling constants have been halved.

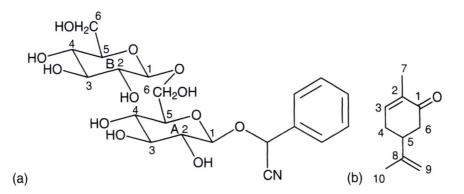

Figure 9.7 The structures of (a) D-amygdalin and (b) carvone.

9.1 *J*-RESOLVED TWO-DIMENSIONAL NMR SPECTROSCOPY

9.1.1 Homonuclear *J*-resolved two-dimensional NMR spectroscopy

This experiment separates chemical shift information from coupling constant information. The result is a ^1H NMR spectrum with all the first-order ^1H coupling removed and cross-sections giving the coupling patterns.

This technique is based on the Carr–Purcell pulse sequence described in section 6.1.2.2 for the determination of T_2, which only works if the resonances studied are singlets, since the refocusing of doublets caused by spin–spin coupling modulates the output intensity as a function of the waiting time. This is easily transformed to advantage into a two-dimensional experiment. The pulse sequence is shown in Fig. 9.8.

This two-dimensional NMR experiment was applied to carvone. After transformation in both the t_2 and t_1 dimensions, the two-dimensional NMR spectrum shown in Fig. 9.9 is obtained. The chemical shift information is in the f_2 dimension and the coupling information is spread out in the f_1 dimension. It is then normal to 'tilt' the spectrum so that all the lines of each multiplet line up in the f_1 dimension. This is a very simple operation which is done automatically by the computer. Each row of the two-dimensional matrix is shifted by an amount proportional to its distance from the centre of the spectrum. Thus the row corresponding to a 15 Hz offset in the f_1 dimension is shifted to a 15 Hz increase in frequency, and this produces a shift to the left, higher frequency. The result is shown in Fig. 9.10. If all the individual spectra are now added together, the result is a ^1H decoupled ^1H NMR spectrum, which is shown above the two-dimensional spectrum. It is possible to clean up the spectrum by symmetrizing it about the $f_1 = 0$ Hz row. This consists of comparing the intensities in the spectra equally spaced above and below the $f_1 = 0$ Hz row and taking the smallest value. This is shown in Fig. 9.11 and the improved ^1H decoupled ^1H NMR spectrum is shown above the two-dimensional spectrum. A section taken along the length of a

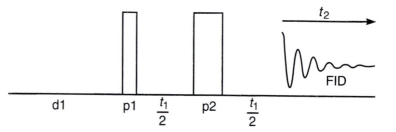

Figure 9.8 The *J*-resolved pulse sequence. d1 is for relaxation, p1 is a 90° pulse, $t_1/2$ and t_2 are the two time dimensions for the two-dimensional experiment, and p2 is a 180° pulse.

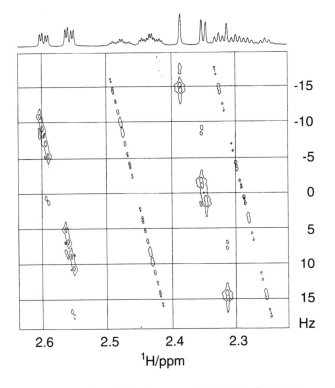

Figure 9.9 A contour plot of the *J*-resolved 400 MHz ^{1}H NMR partial spectrum of carvone in $CDCl_3$ before tilting. A projection of the spectrum is shown above.

multiplet, parallel with the f_1 axis (a cross-section) gives the coupled spectrum for each proton of that chemical shift, and an example of one such section is given in Fig. 9.12. This technique is very valuable for the separation of overlapping multiplets and producing ^{1}H broad-band 'decoupled' ^{1}H NMR spectra.

A model compound such as carvone used to illustrate the experimental aspects of the technique does not fully show its power. In a much more congested spectrum, such as that for D-amygdalin, the signals at δ 3.9 and between δ 3.3 and 3.6 are much more difficult to disentangle (Fig. 9.13), but the *J*-resolved experiment separates the multiplets. A typical plot is shown in Fig. 9.14 for the low frequency sugar resonances of D-amygdalin, where most spectral congestion lies. The chemical shifts are plotted on the horizontal scale, whose length is equivalent to 800 Hz, and coupling constants are in the vertical axis. The plot below the chart is the normal one-dimensional spectrum. That at the top is the projection of the contour plots onto the chemical shift axis and is, in fact, a spectrum in which all the protons have been fully decoupled from one another, i.e. the experimentally unachievable homonuclear broad-band decoupling experiment. Clearly, this plot helps to distinguish resonances that overlap in the one-dimensional

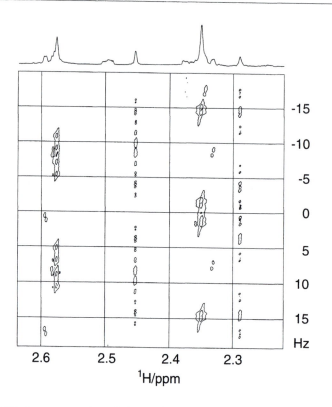

Figure 9.10 A contour plot of the *J*-resolved 400 MHz ¹H NMR partial spectrum of carvone in CDCl₃ after tilting. A projection of the spectrum is shown at the top.

trace. There are, for instance, two 1 : 1 : 1 : 1 quartets at δ 4.6 and 3.73, and we can see immediately that the former consists of two equal overlapping doublets whereas the latter is a doublet of doublets involving only one chemical shift value. The multiplet structure appears in the coupling dimension. 128 values of τ were used and 128 FIDs were obtained, which limits the digital resolution to 0.39 Hz. It is a simple matter to obtain all the coupling constants to within this limit. What can we deduce about the spectrum of D-amygdalin from this chart? The lines are already assigned on the figure but we will attempt to proceed as if this had not already been done. Four lines are marked as originating from proton-6. They are all doublets of doublets, but one of the couplings is rather large at about 11 Hz and is found nowhere else in the spectrum, and we can think immediately in terms of a possible geminal pair. The two large couplings are slightly different, sufficiently so that it is possible to separate the four resonances into two pairs linked by identical coupling constants. These are then the CH₂ protons linked to the two rings, although we do not actually know yet which is which. Each proton of these geminal pairs is further coupled to proton 5, though the responses from both of these

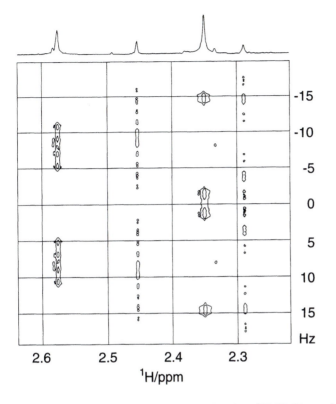

Figure 9.11 A contour plot of the *J*-resolved 400 MHz ^{1}H NMR partial spectrum of carvone in CDCl$_3$ after tilting and symmetrizing about a horizontal running through the middle of the spectrum at 0 Hz. A projection of the spectrum is shown at the top.

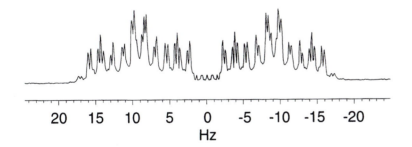

Figure 9.12 The cross-section of the 400 MHz ^{1}H NMR spectrum of the signal at δ 2.46 of carvone in CDCl$_3$.

is rather weak and we cannot reliably proceed to further assignment. The two doublets near δ 4.6 have a chemical shift that is characteristic of the anomeric proton H1 of β-pyranosides of glucose. Further, these protons are the only ones that will be split into simple doublets, and so the assignment is confirmed. Finally, we can see that all the

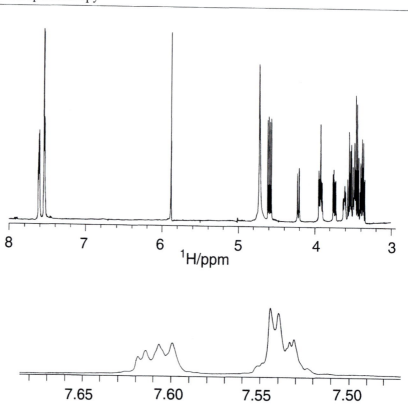

Figure 9.13 The normal, i.e. one-dimensional, 500 MHz ¹H NMR spectrum of D-amygdalin. There is a pattern in 2 : 3 intensity ratio at 7.6 ppm due to phenyl, a CHCN singlet at 5.88 ppm, HOD at about 4.75 ppm and a complex pattern to low frequency due to the remaining 14 sugar protons. (From Ribiero (1990) *Magn. Reson. Chem.*, **28**, 765–7, copyright (1990) John Wiley and Sons Ltd, reprinted with permission.)

coupling constants other than those between proton-6 are about 7–8 Hz, so that we have axial–axial stereo-chemistry rather than equatorial–equatorial or equatorial–axial.

9.2 HOMONUCLEAR COSY NMR SPECTROSCOPY

COSY stands for **CO**rrelation **S**pectroscop**Y**, and was the first two-dimensional technique to be proposed by Jeener in 1971. As we shall see, it has given rise to many variants, which depend upon the existence of spin–spin coupling between nuclei to provide supplementary responses that relate the chemical shift positions of the coupled nuclei. It is equivalent to carrying out simultaneously a series of double-resonance experiments at each multiplet in the spectrum and looking for the part of the spectrum where a perturbation has occurred. It is thus rapid and avoids the need to know what irradiation frequency to use

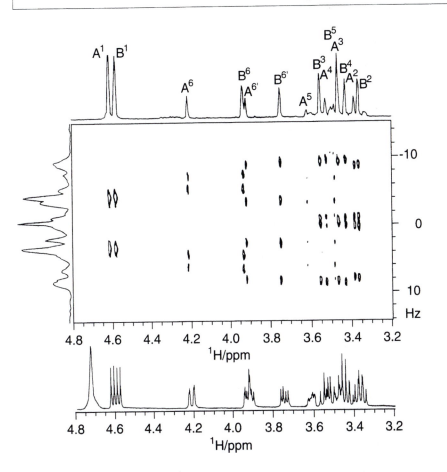

Figure 9.14 Partial homonuclear *J*-resolved two-dimensional spectrum of D-amygdalin. The responses are shown as contour plots on the *J/δ* chart. The normal one-dimensional spectrum is below and the fully decoupled version of this is above the chart. All coupling constants can be read off on the right-hand scale. (From Ribiero (1990) *Magn. Reson. Chem.*, **28**, 765–73, copyright (1990) John Wiley and Sons Ltd, reprinted with permission.)

for each experiment, information that in any case is often difficult to obtain for a complex or crowded spectrum. Figure 9.15 shows the basic pulse sequence used. There are many variations on this pulse sequence and some of the more important ones will be illustrated in the following sections.

9.2.1 COSY-90 NMR spectroscopy

The simplest and most sensitive version of the COSY experiment is the COSY-90 experiment where p1 is a 90° pulse. P1 produces a normal FID and p2 distorts this in a way which depends upon the delay t_1. Treatment of the set of distorted spectra which result, gives a two-dimensional diagram with the proton chemical shift along both

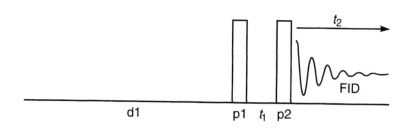

Figure 9.15 The basic pulse sequence for the COSY experiment. d1 is a relaxation delay, p1 is a variable pulse, fixed for a given experiment, p2 is a 90° pulse, and t_1 is the variable delay for the second dimension.

axes and the spectrum along the diagonal, which projects onto either axis as the normal spectrum. The distortion introduced by the pulse sequence introduces extra responses, off the diagonal, for those protons which are spin coupled and none for those which are not. These extra resonances appear on the two-dimensional map at the points which describe the two chemical shifts of the coupled protons and so enable the shifts of coupled resonances to be discovered, or correlated. These peaks are often called cross-peaks.

A partial COSY-90 spectrum of D-amygdalin is shown in Fig. 9.16, concentrating on the low frequency part of the spectrum where all the interest lies. The diagonal runs from top left to bottom right of the map and when projected onto a chemical shift axis is seen to be the one-dimensional spectrum, in this case shown at the top of the figure. Several off-diagonal responses are evident, disposed symmetrically about the diagonal, and it is these we must examine for information about correlation. The anomeric signals at about 4.6 ppm can be used as a point of entry for analysis of the spectrum, and the connectivities to the H^2 signals can easily be traced. Unfortunately, the diagonal is very cluttered in the 3.4 to 3.6 ppm region, and we cannot move on with any degree of certainty to the connectivities to protons further around the ring. This is a disadvantage of COSY-90, which we will see how to circumvent shortly. The doublet of doublets at 4.22 ppm has a cross-peak to the complex multiplet at 3.92 ppm and to the multiplet at 3.62 ppm, the latter also being correlated with the 3.92 ppm resonance. This pattern is that of an ABX spin system and so has to be due to H^6, $H^{6'}$ and H^5 of one ring. The multiplet at 3.62 ppm is an octet and so has to be H^5, in agreement with the J-resolved data, which has already allowed us to locate the protons 6. It follows from the cross-peak near the diagonal that H^4 of this ring is at about 3.5 ppm.

A second, similar pattern, connects signals at 3.9, 3.75 and 3.49 ppm and presumably arises from the protons 5 and 6 of the second ring. H^4 cannot be assigned in this case. It remains to assign this and the H^3 protons.

Figure 9.16 shows the spectrum before symmetrization. The symmetrized spectrum is in Fig. 9.17.

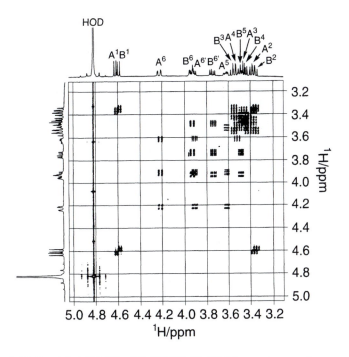

Figure 9.16 Partial 400 MHz ^{1}H COSY-90 NMR spectrum of D-amygdalin before symmetrization showing the sugar resonances.

There is a problem with symmetrization and care is necessary in its use. Examination of the peaks on the diagonal in Fig. 9.16 shows that the peaks are not round or oval, but have horizontal and vertical 'tails'. When peaks are close in chemical shift, the 'tails' can reinforce each other, giving an apparent cross-peak. This is obvious before symmetrization but after, it is easy to be fooled.

COSY is not restricted to ^{1}H NMR spectroscopy or indeed to nuclei where coupling is resolved. As long as the coupling contributes to the linewidth of the signals, then COSY can be used. For example, COSY is very valuable in elucidating the structure of boron hydrides and carboranes. This is illustrated here for 9-SMe$_2$-7,8-*nido*-C$_2$B$_9$H$_{11}$, (**9.1**) (Fig. 9.18). Examination of the COSY spectrum shows strong correlations, as shown by the large number of contour lines for each cross peak, acting through $^1J(^{11}B^{11}B)$ and weak correlations through $^2J(^{11}B^{11}B)$ and/or $^3J(^{11}B^{11}B)$. This spectrum then allows the resonances to be assigned to the nearest neighbour boron atoms and to confirm this via the longer range correlations.

9.2.2 COSY-45 NMR spectroscopy

The COSY-45 NMR spectrum only differs from the COSY-90 NMR spectrum by using a 45° pulse for p1 in the pulse sequence in Fig. 9.15. There are three reasons for measuring a COSY-45 NMR spectrum in

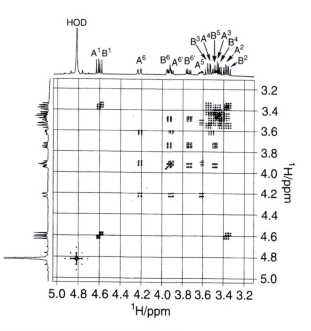

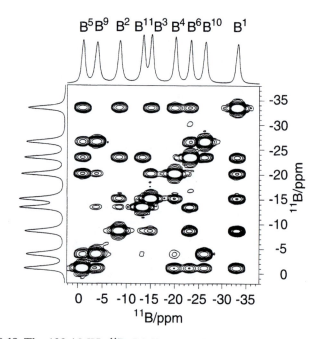

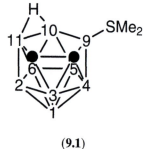

(9.1)

Figure 9.17 Partial 400 MHz ^{1}H COSY-90 spectrum of D-amygdalin after symmetrization, showing the sugar resonances.

Figure 9.18 The 128.4 MHz ^{11}B COSY-90 NMR spectrum of 9-SMe$_2$-7,8-*nido*-C$_2$B$_9$H$_{11}$, (9.1), in CDCl$_3$. Numbers are used to identify the boron atoms and ● to indicate the carbon atoms. (Reproduced from Rosair *et al.* (1997) *J. Organomet. Chem.*, **536**, 299, copyright (1997), with permission from Elsevier Science.)

preference to the more sensitive COSY-90 NMR spectrum. First, the peaks on the diagonal are weaker, so that it is easier to identify correlations close to the diagonal. Second, it is easier to decide which coupling constant is operating for a particular correlation, and third, the relative signs of coupling constants can be determined. As noted earlier in sections 3.1 and 3.7.3, coupling constants have sign. If we only measure the magnitude of a coupling constant, then we are throwing away valuable information. The most common case where sign is important is in $-CH_2CH<$ fragments in molecules. $^2J(^1H^1H)$ normally lies between -12 and -14 Hz, while $^3J(^1H^1H)$ can lie between 0 and $+16$ Hz depending on the torsion angle and substituents. It is quite common to have situations where in order to distinguish between assignments, we need to know which coupling is positive and which is negative.

A COSY-45 ^1H NMR spectrum of D-amygdalin is shown in Fig. 9.19. Examination of the spectrum shows that the diagonal is cleaner making it easier to observe cross peaks close to the diagonal and cross-peaks are no longer oblong, like the COSY-90 ^1H NMR spectrum, but lean. In order to compare a COSY-90 and a COSY-45 NMR spectrum two partial spectra are given in Fig. 9.20. The peak on the diagonal at δ 3.5 in the COSY-90 spectrum also consists of off-diagonal signals which map out a square for both protons B^6 and $A^{6'}$. Comparing this with the same peak in the COSY-45 NMR

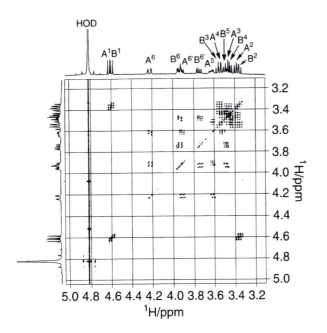

Figure 9.19 Partial 400 MHz ^1H COSY-45 spectrum of D-amygdalin in D_2O showing the sugar resonances.

spectrum shows that the off-diagonal signals are not present. The result is that the spectrum close to the diagonal is also much cleaner, and it is easier to identify cross peaks in crowded regions of the spectrum, for example between δ 3.3 and 3.6, compare Figs 9.16 and 9.19.

Examination of the cross peak at δ 3.5/3.9 shows that in the COSY-90 NMR spectrum the responses map out a rectangle (Fig. 9.20(a)), while in the COSY-45 NMR spectrum, half the responses are missing (Fig. 9.20(b)). The result is that the cross peak appears to 'lean'. This provides us with useful information. If we compare the two B^6 cross peaks at δ 3.5 and 3.74, they lean in opposite directions. This is because the cross peak at δ 3.74 is to protons $B^{6'}$ and B^5. $^2J(B^6,B^{6'})$ is a geminal coupling constant and is negative, while $^3J(B^6,B^5)$ is a vicinal coupling

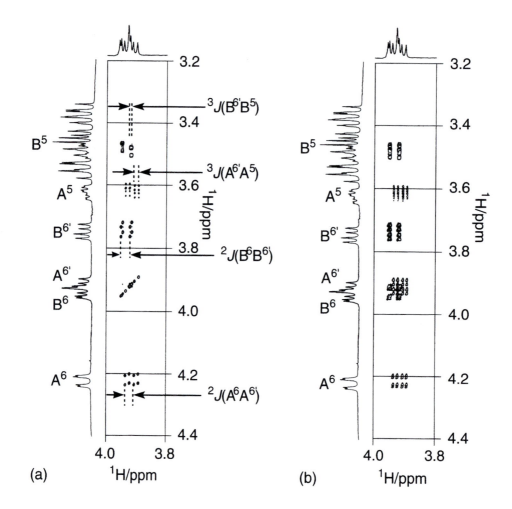

Figure 9.20 Partial 400 MHz ^{1}H (a) COSY-45 and (b) COSY-90 NMR spectrum of D-amygdalin in D$_2$O. The active coupling constant is indicated by dotted lines and arrows. The diagonal resonances are near δ 3.95.

constant and is positive. The direction of 'lean' gives the relative sign of the coupling constant that is operating. The missing peaks also tell us which coupling constant is operating between a pair of protons. It is the one where both components are present.

9.2.3 Phase-sensitive COSY NMR spectroscopy

The COSY-90 and COSY-45 NMR spectra described above are transformed in the magnitude mode, i.e. the intensity of each data point is calculated as $\sqrt{(u^2 + v^2)}$, where u and v are the intensities of the real and imaginary parts of the transform spectra. This removes the problems of phasing and makes all the signals positive. If high resolution is used in both the f_1 and f_2 directions, then it is possible to obtain a phase-sensitive COSY spectrum. The spectra have to be phased in both the f_1 and f_2 directions, and both positive and negative peaks are obtained. The peaks on the diagonal are out-of phase and give long tails in the f_1 and f_2 directions leading to very messy looking spectra. However, if cross-sections are taken, then the resulting spectrum makes it easy to determine in the cross peak the coupling which connects the two resonances (Figs 9.21, 9.22). There is a 180° phase change between lines associated with the active coupling constant which can then be identified.

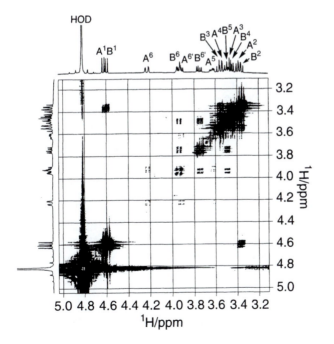

Figure 9.21 Partial phase-sensitive 400 MHz ^{1}H COSY spectrum of D-amygdalin showing the sugar resonances.

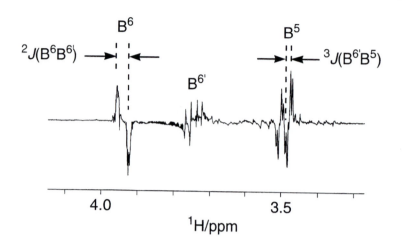

Figure 9.22 A cross-section through a signal at $\delta\,3.74$ from the partial phase-sensitive 400 MHz ^{1}H COSY spectrum of D-amygdalin shown in Fig. 9.21. The cross-section is through $B^{6'}$, which gives the out-of-phase signal at $\delta\,3.74$, and shows the coupling to B^6 at $\delta\,3.94$ and B^5 at $\delta\,3.49$.

9.3 HETERONUCLEAR COSY NMR SPECTROSCOPY

The next stage in understanding the spectroscopic properties of our model molecule is to assign the ^{13}C resonances. This is done by investigating the connectivities imparted by ^{1}H–^{13}C coupling paths. The task is facilitated by the fact that the one-bond CH coupling constants are much larger than the two- or three-bond (CCH or CCCH) coupling constants, so that the pulse sequence used can differentiate between different types of coupling path. Decoupling of protons from ^{13}C is achieved in the usual way by irradiating the protons in t_2, while accumulating the ^{13}C FID. The sequences used also illustrate how broadband decoupling of the protons may be achieved simultaneously by using a specially adapted pulse sequence. The pulse sequences needed are illustrated in Fig. 9.23, and are shown in order of increasing complexity so that the experimental technique can be built up in stages. The first sequence shown produces ^{1}H–^{13}C connectivities via the one-bond coupling. The first 90° pulse at the ^{1}H frequency produces transverse y magnetization. This then evolves with t_1 and the position of the magnetization of a given proton depends upon its chemical shift. Those protons attached to carbon-13 will also have two magnetization components, which precess at different frequencies, and these are refocused by inverting the carbon magnetization half-way through the t_1 period. In principle, then, the proton magnetization at the end of t_1 is a function of t_1, and this can be transferred into the carbon magnetization by a second proton 90° pulse, which places the y magnetization into the z direction and permits polarization transfer. Unfortunately, the inversion of the ^{13}C nuclei also causes the proton lines of the ^{1}H–^{13}C doublet to be of opposed phases with zero resultant. It is

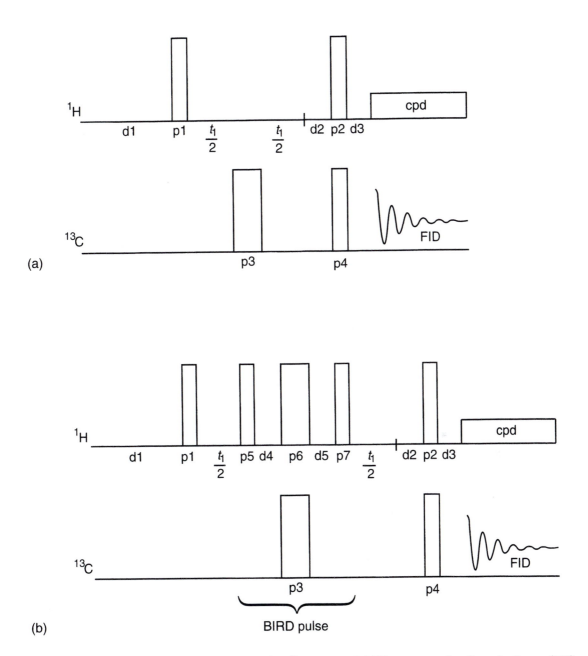

Figure 9.23 Pulse sequences used to produce ^{1}H–^{13}C hetero COSY spectra. p1, p2, and p5 are $(90°)_x$ pulses, p7 is a $(90)°_{-x}$ pulse, p4 is a 90° pulse, p6 is a $(180°)_x$ pulse and p3 is a 180° pulse. d1 is the relaxation delay, t_1 is the time for the second dimension. (a) Upper sequence is the one used for detecting correlations through $^1J(^{13}C^1H)$ as used for Fig. 9.24. d2 and d3 are $1/2J$ for $^1J(^{13}C^1H)$ and 0.0036 s was used. (b) The lower sequence is used to produce Fig. 9.25 where the correlation is through $^2J(^{13}C^1H)$ and $^{3}J(^{13}C^1H)$. d2 and d3 are $1/2J$ for $^{2,3}J(^{13}C^1H)$ and 0.14 s was used. d4 and d5 are $1/2J$ for $^1J(^{13}C^1H)$, 0.0036 s. cpd represents composite pulse decoupling.

therefore necessary to wait a period of d1 for the two components to come into phase before the 90° pulse is applied. The delay d1, which is fixed, is made to be $1/2J(CH)$. A 90° pulse is applied at the carbon frequency at the same time as the second proton pulse, which creates transverse carbon magnetization and produces an output. It is convenient to remove the multiplicity due to coupling to the protons from the carbon signals and this is done using broad-band decoupling. However, again because of the different phases of the multiplet signals, it is necessary to introduce a refocusing delay d2, which often is of the same length as d1, before decoupling and signal acquisition are initiated. The experiment is repeated for many values of t_1 and a stack of FIDs obtained, which contain both carbon and proton chemical shift information for all the C–H pairs in the molecule. The pulse sequence is tuned to $^1J(CH)$ because of the values chosen for d1 and d2 and which also suppresses any responses due to the much smaller two- and three-bond couplings. It follows that, in this type of hetero COSY spectrum, quaternary carbon atoms will not give a response. Numerical treatment of the data produces a two-dimensional chart in which there is a response for each CH pair at coordinates $(\delta C),(\delta H)$.

A spectrum obtained in this way for D-amygdalin is shown in Fig. 9.24, with the ^{13}C spectrum along the horizontal axis and proton on the vertical. The response was tuned to a $^{13}C–^1H$ correlation through $^1J(^{13}C^1H)$ by choosing values of d2 and d3 of 0.0036 s corresponding to an average coupling constant of 139 Hz. The lack of significant response

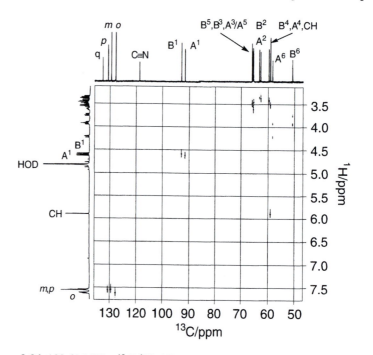

Figure 9.24 100.62 MHz $^{13}C–^1H$ NMR correlation plot for hydrogen and carbon atoms that are directly bonded in the molecule D-amygdalin in D_2O.

for the phenyl quaternary carbon and CN carbon atoms is evident. It is also possible to read off the carbon chemical shifts corresponding to the now known proton assignments. Also note that the CH$_2$ groups due to A^6 and B^6 give two responses in the ^{1}H dimension corresponding to the two different hydrogen environments. This can be a valuable tool in identifying chemically inequivalent CH$_2$ protons.

A correlation through $^1J(^{13}C^1H)$ is very valuable for assigning individual ^{13}C NMR signals. However, no correlation is observed for ^{13}C nuclei which do not bear protons. Such nuclei can be detected through $^2J(^{13}C^1H)$ and $^3J(^{13}C^1H)$. In principle, a correlation can be achieved using the pulse sequence given in Fig. 9.23(a), using values of d2 and d3 of about 0.14 s, appropriate for the much smaller values of $^2J(^{13}C^1H)$ and $^3J(^{13}C^1H)$. This can also result in correlations through $^1J(^{13}C^1H)$. This problem can be reduced by lengthening the pulse sequence. This sequence is shown in the lower part of the Fig. 9.23(b) and gives an idea of the way it is possible to manipulate spin systems to obtain particular objectives. Three pulses are added to the proton sequence and one to the carbon sequence, with a particular relationship between them, which

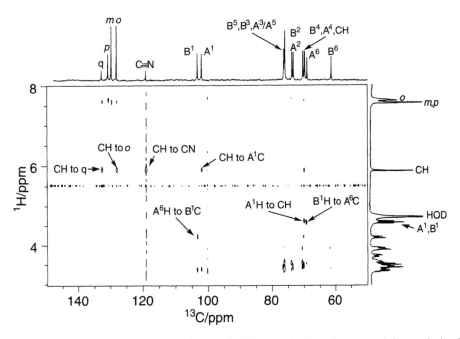

Figure 9.25 Heteronuclear proton–carbon chemical shift correlation of D-amygdalin optimized for ^2J(CH) and ^3J(CH) long-range couplings (6.25 Hz). Direct responses due to ^1J(CH) were suppressed with the use of a BIRD pulse. Major inter-residue, long-range responses have been labelled. Additional intra-residue, long-range correlations include:
Ring A: AH2 → AC 1; AH2 → AC3; AH3 → AC2; AH3 → AC4; AH4 → AC3; AH4 → AC5; AH4 → AC6; AH5 → AC3.
Ring B: BH2 → BC3; BH3 → BC2; BH3 → BC4; BH4 → BC5 and BH6 → BC4.
Aryl ring: *ortho* H → *para* C; *meta* H → quaternary; *meta* H → *ortho* C; *para* H → *ortho* C; *para* H → *meta* C. (From Ribiero (1990) *Magn. Reson. Chem.*, **28**, 765–73, copyright (1990) John Wiley and Sons Ltd, reprinted with permission.)

is called a BIRD (**Bi**linear **R**otation **D**ecoupling) pulse, see section 6.3.2. d4 and d5 in the BIRD pulse are chosen as 0.0036 s to optimize the rejection of correlations through $^1J(^{13}C^1H)$. The resulting spectrum is shown in Fig. 9.25 and provides a great deal of extra data, which gives unequivocal assignments. For instance, the B_1H to A_6C and A_6H to B_1C responses show the connectivity between the two rings.

We have thus used a series of two-dimensional experiments to make a full assignment of the proton and carbon spectra of D-amygdalin and obtain unequivocal evidence about its structure, despite the very congested and unpromising appearance of the one-dimensional spectrum as a source of useful information. We have described five two-dimensional techniques. A recent review (see Bibliography) lists some 40 pulse sequences, and there are said to be perhaps 500 variations of two-dimensional spectroscopy proposed currently. There exists, then, a whole battery of techniques that permit the successful examination of molecules much more complex than D-amygdalin, and two-dimensional NMR is being extensively used by biochemists to understand the properties (and, indeed, structures in solution) of many molecules found to regulate the behaviour of living systems.

9.4 HOHAHA OR TOCSY

In many molecules there are groups of independent coupled protons, for example in a peptide where each amino acid residue gives a coupled spin system, but there is negligible coupling between each residue. Amygdalin is another example, where there is coupling within each ring, but not between them. HOHAHA (**HO**monuclear **HA**rtman **HA**hn), or TOCSY (**TO**tal **C**orrelation **S**pectroscop**Y**) provide a method to separate the various spin systems and hence identify the single residues. The technique uses a spin-lock using the MLEV-17 pulse sequence. Spin-locking has been described in section 6.1.2.3. The MLEV-17 pulse sequence is similar to the WALTZ sequence and is based on a sequence of phase-cycled composite 180° pulses, $(90°)_x(180°)_y(90°)_x$. The pulse sequence is given in Fig. 9.26 and its application to D-amygdalin in Fig. 9.27.

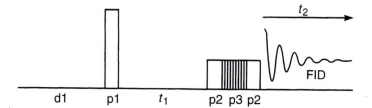

Figure 9.26 The basic pulse sequence for the HOHAHA or TOCSY experiment. d1 is a relaxation delay, p1 is a 90° pulse, t_1 is the variable delay for the second dimension, p2 is a pulse, typically 2.5 ms, attenuated to the spin lock level, and p3 is the spin lock pulse sequence, where the transmitter power has been attenuated so that the 90° pulse is of the order of 40 μs.

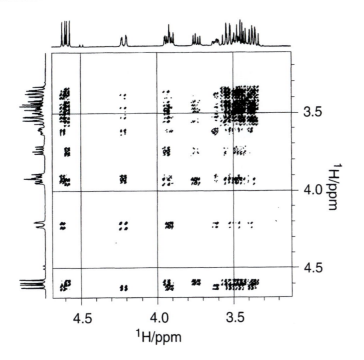

Figure 9.27 A partial 400 MHz ^{1}H HOHAHA or TOCSY experiment applied to D-amygdalin in D$_2$O. Note the two separate sets of signals for the protons of the two sugar residues.

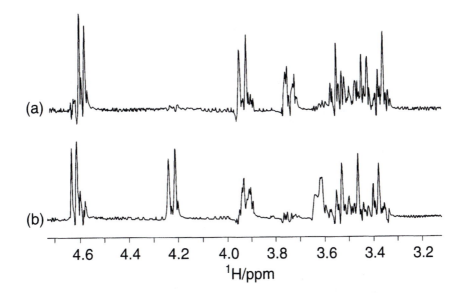

Figure 9.28 Two cross-sections from the 400 MHz ^{1}H HOHAHA or TOCSY experiment shown in Fig. 9.27. (a) A cross-section though δ 4.58. (b) A cross-section through δ 4.63. Note the two separate sets of signals for the protons of the two sugar residues.

9.5 TWO-DIMENSIONAL INADEQUATE

INADEQUATE (**In**credible **N**atural **A**bundance **D**oubl**E** **QUA**ntum **T**ransfer **E**xperiment) is the ultimate experiment to determine connectivity in the carbon skeleton of a molecule. It uses $^1J(^{13}C^{13}C)$ to map out which carbon atoms are attached to each other. The problem is that ^{13}C is only 1.1% abundant. In general, this means that only 0.012% of the molecules have two connected ^{13}C atoms. The result is that very concentrated solutions are required. The use of field gradients improves the sensitivity of the experiment by at least a factor of ten. COSY will produce the same results in theory, but the molecules containing isolated ^{13}C nuclei give very strong signals which obscure the correlations. INADEQUATE is a double quantum transfer experiment and selects AX spin systems, and hence selects coupled nuclei.

The pulse sequence is given in Fig. 9.29. An INADEQUATE ^{13}C NMR spectrum is shown in Fig. 9.30.

INADEQUATE can be applied to detect homocoupling between any pair of low abundance nuclei. An example is given in Fig. 9.31 where the technique is applied to ^{183}W in $[SiVW_{11}O_{40}]^{5-}$. This anion has a structure closely related to $[AlO_4Al_{12}(OH)_{24}(H_2O)_{12}]^{7+}$ (Fig. 7.30), where the central tetrahedral aluminium has been replaced by silicon, eleven octahedral aluminium atoms have been replaced by tungsten, and one octahedral aluminium atom has been replaced by vanadium so that the majority of the tungsten atoms are in distinguishable sites.

9.6 OVERHAUSER AND MAGNETIZATION TRANSFER BASED TWO-DIMENSIONAL NMR SPECTROSCOPY

It remains to be ascertained which of rings A and B of D-amygdalin is attached directly to the CHCN group. This is done simply in this case by irradiating the CHCN proton at 5.88 ppm and observing which of the doublets around 4.6 ppm (the H1 protons) show an NOE enhancement. It proved to be the one slightly to high frequency at

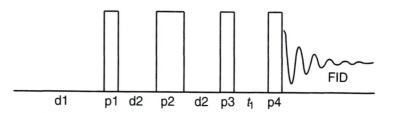

Figure 9.29 The INADEQUATE pulse sequence. d1 is for relaxation, p1, p3, and p4 are 90° pulses, d2 is 1/4J', where J is the coupling constant between the nuclei it is wished to detect, usually, $^1J(^{13}C^{13}C)$, t_1, is the second dimension for the two-dimensional experiment, and p2 is a 180° pulse.

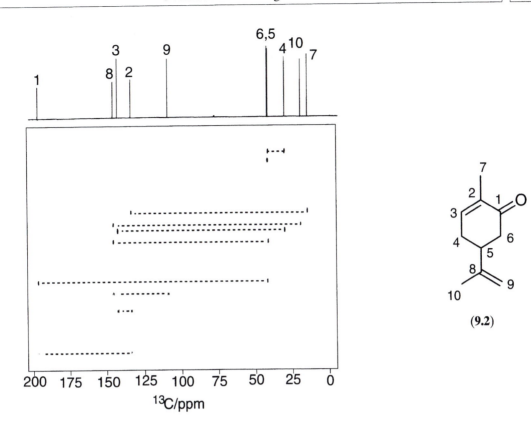

Figure 9.30 100.62 MHz ^{13}C INADEQUATE experiment applied to carvone, (**9.2**), in CDCl$_3$. There is a lone signal at δ 43 which arises from the close chemical shifts of C^5 and C^6, giving an AB spin system.

4.61 ppm, and this is then in the A ring, as, conveniently, we have labelled the peaks throughout. The inverse experiment is also possible, in which the doublets at 4.61 and 4.58 ppm are irradiated in turn and the effect observed on the CHCN proton. This leads to the same conclusion, though with less clarity, since it is not possible to irradiate such close resonances with sufficient selectivity so that one gives an NOE and the other none. A weaker effect is detected in the latter case due to spill-over of the irradiation power into the 4.61 ppm doublet region.

It is, of course, possible to carry out a two-dimensional version of this experiment on D-amygdalin. Such experiments are very important in the case of large, flexible biomolecules such as peptides. In solution, it is possible, once the proton resonances of identifiable residues have been assigned, to determine which are in close proximity in space. Thus the way the chains of such molecules are folded can be ascertained, and the data currently being obtained in solution studies of large molecules are comparable in accuracy with crystallographic data on the same molecules. This experiment is called **Nuclear Overhauser**

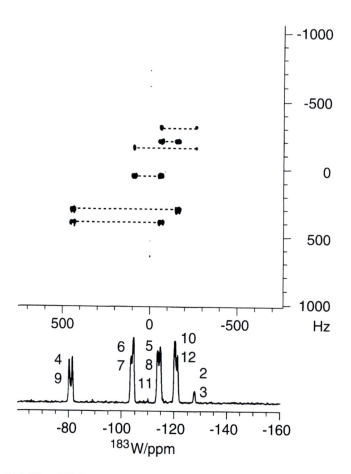

Figure 9.31 The 15 MHz two-dimensional INADEQUATE ^{183}W NMR spectrum of ^{51}V decoupled $Li_5[SiVW_{11}O_{40}]$, 0.6M, 30°C, 1536 transients, $128 \times 2k$ files, 115 h, 20 mm sideways tube. Note that the signals due to W(2) and W(3) are weak. (Reproduced by permission of the American Chemical Society from Domaille (1984) *J. Am. Chem. Soc.*, **106**, 7677.)

Effect **S**pectroscopy (NOESY), and while it will not be illustrated in detail here, we will shortly discuss the investigation of chemical exchange by two-dimensional spectroscopy, which uses essentially the same technique.

9.6.1 NOESY

The NOESY (**N**uclear **O**verhauser **E**nhancement **S**pectroscop**Y**) experiment is the two-dimensional version of NOE difference NMR spectroscopy, see section 8.2. The NOESY pulse sequence is given in Fig. 9.32.

The NOE builds up during the mixing time d2. This presents an experimental problem. The individual contributions to T_1 from an

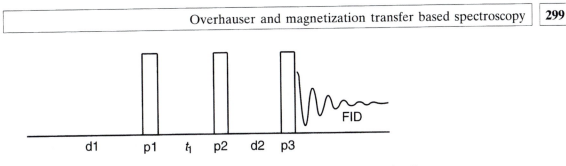

Figure 9.32 The NOESY pulse sequence. d1 is the relaxation delay, p1, p2, and p3 are 90° pulses, t_1 is the time for second dimension, and d2 is the mixing delay for the NOE to build up.

individual H–H dipole–dipole interaction are always longer than the overall T_1 and T_2. The result is that the signal is decaying while the NOE is building up. The result is that the NOESY experiment is less sensitive that the NOE difference experiment. It can be advantageous to use the experiment if many correlations are required due to the time that would be required to carry out all the NOE difference experiments, but if only a few correlations are required it is better to use NOE difference NMR spectroscopy. The application of NOESY to D-amygdalin is given in Fig. 9.33. As with the one-dimensional experiment, a correlation is found between the CH and A^1 protons providing the relative assignments of A^1H and B^1H.

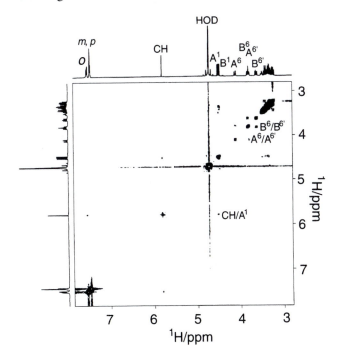

Figure 9.33 A symmetrized 400 MHz ^1H NOESY NMR experiment applied to D-amygdalin in D_2O. Note particularly the NOE observed between the CH and A^1 protons. This identifies the sugar ring next to the CH group.

9.6.2 Two-dimensional chemical exchange NMR spectroscopy, EXSY

Chemical exchange has perforce to be a homonuclear process and is studied in two dimensions by the pulse sequence shown above for NOESY. If we have two uncoupled spins with different chemical shifts that undergo slow exchange, then the pulse sequence affects them as follows. The first 90° pulse places the magnetization in the xy plane and the two components then precess at their individual frequencies for a time t_1. At the end of t_1, they will each have a particular orientation, and in general both will have x and y components of magnetization. The second pulse then turns this magnetization into the xz plane, where the y components continue to precess about the z axis. A short mixing time (τ_m), typically 0.05 s, is then given during which time there will be an exchange of magnetization. However, the z magnetization of the two components depends upon t_1; indeed, for some ranges of t_1 one of the two components will be inverted and the exchange will cause quite marked changes in magnetization. Thus at the end of τ_m we have the two component frequencies modulated by the exchange process as a function of t_1, and this gives cross-peaks in the transformed two-dimensional trace. A FID is produced by moving the z magnetization back into the xy plane at the end of τ_m. We should also note that, in non-exchanging molecules, if there exists through-space relaxation, then there can also be an exchange of magnetization through the NOE effect, since the second pulse produces

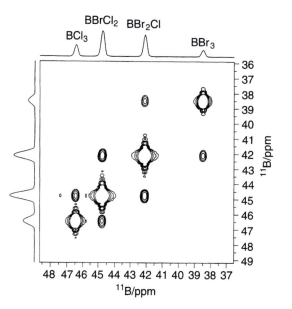

Figure 9.34 128.37 MHz ^{11}B EXSY NMR spectrum of a 1.05 : 1.0 M mixture of BCl_3 and BBr_3 at 400K using a mixing time of 0.05 s. (Reproduced with permission from Derose *et al.* (1991) *J. Magn. Reson.*, **93**, 347.)

non-equilibrium spin distributions. The same pulse sequence is thus also used for NOESY experiments as mentioned above. The acronym EXSY is often used for the exchange experiment.

A relatively simple example of the use of the EXSY pulse sequence to map out exchange is for a mixture of BCl_3 and BBr_3 where the halogens scramble (Fig. 9.34). The cross peaks clearly demonstrate that exchange involves the transfer of only one halogen. Hence BBr_2Cl exchanges with $BBrCl_2$ and BBr_3, but not with BCl_3. This result was obtained by using a mixing time of 0.05 s. At 400K this time is only long enough for a significant number of single exchanges to occur. If a longer mixing time had been used, then cross peaks corresponding to the transfer of two or three halogens would have been detected due to there being time for two or three exchanges to occur.

A more complicated EXSY spectrum is given in Fig. 9.35 for exchange between isomers of $[Cr(CO)_2(CS)\{P(OMe)_3\}_3]$. In solution, the compound exists as three isomers, (**9.3**), (**9.4**), and (**9.5**). All three isomers give AB_2 ^{31}P NMR spectra. It was proposed that the *mer*-isomers undergo interconversion *via* the trigonal twist mechanism (Fig. 9.36).

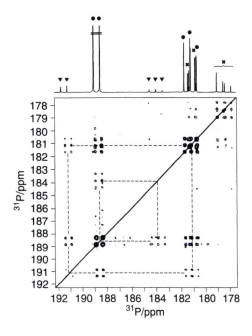

Figure 9.35 121.5 MHz ^{31}P EXSY NMR spectrum of $[Cr(CO)_2(CS)\{P(OMe)_3\}_3]$ in $CD_3C_6D_5$ at 61°C. The three isomers are identified by symbols, ✖, (**9.3**), ●, (**9.4**), and ▼, (**9.5**). Note that the doublet at δ 189 in the one-dimensional NMR spectrum has been truncated. (Reproduced with permission from Ismail *et al.* (1985) *Organometallics*, **4**, 1914, copyright (1985) American Chemical Society.)

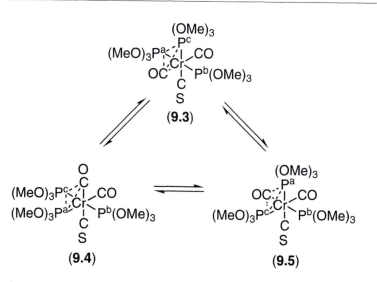

Figure 9.36 The trigonal twist mechanism applied to $[Cr(CO)_2(CS)$ $\{P(OMe)_3\}_3]$. Two of the $P(OMe)_3$ ligands rotate about the pseudo C_3 axis *via* a trigonal bipyramidal indermediate.

9.6.3 CAMELSPIN or ROESY

NOE changes its sign as τ_c increases, see section 8.2. The result is that when the molecular mass is in excess of 1000, NOESY measurements are generally unsuccessful. The problem is removed by working under spin-lock conditions where the effective ω in equations (8.1) – (8.3) is reduced from the NMR frequency range to the kHz range. The result is that τ_c is very short compared with the effective ω and the experiment operates in the positive NOE region. The experiment is called ROESY (**R**otating frame **O**verhauser **E**nhancement **S**pectroscop**Y**) and uses the pulse sequence given in Fig. 9.37.

A typical experiment is shown in Fig. 9.38. In this example, composite 180° spin-locking pulses were used.

ROESY generally gives stronger correlations than NOESY as the irradiation is continued for typically 300 ms. However, it does suffer

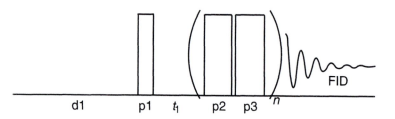

Figure 9.37 The ROESY pulse sequence. d1 is the relaxation delay, p1 is a $(90°)_x$ pulse, t_1 is the time for second dimension, p2 is a $(180°)_{-y}$ pulse and p3 is a $(180°)_y$ pulse and the transmitter is attenuated so that 180° pulses are approximately 180 μs. The 180° pulses make up a train of n pulses taking typically 300 ms.

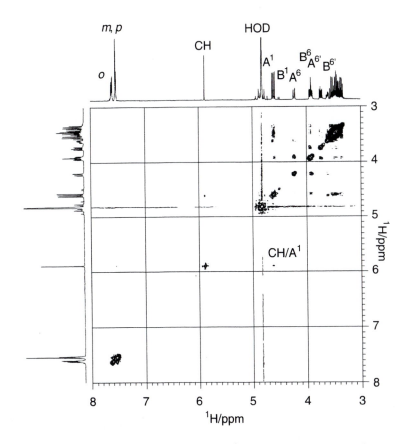

Figure 9.38 The 400 MHz two-dimensional ^{1}H ROESY spectrum of D-amygdalin in D$_2$O.

from the problem that correlations can occur through coupling, arising from the same process as is exploited in TOCSY. Consequently more caution is necessary in interpreting the results. The key observation in Fig. 9.38 of the correlation between the CH group and the A^{1}H proving the relative correlation between these two protons, is safe as no coupling pathway exists between them.

9.6.4 HOESY

Two-dimensional NOE can be used to link different species of nuclei, when the technique is called HOESY (**H**eteronuclear **O**verhauser **E**ffect **S**pectroscop**Y**). The technique works well whenever there is a significant dipole–dipole relaxation process operating between the nuclei. This normally means pairs of nuclei such as ^{1}H and ^{13}C or ^{31}P, but in favourable cases quadrupolar nuclei can be involved. The example chosen here is of ^{6}Li–^{1}H HOESY. Due to the low quadrupole moment of ^{6}Li, dipole–dipole relaxation is an important relaxation pathway, and hence the Overhauser effect is observed. The compounds

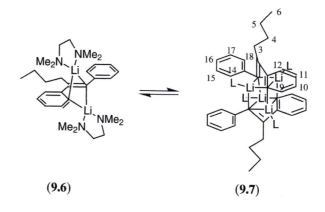

(9.6)   (9.7)

(9.6) and (9.7) (where L is monodentate $Me_2NCH_2CH_2NMe_2$) are in equilibrium in solution and so their proton spectra are always seen together. In this case we use the Overhauser effect between protons and ^6Li to show which protons are close to which lithium. The pulse sequence is given in Fig. 9.39 and it will be seen that the inversion of the ^6Li magnetization will perturb the magnetization of proximate protons. The two-dimensional plot is shown in Fig. 9.40 where it will be seen that the monomer shows a strong response to the bidentate ligand and also to protons 9, 14 and 18, whereas in the more tightly arranged dimer, the pendant ligands show no interaction and interaction is seen with almost all the protons of the unsaturated ligands.

9.7 INVERSE DETECTION

Inverse detection is a very powerful way of detecting insensitive nuclei attached to sensitive nuclei. The method uses the sensitivity of the

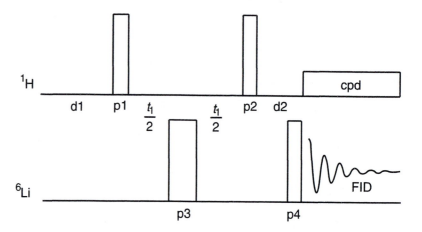

Figure 9.39 The HOESY pulse sequence. d1 is for relaxation, d2 is the mixing time for the NOE to develop, p1, p2, and p4 are 90° pulses, $t_1/2$ is the second dimension for the two-dimensional experiment, and p3 is a 180° pulse.

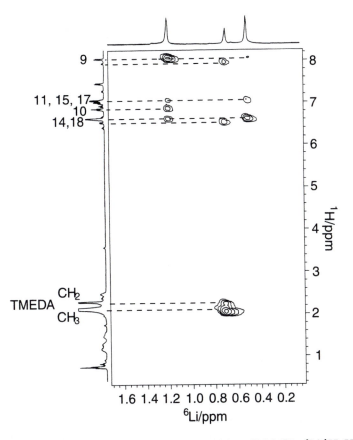

Figure 9.40 A contour plot of a phase-sensitive 58.8 MHz ^{6}Li,^{1}H HOESY NMR spectrum of an equilibrium mixture of (**9.6**) and (**9.7**) in THF-d_8 at –64°C. (Redrawn by permission of John Wiley and Sons Ltd, from Bauer and Schleyer (1988) *Magn. Reson. Chem.*, **26**, 827.)

sensitive nucleus, frequently ^{1}H, to observe other spin-coupled nuclei. Inverse detection through two-dimensional correlation spectroscopy can, in theory, give a gain in sensitivity of $(\gamma_H/\gamma_X)^{5/2}$ over straight detection. For ^{13}C this gives an increase in signal strength of 31.6 compared with no enhancement or 7.9 compared with an INEPT or DEPT spectrum. For lower γ nuclei the gain is even more dramatic (Table 9.1). These gains are actually overstated as the ^{1}H NMR spectra are ^{1}H–^{1}H coupled multiplets and consequently the full gain is not achieved. However, the gains are useful and inverse detection is generally preferred.

There are problems associated with inverse detection. Firstly, low frequency nuclei such as ^{103}Rh have relatively long 90° pulses, with the consequence that, as 1/4PW is small, they can only be detected over a fairly narrow frequency range. However, the correct frequency can be estimated by using a BIRD type pulse sequence. Secondly,

Table 9.1 The intensity gains on using inverse detection of several nuclei, X, by ^1H. Two gains are given. $(\gamma_H/\gamma_X)^{5/2}$ is the gain relative to simple detection and $(\gamma_H/\gamma_X)^{3/2}$ is the gain relative to an INEPT or DEPT spectrum.

| Observe nucleus | ^1H | | | | ^{31}P | |
Inverse detect	^{13}C	^{15}N	^{29}Si	^{103}Rh	^{15}N	^{103}Rh
$(\gamma_H/\gamma_X)^{5/2}$	31.6	306	56.8	5634	31.9	587
$(\gamma_H/\gamma_X)^{3/2}$	7.9	31	11.3	178	8.0	45.8

nuclei such as ^{13}C have shifts spread over a wide frequency range. In order to obtain reasonable resolution in the t_1/f_1 dimension, a large number of spectra have to be acquired.

We will describe two inverse detection experiments, HMQC or **H**eteronuclear **M**ultiple **Q**uantum **C**oherence and HMBC or **H**eteronuclear **M**ultiple **B**ond **C**orrelation. HMQC is the simplest $^1H,X$ correlation experiment and is normally combined with BIRD selection (see section 6.3.2) of the magnitude of the coupling to be observed and GARP decoupling of the X nuclei (see section 6.1.5) and it is this combination which is given here. The pulse sequence is given in Fig. 9.41. Delay d2 is $1/nJ(XH)$ and d3 is set so the ^1H spin vectors are out of phase. HMQC is an alternative to heteronuclear COSY (see section 9.3) but is generally more sensitive.

HMBC is used to study the two or three bond coupling correlations. This can be done using HMQC or $^1H,X$ COSY (see section 9.3) but HMBC suppresses more effectively the one bond coupling correlation. The pulse sequence is given in Fig. 9.42. Both sequences start with the BIRD sequence which, for HMQC is tuned to the coupling interaction desired and for HMBC is tuned to $^1J(XH)$, the smaller couplings being selected by d4.

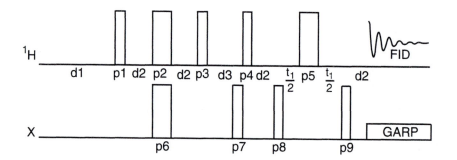

Figure 9.41 The pulse sequence used for the HMQC experiment with BIRD selection and GARP decoupling. d1 is a relaxation delay, p1, p3, p4, p8, and p9 are 90° pulses, p2, p5, and p6 are 180° pulses, d2 is $1/2J$, d3 is the BIRD delay, optimized for minimum ^1H NMR signal, and t_1 is the variable delay for the second dimension.

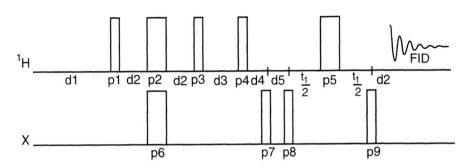

Figure 9.42 The pulse sequence used for the HMBC experiment with BIRD selection. d1 is a relaxation delay, p1, p4, p7, p8, and p9 are $(90°)_x$ pulses, p3 is a $(90°)_{-x}$ pulse, p2, p5, and p6 are 180° pulses, d2 is $1/2^1J'$, d3 is the BIRD delay, optimized for minimum ^1H NMR signal, d4 is $1/2^nJ$ where $n = 2$ or 3, d5 = d2 – d4, and t_1 is the variable delay for the second dimension.

9.7.1 The study of [Ru(η^5-C$_5$Me$_5$)(η^5-2-C$_4$H$_3$S-2',6'-C$_4$H$_2$S-2-C$_4$H$_3$S)] (9.8)

The HMQC plot of this compound, tuned for ^1J(CH), is shown in Fig. 9.43. The CH correlations are easily seen and it is also evident which are the quaternary carbon atoms with no response. The proton

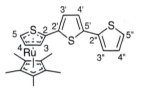

(9.8)

Figure 9.43 A 500 MHz ^1H–^{13}C HMQC NMR spectrum of [Ru(η^5-C$_5$Me$_5$) (η^5-2-C$_4$H$_3$S-2',6'-C$_4$H$_2$S-2-C$_4$H$_3$S)], **(9.8)**, in (CD$_3$)$_2$CO. The experiment permits the assignment of the hydrogen bearing carbon atoms. (Reproduced with permission from Graf *et al.* (1995) *Inorg. Chem.*, **34**, 1562, copyright (1995) American Chemical Society.)

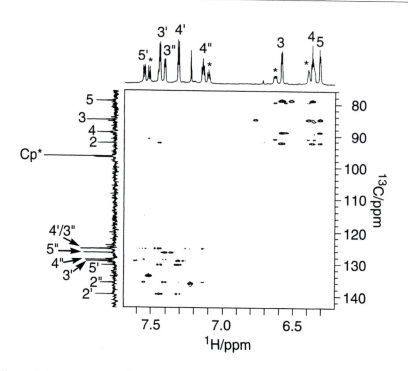

Figure 9.44 A 500 MHz ^{1}H–^{13}C HMBC NMR spectrum of [Ru(η^5-C$_5$Me$_5$)(η^5-2-C$_4$H$_3$S-2′,6′-C$_4$H$_2$S-2-C$_4$H$_3$S)], (**9.8**), in (CD$_3$)$_2$CO. The experiment permits the assignment of the non-hydrogen bearing carbon atoms. (Reproduced with permission of the American Chemical Society, from Graf *et al.* (1995) *Inorg. Chem.*, **34**, 1562, copyright (1988).)

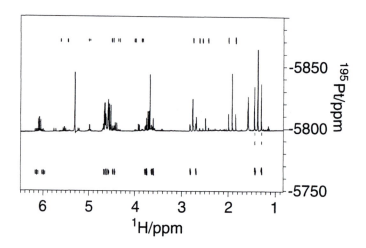

Figure 9.45 400 MHz ^{1}H,^{195}Pt two-dimensional inverse correlation NMR spectrum for the two isomers of [Pt(SnCl$_3$)(η^3-C$_4$H$_7$)(η^2-PhCH=CH$_2$)]. The normal ^{1}H NMR spectrum is superimposed on the spectrum. The ^{195}Pt chemical shifts are referenced to H$_2$PtCl$_6$. (Reproduced by permission of the American Chemical Society, from Pregosin *et al.* (1988) *Organometallics*, **7**, 2130, copyright (1988).)

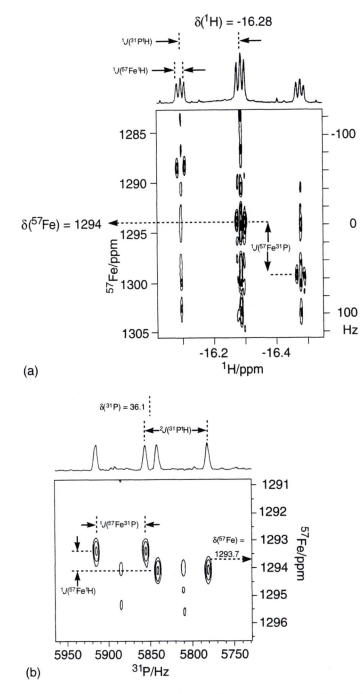

Figure 9.46 Inverse two-dimensional NMR spectra of $[(\eta^5\text{-}C_5H_5)Fe(H)$ $(PMe_3)_2]$ at 300K. (a) $^1H,^{57}Fe$ spectrum and (b) $^{31}P,^{57}Fe$ spectrum with single frequency decoupling of the methyl protons. (Reproduced from Benn, in *Transition Metal Nuclear Magnetic Resonance*, ed. P.S. Pregosin, Elsevier, Amsterdam, 1991, p. 112, copyright (1991), with permission from Elsevier Science.)

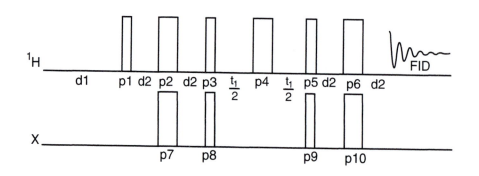

Figure 9.47 The pulse sequence used for the HSQC experiment with BIRD selection. d1 is a relaxation delay, p1, p5, p8, and p9 are $(90°)_x$ pulses, p3 and p5 are $(90°)_y$ pulses, p2, p4, p6, p7, and p10 are 180° pulses, d2 is $1/4^1J$, and t_1 is the variable delay for the second dimension.

spectrum is partially assignable since protons 3′ and 4′ give an AB spectrum and the other two rings give two doublets and a triplet with the ring bonded to ruthenium giving a low frequency chemical shift.

The HMBC plot is shown in Fig. 9.44 with impurity peaks marked with an asterisk. This permits the assignment of the non-proton bearing carbon atoms by the observation of correlations through the longer distance coupling paths. It also helps to sort out ambiguities in the proton spectrum. The quaternary carbon correlated with protons 3, 4, and 5 is also correlated with a resonance of the AB spectrum, which is thus proton 3′, and this leads us on to assign $C^{2'}$ via its weak correlation with $H^{4'}$. Similarly, H^3 is correlated with C^2, and we can assign the whole spectrum in this way.

9.7.2 Other applications of the HMQC method

A typical application of HMQC is shown in Fig. 9.45. The complex $[Pt(SnCl_3)(\eta^3-C_4H_7)(\eta^2-PhCH=CH_2)]$ exists as two isomers in solution and the ^1H spectrum is complex as the two spectra are interleaved. The HMQC spectrum in which X = ^{195}Pt, $[(\gamma_H/\gamma_{Pt})^{5/2} = 44.95]$ shows two platinum resonances at δ −5770 and δ −5875 ppm in the f_1 dimension and the associated proton resonances in the f_2 dimension. The normal ^1H spectrum is superimposed on the two-dimensional NMR spectrum and consists of a series of resonances flanked by their ^{195}Pt satellites which are the only ones to appear in the two-dimensional NMR spectrum, where it is clear which proton resonances are associated with which isomer resonance.

Inverse detection has been used with considerable success to observe very insensitive nuclei such as ^{57}Fe or ^{187}Os either through ^1H or ^{31}P $[(\gamma_H/\gamma_{Fe})^{5/2} = 5263; (\gamma_P/\gamma_{Fe})^{5/2} = 550; (\gamma_H/\gamma_{Os})^{5/2} = 12\ 264; (\gamma_P/\gamma_{Os})^{5/2} = 1282$; the receptivities R^C of the two elements are $^{57}Fe = 4.25 \times 10^{-3}$

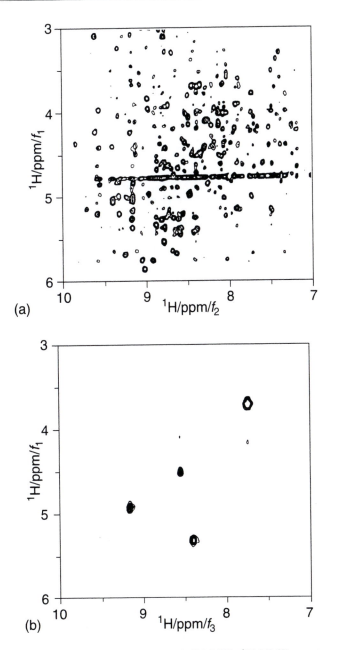

(a)

(b)

Figure 9.48 Two- and three-dimensional 600 MHz ^{1}H NMR spectra of inter-leukin-1β. In both cases, NOESY is used to identify the connection between the NH and the αCH protons of individual amino acid residues. ^{15}N editing was used so that only ^{15}NH groups are observed. (a) The two-dimensional NOESY spectrum showing the large number of responses from the amino acid residues. (b) A three-dimensional NOESY spectrum with the ^{15}N chem-ical shift being used as the f_2 direction. One slice is plotted corresponding to a δ(^{15}N) = 123.7. The introduction of the third dimension has considerably simplified the problem. (Reproduced with permission from Bax *et al.* (1989) *Biochemistry*, **28**, 6150, copyright (1989) American Chemical Society.)

and $^{187}Os = 1.15 \times 10^{-3}$]. An example is shown in Fig. 9.46 where the ^{57}Fe signal of $[Fe(\eta^5\text{-}C_5H_5)H(PMe_3)_2]$ is observed through (a) ^1H or (b) ^{31}P. The ^1H spectrum of the ^{57}Fe containing species is a triplet of doublets though some intensity shows through from the non ^{57}Fe complex, which is almost 50 times more abundant, to give a triplet of triplets. The ^{57}Fe resonance is also a triplet of doublets. In (b) only the couplings to P and H show.

9.7.3 HSQC

The HSQC (**H**eteronuclear **S**ingle **Q**uantum **C**oherence) experiment produces similar correlations to those obtained using the double quantum experiment, HMQC, but has the advantage of removing $^1H^1H$ coupling leading to sharper signals. This offers considerable advantages for larger molecules with crowded spectra. The pulse sequence is given in Fig. 9.47.

9.8 THREE-DIMENSIONAL NMR SPECTROSCOPY

It will come as no surprise that the techniques of two-dimensional NMR can be combined to produce experiments into which further dimensions are introduced. Possibilities include NOESY–COSY spectra or correlations between ^{15}N in ^{15}N-enriched molecules with the proton–proton COSY spectrum. While such techniques are in their infancy, they should assist in improving the resolution of the spectra of complex molecules, and are likely to take up an increasing amount of the time of NMR spectroscopists. The example chosen here is where a ^1H–^1H NOESY spectrum is greatly simplified. The spectrum is of interleukin-1β (Fig. 9.48). The two-dimensional NMR spectrum was first simplified so that only NH groups are observed by using ^{15}N filtering via a BIRD like sequence. NOEs are observed to the α-CH protons. Due to the very large number of amino acid residues in the molecule, the spectrum is still very complex. A third dimension, that of the ^{15}N chemical shift, is then added and the signals are sorted in this third dimension giving sets of much simpler spectra.

9.9 QUESTIONS

The following selection of problems is chosen to test you on the knowledge that you have acquired from the first nine chapters of this book. They do not refer exclusively to the information in Chapter 9.

9.1. Figure 9.49 shows the ^{119}Sn NMR spectrum of a mixture of $[Bu^t(F)Si(OSnBu^t_2O)_2Si(F)Bu^t]$ and $[Bu^t(F)Si(OSnBu^t_2)_2O(\mu\text{-}F)_2 SnBu^t_2]$ in $CDCl_3$. The two compounds in the solution are:

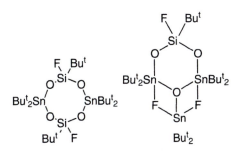

Assign the ^{119}Sn signals and give the chemical shifts of each different type of tin. Account for the multiplicity of the signals. Derive the $J(^{119}Sn^{19}F)$ values, where observed.

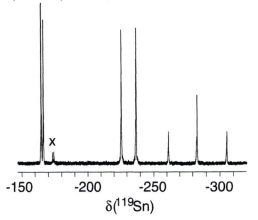

Figure 9.49 149.2 MHz ^{119}Sn{^{1}H} NMR spectrum of a mixture of [Bu^{t}(F)Si(OSnBu$^{t}_{2}$O)$_{2}$Si(F)Bu^{t}] and [Bu^{t}(F)Si(OSnBu$^{t}_{2}$O)$_{2}$(μ-F)$_{2}$ SnBu$^{t}_{2}$] in CDCl$_{3}$. x marks an impurity. (Reproduced with permission from Jurkschat *et al.* (1998) *Organometallics*, **17**, 5697, copyright (1998) American Chemical Society.)

9.2. Figure 9.50 shows the ^{31}P and ^{195}Pt NMR spectra of a mixture of two products formed by the addition of H$^-$ to

$$\left[(Ph_3P)_2Pt\text{---}\underset{Mo(CO)_3}{\bigcirc} \right]^+$$

Given that [(η^7-C$_7$H$_7$)Mo(CO)$_3$]$^+$ reacts with H$^-$ to give [(η^6-C$_7$H$_8$)Mo(CO)$_3$], suggest two products which would account for the NMR spectra and derive the $^1J(^{195}Pt^{31}P)$ values.

Given that [PtCl$_4$]$^{2-}$ is at δ 2887 relative to Ξ(^{195}Pt) = 21.4 MHz, re-reference the signals in Fig. 9.50(a) to Ξ(^{195}Pt) = 21.4 MHz.

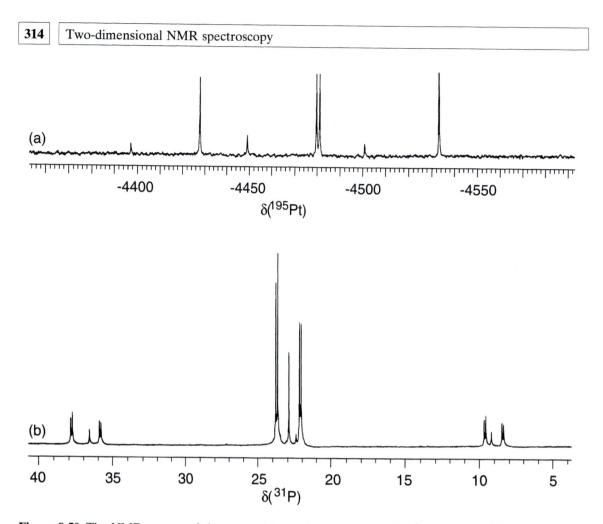

Figure 9.50 The NMR spectra of the two products of the reaction of $[\{\eta^2\text{-}(Ph_3P)_2Pt\}\{\eta^6\text{-}Mo(CO)_3\}C_7H_5]^+$ with H^-. (a) 64 MHz ^{195}Pt NMR spectrum, referenced to K_2PtCl_4 in D_2O. (b) 121 MHz ^{31}P NMR spectrum. (Reproduced with permission from Jones *et al.* (1996) *Organometallics*, **15**, 596, copyright (1996) American Chemical Society.)

9.3. Figure 9.51 shows the cyclopentadienyl region of the ^1H and ^{13}C NMR spectra of $[\{\eta^5\text{-}C_5H_4P(OC_6H_4NMe)_2\}Fe(CO)_2Me]$. Account for the number of signals and derive the chemical shifts of the signals and in the case of ^{13}C the $J(^{31}P^{13}C)$ values.

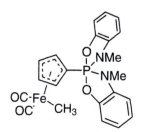

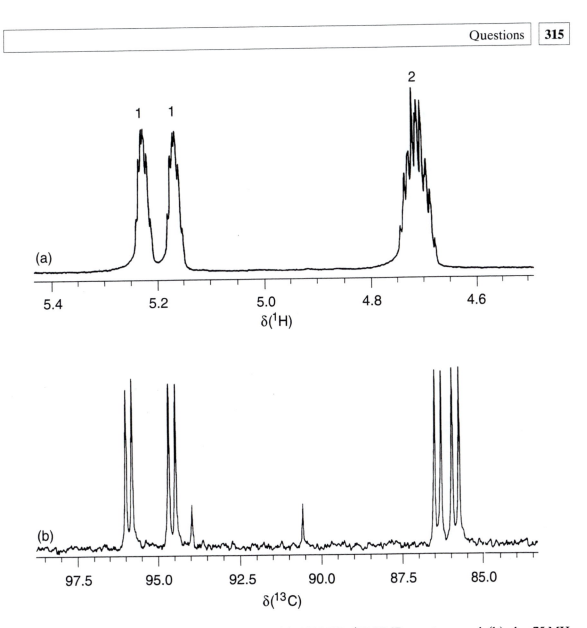

Figure 9.51 The cyclopentadienyl region in the (a) 300 MHz ^{1}H NMR spectrum and (b) the 75 MHz ^{13}C{^{1}H} NMR spectrum of [{η^5-C$_5$H$_4$P(OC$_6$H$_4$NMe)$_2$}Fe(CO)$_2$Me] in CDCl$_3$. (Reproduced with permission from Nakazawa *et al.* (1998) *J. Am. Chem. Soc.*, **120**, 6715, copyright (1998) American Chemical Society.)

9.4. Figure 9.52 shows the ^{31}P{^{1}H} NMR spectrum of 3-Cl$_2$P-5-Ph$_2$P-1,2-P$_2$C$_3$H. Explain how the coupling pattern arises.

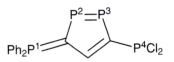

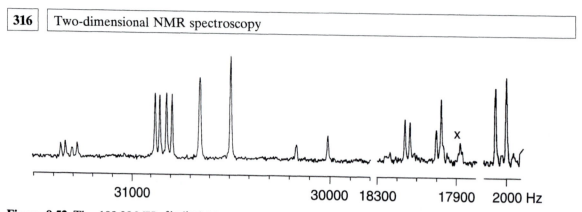

Figure 9.52 The 109.38 MHz ^{31}P{^{1}H} NMR spectrum of 3-Cl$_2$P-5-Ph$_2$P-1,2-P$_2$C$_3$H in pyridine. x marks an impurity signal. (Reproduced with permission from Schmidpeter *et al.* (1998) *Eur. J. Inorg.Chem.*, 1907, copyright Wiley-VCH.)

9.5. Figure 9.53 shows the ^{1}H NMR spectrum of the SMe$_2$ signal of [Pt$_2$(C$_6$H$_4$CH=N-1-C$_6$H$_{10}$-2-N=CHC$_6$H$_4$)Me$_4$Br$_2$(μ-SMe$_2$)], complete with ^{195}Pt satellites. Account for the coupling pattern.

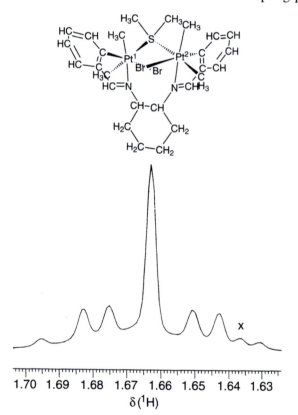

Figure 9.53 The 300 MHz ^{1}H NMR spectrum of the SMe$_2$ group of [Pt$_2$(C$_6$H$_4$CH=N-1-C$_6$H$_{10}$-2-N=CHC$_6$H$_4$)Me$_4$Br$_2$(μ-SMe$_2$)] in CDCl$_3$. x marks an impurity. (Reproduced with permission from Puddephatt *et al.* (1998) *Organometallics*, **17**, 32, copyright (1998) American Chemical Society.)

9.6. Figure 9.54 shows the $^{31}P\{^1H\}$ NMR spectrum of [Pd(*cis*-Ph$_2$ PCMe=CHPPh$_2$)$_2$]$^{2+}$ Assign the signals and account for the coupling pattern. {Remember that $^2J(^{31}P^{31}P)$ *trans* across platinum is several hundreds of Hz, while *cis* is typically 10–20 Hz.}

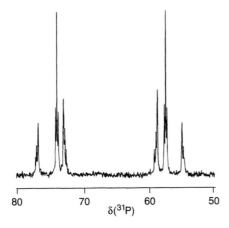

Figure 9.54 The 101 MHz ^{31}P NMR spectrum of [Pd(*cis*-Ph$_2$PCMe= CHPPh$_2$)$_2$]$^{2+}$ in CD$_3$CN. (Reproduced with permission from Higgins *et al.* (1998) *J. Chem. Soc., Dalton Trans.*, 1787.)

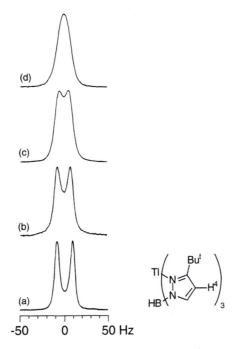

Figure 9.55 The magnetic field dependence of the ^1H NMR signal of H^4 of [HB(5-Bu^t-pyrazolyl)$_3$Tl] in CDCl$_3$ at room temperature. (a) 200 MHz. (b) 300 MHz. (c) 400 MHz. (d) 500 MHz. (Reproduced with permission from Parkin *et al.* (1998) *J. Am. Chem. Soc.*, **120**, 10416, copyright (1998) American Chemical Society.)

9.7. Figure 9.55 shows the effect of magnetic field on H⁴ of [HB(5-Buᵗ-pyrazolyl)₃Tl]. Explain why there is a doublet observed at 200 MHz, but a broad singlet is observed at 500 MHz.

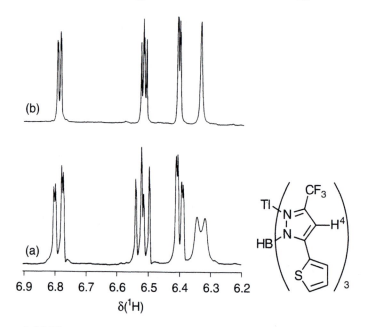

Figure 9.56 The magnetic field dependence of the ¹H NMR signal of H⁴ of [HB{3-(2-thienyl)-3-CF₃-pyrazolyl}₃Tl] in CDCl₃ at room temperature. (a) 200 MHz. (b) 500 MHz. (Reproduced with permission from Parkin *et al.* (1998) *J. Am. Chem. Soc.*, **120**, 10416, copyright (1998) American Chemical Society.)

Figure 9.56 shows the effect of magnetic field on the ¹H NMR spectra of [HB{3-(2-thienyl)-3-CF₃-pyrazolyl}₃Tl]. Explain the differences in the appearance of the spectra between 200 and 500 MHz.

9.8. Figure 9.57 shows the ³¹P{¹H} NMR spectrum of [{(dppe)(2,6-Me₂C₆H₃NC)Pt}₂Hg].

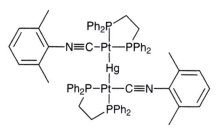

Analyse the spectrum and derive as many coupling constants as you can.

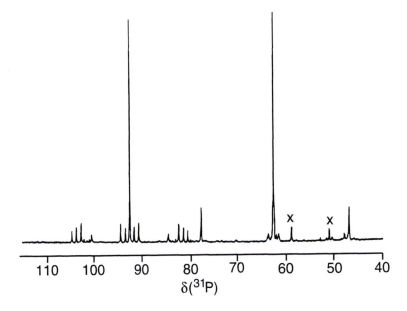

Figure 9.57 The 100.6 MHz ^{31}P{^{1}H} NMR spectrum of [{(dppe)(2,6-Me$_2$ C$_6$H$_3$NC)Pt}$_2$Hg] in CD$_2$Cl$_2$. x marks impurities. (Reproduced with permission from Puddephatt *et al.* (1996) *Organometallics*, **15**, 1502, copyright (1996) American Chemical Society.)

9.9. Partial ^{1}H NMR spectra of [Ru{1,5-(C$_3$H$_3$N$_2$CH$_2$)$_2$-2,4-Me$_2$-C$_6$H$_4$} (terpy)]$^{2+}$ are shown in Fig. 9.58. Use the NOE difference spectra to assign the ^{1}H NMR signals due to the CH$_2$ protons, H^2, H^5, H$^{5'}$, H$^{6''}$, and H$^{6'''}$. Explain why the CH$_2$ protons are inequivalent. Explain the marked difference in chemical shifts of H$^{6''}$, and H$^{6'''}$. What extra experiment(s) would you do to assign the remaining protons?

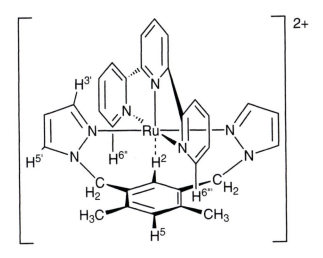

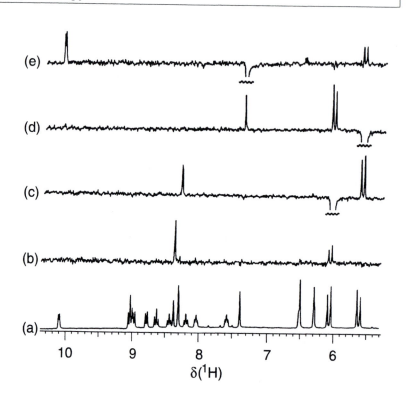

Figure 9.58 Partial 300 MHz ^{1}H NMR spectra of [Ru{1,5-(C$_3$H$_3$N$_2$CH$_2$)$_2$ -2,4-Me$_2$-C$_6$H$_4$}(terpy)]$^{2+}$ in (CD$_3$)$_2$CO. The only signal omitted is that of the methyl groups. (a) The simple spectrum. (b)–(e). NOE difference spectra with presaturation at (b) CH$_3$ (c) δ 6.0; (d) δ 5.6; (e) δ 7.4. (Reproduced with permission from Steel *et al.* (1998), *Organometallics*, **17**, 3487, copyright (1998) American Chemical Society.)

9.10. The ^{15}N NMR spectrum of [Ru(NH$_3$)$_4$(2-bzpy)]$^{2+}$ in Fig. 9.59 shows the three different types of ^{15}NH$_3$ groups. Use stick diagrams to show the three separate ^{15}NH$_3$ INEPT signals.

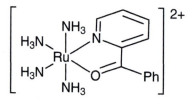

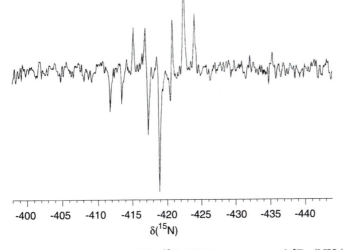

Figure 9.59 The partial 40.56 MHz ^{15}N NMR spectrum of $[Ru(NH_3)_4 (2\text{-}bzpy)]^{2+}$ in d_6-DMSO at room temperature recorded using the INEPT pulse sequence. (Reproduced with permission from de Paula *et al.* (1999) *Polyhedron*, **18**, 2017, copyright (1999), with permission from Elsevier Science.)

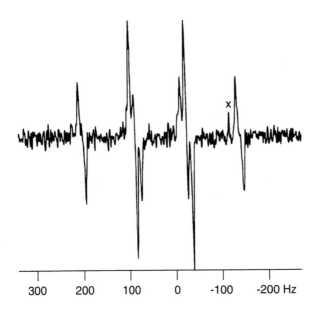

Figure 9.60 The 7.91 MHz ^{103}Rh INEPT NMR spectrum of $[Rh\{C(C_6H_4\text{-}2\text{-}OH)=N\text{-}2\text{-}6\text{-}Me\text{-}C_5H_4N\}H\{P(OMe)_3\}(PPh_3)_2]^+$ in $CDCl_3$ recorded using polarization transfer from ^1H. x marks an impurity. (Reproduced with permission from Pregosin *et al.* (1987) *Magn. Res. Chem.*, **25**, 158, copyright John Wiley and Sons Ltd.)

9.11. Figure 9.60 shows the ^{103}Rh INEPT NMR spectrum of [Rh{C(C$_6$H$_4$-2-OH)=N-2-6-Me-C$_5$H$_4$N}H{P(OMe)$_3$}(PPh$_3$)$_2$]$^+$. Analyse the coupling pattern and derive $^1J(^{103}$Rh^{1}H), ^1J{^{103}Rh(^{31}PPh$_3$)$_2$} and ^1J{^{103}Rh^{31}P(OMe)$_3$}. Suggest a structure which is consistent with this spectrum, given that the hydride is *trans* to nitrogen.

9.12. Figure 9.61 shows the variable temperature ^{1}H NMR spectrum of the methyl region of [AsH(SiBu^{t}Me$_2$)$_2$]. Account for the observation of two methyl signals at 10°C. On warming, they exchange. Estimate $\Delta G^{\ddagger}_{303}$. Identify sources of error in $\Delta G^{\ddagger}$. Suggest a mechanism for the exchange.

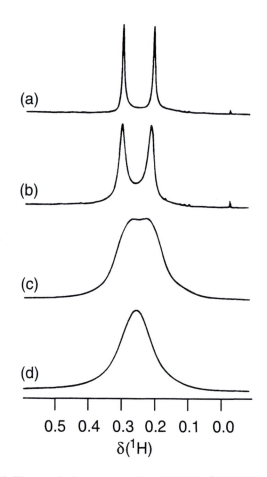

Figure 9.61 The methyl region in the 400 MHz ^{1}H NMR spectrum of [AsH(SiBu^{t}Me$_2$)$_2$] in CDCl$_3$. The spectra are at (a) 10°C, (b) 20°C, (c) 30°C, (d) 40°C. (Reproduced with permission from Westerhausen *et al.* (1998) *Inorg. Chem.*, **37**, 619, copyright (1998) American Chemical Society.)

9.13. Figure 9.62 shows the variable temperature ^{1}H NMR spectrum of the CH$_2$ protons of [Mo(CO)$_3${MeN(CH$_2$-2-C$_5$H$_3$N-6-CH$_2$)$_2$ NMe}]. Explain why there are four signals. Explain why they broaden equally at 280K. Explain why there is only one doublet observed at 310K. Estimate $\Delta G^{\ddagger}_{310}$. Identify sources of error in $\Delta G^{\ddagger}$. Suggest a dynamic process which accounts for the changes in the ^{1}H NMR spectra.

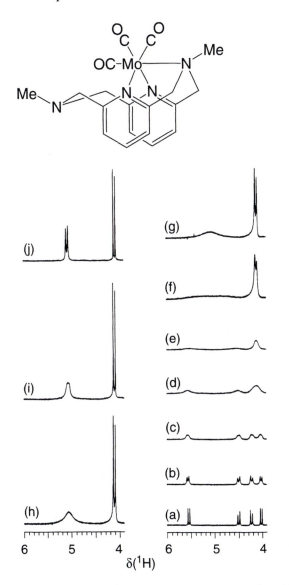

Figure 9.62 The 360 MHz ^{1}H NMR spectrum of [Mo(CO)$_3${MeN (CH$_2$-2-C$_5$H$_3$N-6-CH$_2$)$_2$NMe}] in (CD$_3$)$_2$NDO at (a) 250K, (b) 270K, (c) 280K, (d) 293K, (e) 300K, (f) 310K, (g) 320K, (h) 330K, (i) 350K, (j) 380K. (Reproduced with permission from Kelm, and Krüger (1998) *Eur. J. Inorg. Chem.*, 1381, copyright Wiley-VCH.)

9.14. Figure 9.63 shows the central signals of the variable temperature $^{29}Si\{^1H\}$ NMR spectra of cis-$[Pt(PEt_3)_2(SiFMe_2)_2]$. The ^{195}Pt satellites are outside the spectral range. Explain the multiplicity of the spectra at –80°C and 40°C. Account for the changes in the spectra with temperature. Suggest a mechanism of fluxionality which is consistent with the spectra.

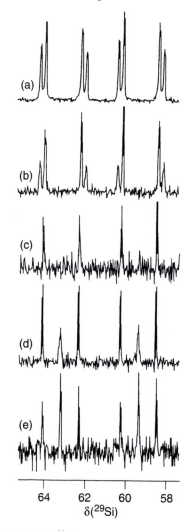

Figure 9.63 The 79.3 MHz ^{29}Si NMR spectrum of cis-$[Pt(PEt_3)_2(SiFMe_2)_2]$ in d_8-toluene at (a) –80°C, (b) –50°C, (c) –30°C, (d) 20°C, (e) 40°C. (Reproduced with permission from Tsuji *et al.* (1998) *Organometallics*, **17**, 507, copyright (1998) American Chemical Society.)

9.15. At 25°C, the ^1H NMR spectrum of the phenyl group of $[1$-Ph-3-$(\eta^5$-$C_5Me_5)$-7-SMe_2-3,1,2-$closo$-$RuC_2B_9H_9]$ shows two sharp triplets at δ 7.12 and 6.95 with further fine structure and a broad signal at δ 6.80, with relative intensities 2 : 1 : 2. On cooling to

–60°C, the signal at δ 7.12 appears to be a quartet, the signal at δ 6.95 does not change significantly, and two doublets appear at δ 6.85 and 6.54. The relative intensities of the four signals are 2 : 1 : 1 : 1. See Fig. 9.64.

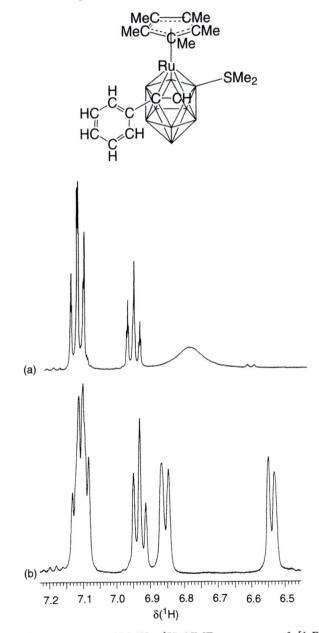

Figure 9.64 A partial 400 MHz ^{1}H NMR spectrum of [1-Ph-3-(η^5-C$_5$Me$_5$)-7-SMe$_2$-3,1,2-$closo$-RuC$_2$B$_9$H$_9$] in CDCl$_3$ showing the phenyl protons. (a) At 25°C. (b) At –60°C. (Reproduced with permission from Welch *et al.* (1998) *Organometallics*, **17**, 3227, copyright (1998) American Chemical Society.)

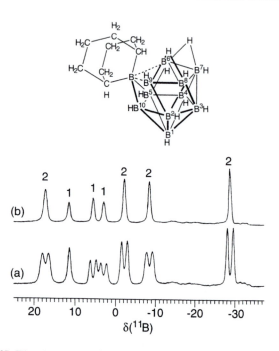

Figure 9.65 The 80.2 MHz ^{11}B NMR spectrum of $[(9\text{-BBN})B_{10}H_{12}]^-$ in d_7-DMF. (a) The ^{1}H coupled spectrum showing $^1J(^{11}B^1H)$. (b) The $^{11}B\{^1H\}$ NMR spectrum. (Reproduced with permission from Shore *et al.* (1998) *Inorg. Chem.*, **37**, 3276, copyright (1998) American Chemical Society.)

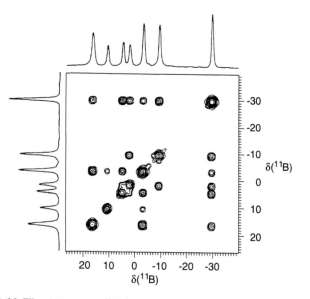

Figure 9.66 The 80.2 MHz ^{11}B$\{^1$H$\}$ COSY NMR spectrum of $[(9\text{-BBN})B_{10}H_{12}]^-$ in d_7-DMF. (Reproduced with permission from Shore *et al.* (1998) *Inorg. Chem.*, **37**, 3276, copyright (1998) American Chemical Society.)

(i)   Assign the signals as far as possible.
(ii) Explain the observation of four signals at –60°C,
(iii) Explain why one signal is broad and the other two signals are sharp at 25°C.

9.16. Assign the ^{11}B NMR signals of $[(9\text{-BBN})B_{10}H_{12}]^-$ using the spectra in Figs 9.65 and 9.66. Note that $^1J(^{11}B^1H)$ coupling is only observed between ^{11}B and **terminal** hydrogens. Only $^1J(^{11}B^{11}B)$ is observed in the COSY spectrum.

9.17. The ^1H NMR spectra in Figs 9.67–9.69 show the ^1H NMR spectra of $[Y(Ph_2CCPh_2)]$. Assign all the ^1H NMR signals.

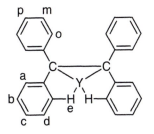

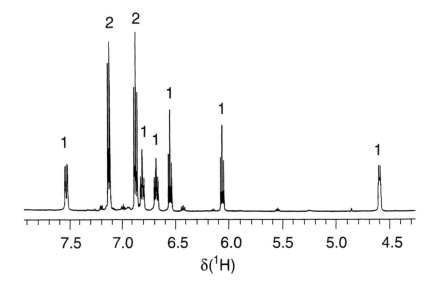

Figure 9.67 The 500 MHz ^1H NMR spectrum of $[Y(Ph_2CCPh_2)]$ in d_8-THF. The numbers above each signal give their relative intensities. (Reproduced with permission from Roitershtein *et al.* (1998) *J. Am. Chem. Soc.*, **120**, 11342, copyright (1998) American Chemical Society.)

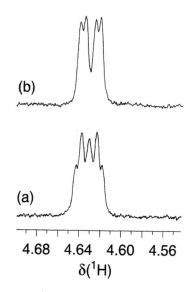

Figure 9.68 The 500 MHz ^{1}H NMR partial spectrum of [Y(Ph$_2$CCPh$_2$)] in d_8-THF. (a) Normal spectrum. (b) ^{1}H{^{89}Y} NMR spectrum. (Reproduced with permission from Roitershtein *et al.* (1998) *J. Am. Chem. Soc.*, **120**, 11342, copyright (1998) American Chemical Society.)

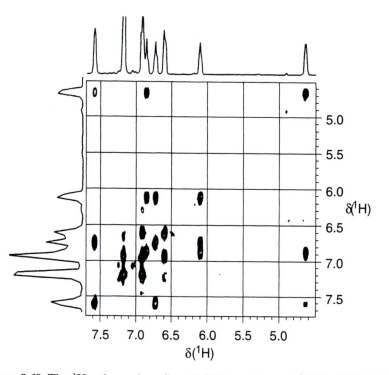

Figure 9.69 The ^{1}H enhanced gradient COSY spectrum of [Y(Ph$_2$CCPh$_2$)] in d_8-THF. (Reproduced with permission from Roitershtein *et al.* (1998) *J. Am. Chem. Soc.*, **120**, 11342, copyright (1998) American Chemical Society.)

Fig. 9.70 shows the variable temperature ^{1}H NMR spectra of [Y(Ph$_2$CCPh$_2$)]. Explain the changes in the spectra as the temperature is increased.

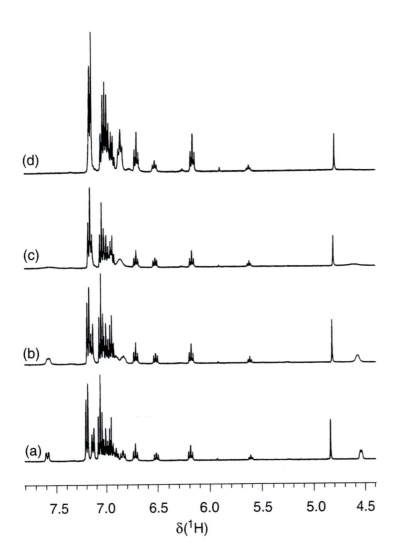

(d)

(c)

(b)

(a)

7.5 7.0 6.5 6.0 5.5 5.0 4.5

$\delta(^1H)$

Figure 9.70 The 400 MHz ^{1}H variable temperature spectrum of [Y(Ph$_2$CCPh$_2$)] in d_8-dioxan. (a) 295K, (b) 323K, (c) 348K, (d) 373K. Note that the sample is impure. (Reproduced with permission from Roitershtein *et al.* (1998) *J. Am. Chem. Soc.*, **120**, 11342, copyright (1998) American Chemical Society.)

9.18. The ^{1}H NMR spectra of the hydrides of $[Ir_2H_4(N_2C_3H_3)_2$ $(NCMe)(PPr^i_3)_2]$ are shown in Fig. 9.71. Selectively decoupling one ^{31}P affects the signals at δ −21.6, −23.3, and −24.15. Selectively decoupling the other ^{31}P affects the signals at δ −20.3 and −24.15. Assign the hydride signals.

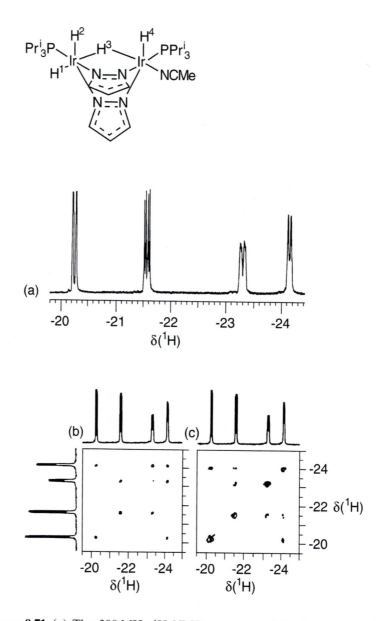

Figure 9.71 (a) The 300 MHz ^{1}H NMR spectrum of the hydride signals of $[Ir_2H_4(N_2C_3H_3)_2(NCMe)(PPr^i_3)_2]$. (b) The COSY spectrum. (c) The NOESY spectrum. (Reproduced with permission from Oro *et al.* (1998) *Organometallics*, **17**, 683, copyright (1998) American Chemical Society.)

9.19. Partial ^{1}H NMR spectra of [OsH$_3$(NH=CPhC$_6$H$_4$))(PPr$^i_3$)$_2$] are shown in Fig. 9.72. Use the NOESY spectrum to assign the hydride signals. Explain the variable temperature ^{1}H NMR spectrum. Derive two values of $\Delta G^{\ddagger}$ from the variable temperature ^{1}H NMR spectrum and identify sources of error in $\Delta G^{\ddagger}$. Predict the appearance of the hydride ^{1}H NMR spectrum at 263K if the spectrum was measured at (i) 100 MHz, (ii) 600 MHz.

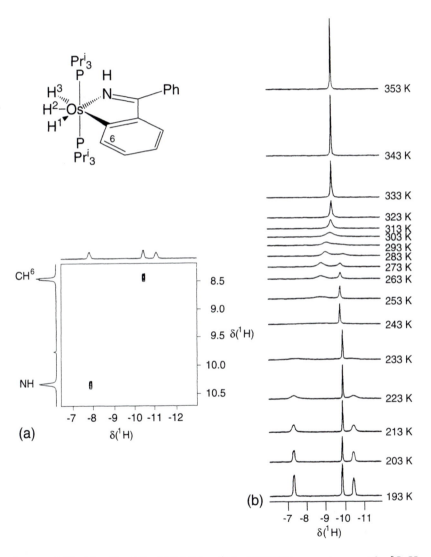

Figure 9.72 (a) Partial 300 MHz ^{1}H NOESY spectrum of [OsH$_3$ (NH=CPhC$_6$H$_4$) (PPr$^i_3$)$_2$] in CD$_2$Cl$_2$ at 193K. (b) Variable temperature partial 300 MHz ^{1}H NMR spectrum of [OsH$_3$(NH=CPhC$_6$H$_4$) (PPr$^i_3$)$_2$] in d_8-toluene. (Reproduced with permission from Esteruelas *et al.* (1998) *Organometallics*, **17**, 4065, copyright (1998) American Chemical Society.)

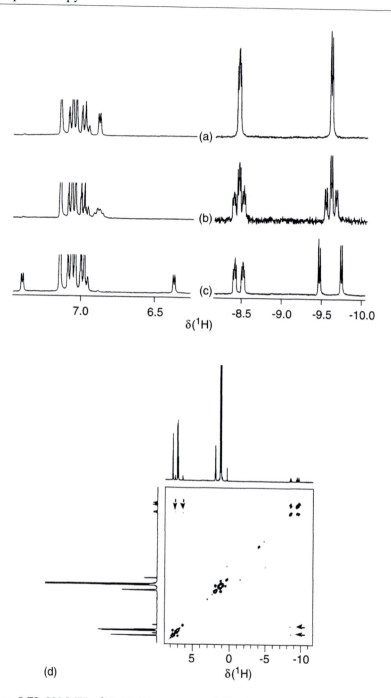

Figure 9.73 300 MHz ^{1}H NMR spectra of $[Ru(CO)H_2(PHPh_2)(PPr^i_3)_2]$ in C_6D_6. (a) A partial spectrum with broad band ^{31}P decoupling. (b) A partial spectrum with continuous wave ^{31}P decoupling centred on the PHPh$_2$ resonance. (c) A partial spectrum with continuous wave ^{31}P decoupling centred on the PPr$^i_3$ resonance. (d) A ^{1}H COSY NMR spectrum. Note the arrowed cross peaks. (Reproduced with permission from Esteruelas *et al.* (1998) *Organometallics*, **17**, 3346, copyright (1998) American Chemical Society.)

9.20. ^1H NMR spectra of $[Ru(CO)H_2(PHPh_2)(PPr^i_3)_2]$ are shown in Fig. 9.73. The $^{31}P\{^1H\}$ NMR spectrum is AX_2. Deduce the stereochemistry of the compound.

There are two hydride signals at δ –8.5 and –9.6 in the $^1H\{^{31}P\}$ NMR spectrum (a) Explain why the signal at δ –8.5 is a triplet or a double doublet and the signal at δ –9.6 is a doublet.

Spectra (a) to (c) show the $^1H\{^{31}P\}$, $^1H\{^{31}PHPh_2\}$, and $^1H\{^{31}PPr^i_3\}$ spectra. Deduce the appearance of ^1H NMR signals due to PH, and the two hydrides in the ^{31}P coupled NMR spectrum. Estimate the values of $J(^{31}P^1H)$ and $J(^1H^1H)$ for these three hydrogens. Account for the broadness of the apparent quartet at δ 6.9 in spectrum (b).

Magnetic resonance imaging and biomedical NMR

10

Magnetic resonance imaging has made immense strides over the last decade. Initially, it was applied to medicine where it has proved to give an extremely useful extra diagnostic method to radiographers and has now become in addition, a medical research tool of increasing usefulness. If these were its only uses, imaging might not be an appropriate subject for a book aimed at chemists but applications in biology and technology are now well established and a basic understanding at least of the subject is necessary in today's world for the aspiring NMR spectroscopist. Note that the word 'nuclear' has been quietly dropped from the description of the technique. This is because the public has difficulty in understanding the difference between stable and unstable isotopes. It also avoids any confusion with Nuclear Medicine, which discipline does use unstable isotopes. We will describe briefly how images are obtained, remembering that there now exist a multitude of RF pulse/magnetic field gradient sequences designed to obtain various ends, and then look at a number of more chemically oriented applications. Biomedical NMR looks at the chemistry taking place in living matter and, apart from straightforward spectroscopy, may use imaging with spatially resolved spectra or simply place a coil on the surface of a sample near an organ of interest, and watch what happens when various constraints are imposed on the system.

10.1 PRODUCING AN IMAGE

The spectra that we have been considering so far contain resolved structure, which arises from chemical differentiation of the nuclei observed and which are obtained in an accurately defined, homogeneous, fixed magnetic field. If, on the other hand, we were to use a magnetic field that had a gradient in one direction and a sample with only one type of nucleus, say, ^1H in H_2O, then the frequency of the water resonance would vary across the sample, and the output would thus contain information about the gradient and the disposition of the sample within it; in other words, the gradient encodes spatial information for us. Since the human body contains a large proportion of water, it became evident that, if one could map its spatial distribution, then one might be able to investigate the nature of the soft body tissue. Water is present as about 55% of body weight and the proportion varies widely in different parts of the body, as does the water mobility, and so its relaxation rates. In general, in body fluids $T_2 < T_1$. We will describe first a basic experiment to indicate the general principles used in obtaining an image following Fig. 10.1.

The basic idea behind producing an image can be understood as follows. An object is three-dimensional but the display of its image has to be two-dimensional. It is necessary therefore to first define a plane whose two-dimensional image is to be reproduced. This is done by subjecting the object to be imaged to a field gradient (and so frequency gradient) in one direction only, and we will take the z direction, which may be vertical or horizontal in the case of imaging. A 90° pulse is applied, but with a fairly long duration or with a gaussian shape, to give a small frequency spread, so that the nuclei are in resonance and swing into the xy plane over only a short distance along the magnetic field gradient. For instance, if the field gradient

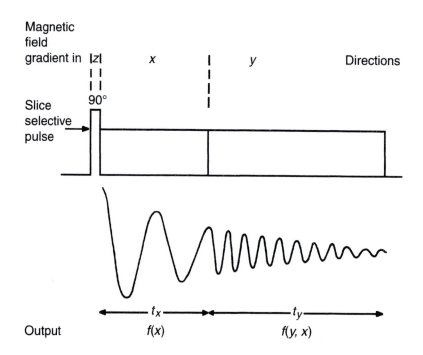

Figure 10.1 Pulse and field gradient timing for obtaining a proton density image of a slice of a subject. The finishing point of the signal produced for each volume element, or voxel, under the x gradient depends on the x coordinate of the resonating fluid and this is encoded into the signal produced under the y gradient as a phase change. The output thus contains information of both x and y coordinates of every part of the subject in the slice defined by the z gradient and RF pulse

$dB_0/dz = 5\,\mu T\ mm^{-1}$ along the main field axis, then the frequency change for ^1H that occurs for a 2 cm displacement along the z axis is 4150 Hz, and a rectangular pulse of the order of 240 μs long will define a slice of the object 2 cm thick. The z field gradient is switched off at the end of the pulse and, say, gradient x is then switched on. The water in the 2 cm slice thus produces a signal, the frequency of which depends upon the x coordinate only so that the rows of spin density perpendicular to the gradient direction each have different frequency. The magnetization evolves with time and, after some time t_x, that of the row at each x coordinate will have a particular phase, which is a function of t_x and x coordinate. This is often called the phase encoding gradient for this reason. The x gradient is then switched off and replaced by a y gradient. The output frequencies now depend upon the y coordinate but with phase determined by the x coordinate. The output is collected in the time domain t_y. Outputs are collected for several values of t_x and the data are then subjected to a two-dimensional Fourier transform, to give a map of proton density in the slice originally defined by the combination of RF pulse and z gradient. The timing of the experiment is summarized in Fig. 10.1. In practice, the technique shown has a number of drawbacks, principally because the relaxation times T_2^* of body fluids are rather short due to the non-homogeneous nature of the subject so that the signal has very little intensity by the end of the collection period and the intensity is also a function of t_x. The example given is however, clear and follows directly from the techniques for two-dimensional spectra already discussed.

10.2 WHOLE BODY IMAGING

Many different sequences have now been developed for whole body imaging which avoid the disadvantages mentioned above and also permit contrast weighting of the image by utilizing the differences in T_1 or T_2 of body tissue. In addition the effects of blood flow and the unavoidable motion of body tissue have to be compensated for and this necessitates adding further complexity to the sequences used. We will describe one basic sequence here which is shown in Fig. 10.2. The first point to note is that the slice selective z gradient (G_z) is followed by a period of z gradient of opposite sign. During the short slice selective period, the spins lose some phase coherence in the xy plane through the depth of the slice and this reversal of gradient compensates for this, refocusing the spins. Secondly, t_x is kept constant and the changes in phase encoding are brought about by varying the intensity of the phase encoding x gradient (G_x), which is shown as a pulse cut by horizontal lines. The y gradient (G_y) is next switched on to give the frequency encoding, though by the end of this time the signal will be of almost zero intensity. A 180° RF pulse is then applied, in conjunction with a z gradient so that it is also slice selective and this produces

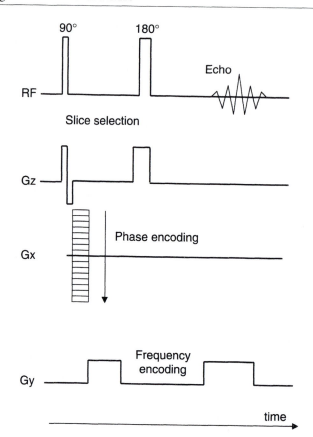

Figure 10.2 A typical sequence used for whole body imaging. The vertical arrow implies that the various phase encoding gradients are applied sequentially. G_z, G_x, G_y, are the magnetic field gradients applied along the three axes of the system.

an echo which is frequency encoded by a further period of y gradient. This procedure is repeated for each value of G_x This is called the spin echo sequence. An alternative approach is to omit the 180° RF pulse and form the echo by applying a negative y gradient then a positive y gradient. This is known as the gradient recalled echo sequence.

The output of such sequences is weighted by the spin density in each voxel of the slice and also by the real T_2 of the voxels. The shorter the sequence, the less the signal intensity lost due to T_2 relaxation. Evidently, by varying the length of the sequence, we can introduce T_2 weighting. The sequences are of quite short duration, insufficient for full T_1 relaxation to take place so that T_1 also affects the intensities, and in this case the weighting can be changed by altering the repetition rate with which spectra are produced. It is equally possible to start the sequence with a 180° RF pulse to invert all the spins and take the image after some recovery period, though this adds to the time needed to take an image, which can be unacceptable to the patient. In summary, spin density, T_1 and T_2 are all manifested in the image but one can be chosen to predominate by choosing appropriate imaging conditions. Typical images of cross sections of a human head are shown in Fig. 10.3. The advantages of the method are that it is completely uninvasive and the contrast between, for instance, the different types of brain tissue is some six times greater than for X-ray radiography. We shall also see below that it is capable of very subtle developments.

One such is based on detecting changes in blood flow in localized parts of an image. If the organism is at rest, then the blood flow will

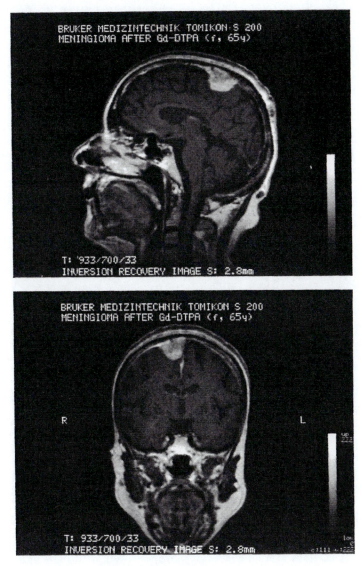

Figure 10.3 Two sections of a human head showing the location of a brain tumour (Images supplied by Bruker Medizintechnik.)

be steady and the ratio between the concentrations of oxyhaemoglobin and deoxyhaemoglobin will have a certain value. Now, oxyhaemoglobin is diamagnetic and the deoxy form is paramagnetic, so that the blood will have a particular value of magnetic susceptibility in the resting state. This will distort the applied magnetic field and so contribute to relaxation via the T_2^* effect. If the organism is stimulated, the blood flow will increase and this will increase the proportion of oxyhaemoglobin present and so change the magnetic susceptibility of the blood. T_2^* will change and if the imaging sequence is chosen so as to be sensitive to T_2^* then the contrast of the image will change in the affected region. The difference between two images taken before and after stimulation, will show clearly where there were changes in the blood flow. This, of course, means that the refocusing pulse cannot be used in the imaging sequence, otherwise inhomogeneity effects would be refocused. An application of this method is shown in Fig. 10.4 where two difference images of a human brain are shown. In each case, images were taken while the subject had periods of repose when nothing was presented to his vision followed by images taken in periods when a visual stimulus was given and the difference was obtained to highlight the changes. If he was asked to look at words then the major part of the activity took place in the left hemisphere

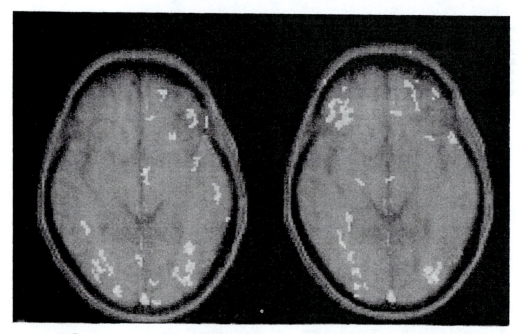

Observe words   **Observe faces**

Figure 10.4 Difference images of thick sections of a human brain showing the different parts of the brain brought into action by the visual stimuli of words (left) and faces (right). (Example supplied by M. Raybaudi, INSERM, Grenoble.)

of the brain, whereas the opposite was the case when faces were in his field of view. The centres of activity are quite dispersed and some of the centres are involved in more than one activity.

10.3 DIFFUSION AND FLOW

Diffusion of molecular species has been studied since the mid-1950s using a method based on the Carr–Purcell $90° - \tau - 180° - \tau -$ echo pulse sequence. Two short periods of magnetic field gradient are added to the sequence as shown in Fig. 10.5 and modify the behaviour of the output as follows. The first 90° pulse turns the magnetization into the xy plane and this then loses phase by the T_2 process and due to inhomogeneity effects. The first field gradient pulse causes further dephasing which depends upon the position of each spin in the sample. The 180° RF pulse inverts the spins and refocuses the inhomogeneity effects but not the effect of the first gradient pulse. This is cancelled by the second field gradient pulse, which after inversion of the spins acts as if it were of opposite sign to the first field gradient pulse. In the absence of translational motion of the spins, we get the usual echo as if the field gradient pulses had not existed. If the spins move however, then a spin which is dephased by an angle ϕ by the first field gradient pulse will be refocused by another angle, say θ, by the second field gradient pulse. Since the two angles are not equal, the magnetization will remain dephased and the intensity of the echo will be reduced, so permitting the estimation of the average spatial displacement of spins between the two magnetic field gradient pulses. In this way, self-diffusion can be estimated (a random process) or flow can be measured in the direction of the magnetic field gradient of the pulses. In the latter case, the result will depend also upon whether the flow is plug flow, turbulent flow or non-turbulent flow, with flow

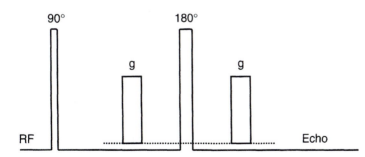

Figure 10.5 The basic pulse sequence used to measure the spatial displacement of spins in short intervals of time. The two RF pulses of 90° and 180° are the usual Carr–Purcell sequence and two short pulses of magnetic field gradient (g) are added equally displaced from the 180° RF pulse. This sequence is followed by an echo whose intensity is reduced if there is displacement of spins between the two pulses g.

faster in the centre of the conduit than at its walls. It will be evident by referring to Fig. 10.2 that this pulse sequence can easily be combined with an imaging sequence and so measure flow or diffusion through an image. Other methods are also possible and are often called time of flight measurements. Thus if the repetition time between pulse sequences is short then stationary spins are saturated and flowing spins are not and so appear with different intensity. Alternatively, if all the spins in a slice are inverted by a 180° pulse and then a waiting period given such that all the stationary spins are at null intensity, new spins flow into the slice wherever there is flow and these appear in the image.

Diffusion weighted images of the brain have been found to be invaluable in the early diagnosis of stroke damage since the diffusion of water is slower in ischemic tissue and, indeed, imaging is the only technique capable of doing this. Other medical applications include the measurement of blood flow in phase contrast magnetic resonance angiography. If flow occurs in a linear gradient there is a phase change induced which is identical for every position along the gradient and allows the flowing fluid to be differentiated from stationary fluid. The magnitude of the phase change depends upon the timing of the pulse sequence, the value of the gradient, and the velocity, which can thus be calculated from the data.

10.4 CHEMICAL SHIFT IMAGING

10.4.1 Water suppression

The techniques discussed so far depend upon the presence of water to provide the signal to generate the image. The water has a strong signal and, with the exception of the signal from fat, obscures the much weaker resonances from metabolites present. In order to carry out spectroscopy of biological samples then it is necessary to eliminate in some way the water resonance, and this holds for one-dimensional spectroscopy, two-dimensional spectroscopy and spectroscopy associated with imaging. One technique is to use a weak continuous wave at the water frequency (long and therefore very selective) during the relaxation delay so as to saturate the water resonance, then produce a spectrum in the normal way. This is useful but not fully effective and resonances near to the water may well be obscured by the remnant water signal. A variety of schemes have been developed to improve water suppression and all are based on the use of magnetic field gradients. One such is WATERGATE (**WATER** suppression by **GrA**dient **T**ailored **E**xcitation). The sequence is shown in Fig. 10.6. The spectrometer frequency is first centred on the water resonance. Transverse magnetization is created by a non-selective $90°_x$ pulse then a gradient is applied which defocuses all the spins. This is followed by a specially designed pulse which consists of two long, selective $90°_{-x}$ pulses

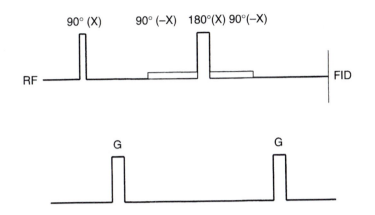

Figure 10.6 The WATERGATE pulse sequence. The two tall RF pulses are non-selective and the two broad, short ones are selective and at the water frequency. The two magnetic field gradient pulses G refocus only the non-water resonances.

separated by a non-selective $180°_x$ pulse. The water magnetization is left unchanged by this combination but the rest of the spins are flipped 180°. A second field gradient pulse refocuses the spins that have been subjected to the 180° flip, but not the water protons. This gradient pair technique is effective because it compensates for imperfections in the $\boldsymbol{B}_1$ pulses.

If large biomolecules are being examined, then we can take advantage of the differences in diffusion rate between them and water, and diffusion weight the spectra using the principles outlined in the previous section. If there is a 20-fold difference in diffusion constants then it is possible to design an experiment such that the signal required is attenuated by only 70% whereas the water intensity is reduced 1000-fold.

In the following section, we will shortly encounter a third water suppression method called CHESS which is used with chemical shift imaging (Fig. 10.8).

10.4.2 Imaging by spectra

By working at high enough magnetic field, it is possible to produce a ^{1}H spectrum from each voxel of the sample. The lines are broad but the significant features of the spectrum can be observed provided the water resonance is attenuated. The basic pulse sequences used to produce such spatially localized spectra are called STEAM (**ST**imulated **E**cho **A**cquisition **M**ode) or PRESS (**P**oint **RES**olved **S**pectroscopy; a spin echo sequence). In STEAM, a series of three 90° pulses are applied, each in the presence of a slice selective gradient along one of the three axes so that a voxel is defined at the intersection of the three slices and a spectrum is obtained from this voxel.

One can do even better than this, however, and produce the spectra of all voxels in a slice using an imaging scan. The slice selective $90° - \tau - 180°$ sequence is used similar to that described above (Fig. 10.2) but gradient pulses are applied along the other two axes, say x and y, just before the $180°$ RF pulse, with many values of G_x and many values of G_y for each value of G_x. Fourier transformation of this data stack gives a spectrum for each voxel of the slice. The sequence has to be preceded by some sort of water suppression sequence such as that included in Fig. 10.8. Figure 10.7 shows such an image. It comprises a matrix of 16×11 spectra at relatively low resolution but within which it is possible to discern major changes over the space occupied by the sample. The sample is the brain of an anaesthetized rat in which a tumor had been implanted. Each spectrum corresponds to a voxel dimension of 1×1 mm. The resonances which are visible are, starting from low frequency, lactate in the tumour, N-acetyl aspartate,

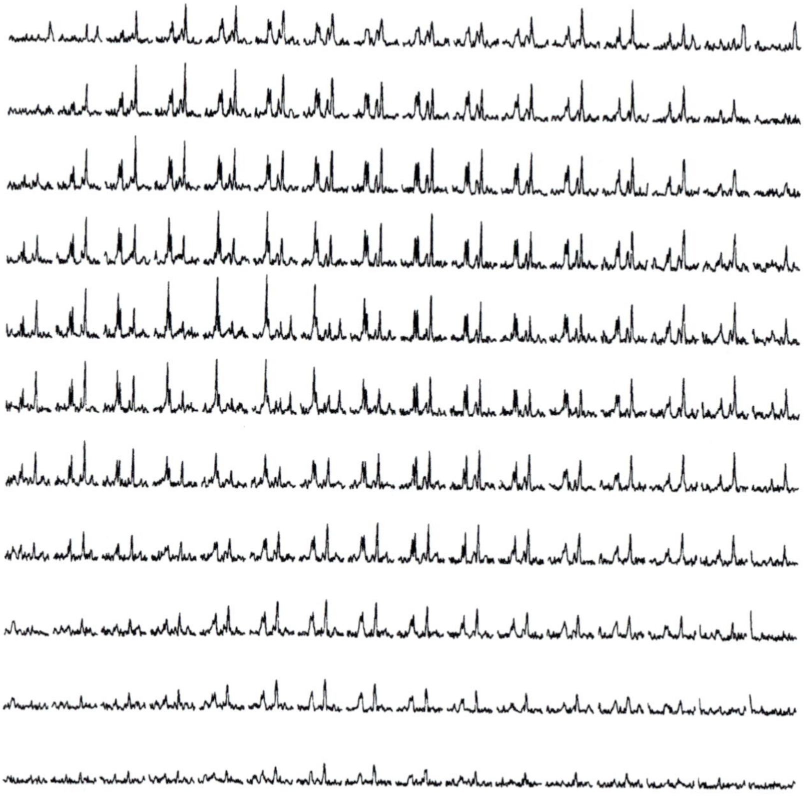

Figure 10.7 ^{1}H spectra of a slice of a rat brain into which a tumour had been implanted. Each spectrum represents a pixel of 1×1 mm. The tumour, which is rich in choline and lactate, is located at the centre left of the figure. Note that the oval shaped brain produces no responses in the lower right and left corners of the matrix. (Figure provided by A. Ziegler, INSERM, Grenoble. To be published in *Magn. Res. in Medicine*)

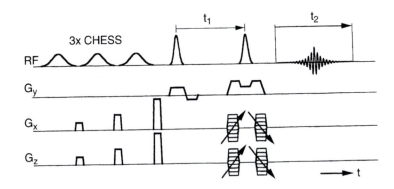

Figure 10.8 ¹H NMR COSY imaging pulse sequence used to obtain the two-dimensional spectra of Fig. 10.9. First the water is subjected to three selective pulses followed by pulses of gradient which dephase its transverse magnetization, so that the water signal is much reduced; this is called the CHESS sequence. The RF sequence that follows is a simple COSY sequence but the magnetic field gradient pulses which are added produce an image. The first 90° selective RF pulse is thus converted into a slice selective pulse, and the second is accompanied by gradients in the three axes, several values of G_x being used for each value of G_z. Note that the gradient pulses can overlap.

glutamate, glutamine, etc., creatine, and choline and related species in the tumour. The normal tissue contains relatively little choline or lactate. The tumour is at the centre left of the figure, and is quite clear in the spectra presented though, naturally, it shows up better in a colour representation based on the intensity of each resonance in each spectrum.

Resolution in such spectra is limited but can be improved by using COSY imaging in which the COSY spectrum is obtained for each voxel and the various compounds present are characterized by the cross peaks on the COSY spectrum of each voxel. A typical pulse sequence is shown in Fig. 10.8 and some representative spectra are shown in Fig. 10.9. The initial CHESS sequence (**CHE**mical **S**hift **S**election imaging) is used to minimize the water resonance and consists of three selective 90° pulses, each followed by a gradient pulse (in the x and z directions in this case, and of varying amplitudes to avoid echo formation). Each RF pulse produces transverse water proton magnetization which is then destroyed by the gradient pulses. Two slice selective 90° pulses are applied next, which produce the COSY spectra as described in Chapter 9 and which are localized in space by the later gradient pulses. Note that the 90° RF pulse/gradient pulse slice selective pair is bracketed by two gradient pulses which compensate for imperfections in the RF pulse. The result of this acquisition is that each voxel produces a COSY spectrum in which there is the diagonal, a still non-negligible water resonance (the vertical response) and a number of cross peaks which arise through the ¹H–¹H correlations in

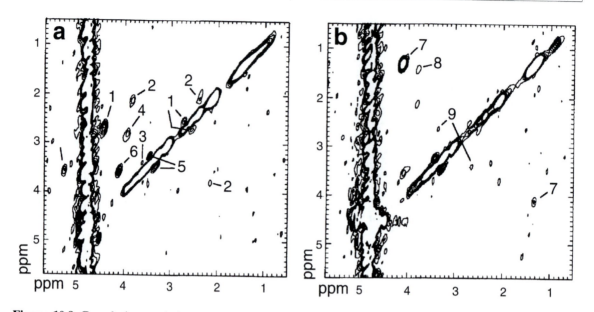

Figure 10.9 Correlation peak imaging of rat brain in which a tumour had been implanted. (a) Normal brain tissue. (b) Tumour tissue. The cross peaks correspond to (1) N-acetyl aspartate, (2) glutamine/glutamate, (3) glucose, (4) aspartate, (5) taurine, (6) inositol, perhaps choline, (7) lactate, (8) alanine and (9) hypotaurine. The nominal voxel volume for each spectrum is 23 μl and slice thickness 5 mm. (Figure provided by A. Ziegler, INSERM, Grenoble.)

each molecule and which are far better resolved than in the simple spectrum. It is thus possible to detect some fourteen compounds in the rat brain, spatially localized (Fig. 10.10). The change in the metabolites present in tumour and normal tissue is very marked. Note that the strong N-methyl resonance of choline is not seen in these COSY plots since there is no coupling. The choline resonance remains in the diagonal.

10.5 BIOLOGICAL USES OF IMAGING – IMAGING MICROSCOPY

Imaging of small biological samples can be carried out in smaller apparatus and the reduction in scale also permits an increase in resolution or decrease in voxel size, typically to 25×25 μm with a slice thickness of 300 to 2000 μm. Otherwise, the techniques are similar to those already described. The technique has been named NMR microscopy. The resolution is less than available with optical microscopy, but different things are imaged by the two techniques and, more importantly, NMR is non-invasive, the sample needs little preparation and studies can be made *in vivo*. In Fig. 10.11, we show the image of a

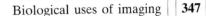

a) NAA b) Glu / Gln c) Glc

d) Lac e) Htau f) Ala

g) Cho / Ino h) Asp i) PE

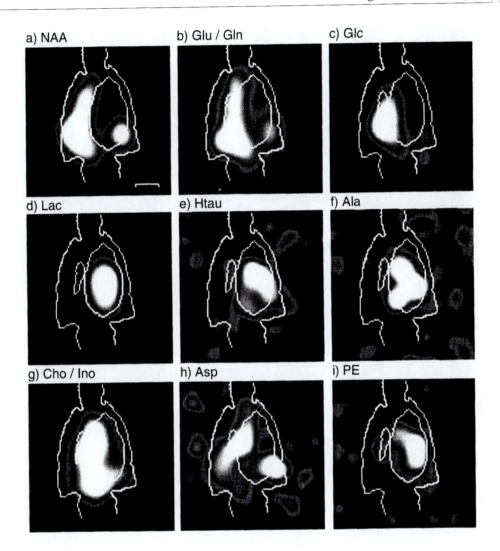

Figure 10.10 Metabolic images for a whole rat brain corresponding to the cross peaks of Fig. 10.9. The contours of brain and tumour are superimposed on the images. (a) shows the distribution of *N*-acetyl aspartate, (b) that of glutamine/glutamate, (c) that of glucose, (d) that of lactate, (e) that of hypotaurine, (f) that of alanine, (g) that of choline/inositol, (h) that of aspartate and (i) that of phosphoethanolamine. (Figure provided by A. Ziegler, INSERM, Grenoble. To be published in *Magn. Res. in Medicine*.)

cross-section of the hypocotyl stem of a castor bean seedling, essentially growing in the probe, in which the structures are well visible. COSY images permitted the localization of metabolites in various parts of the cross-section of the stem and using other techniques, water flow has also been detected moving both up and down the stem. In the latter case, the method used had to be able to differentiate between flow in both directions in a single voxel!

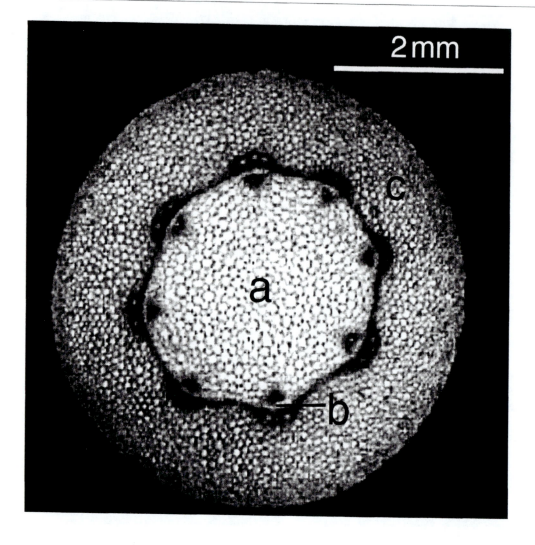

Figure 10.11 Image of a cross-section of the hypocotyl of a castor bean seedling. The eight vascular bundles (b) are connected by the meristem ring. The cellular structures of pith parenchyma (a) and cortex parenchyma (c) are visible. (From Ziegler *et al.* (1996) *J. Magn. Reson.*, **112B**, 141, with permission.)

10.6 INDUSTRIAL USES OF IMAGING TECHNIQUES

The measurement of the distribution of fluid throughout a sample is of interest in a number of other fields. Imaging has, for instance, been used to detect the position of material on a chromatography column, which otherwise was difficult to detect. It has a number of technological uses, such as the study of water in food materials or the cooking of foods, where the non-destructive nature of the technique is useful. The method is also coming to be used in chemical engineering where such things as the distribution of fluid flow in packed bed columns can be visualized or the mixing of fluids of different NMR relaxation times which have contrast in the images.

Some specific examples are the study of packed bed reactors. The fluids present may be gas and liquid when it is easy to determine how the two are distributed because of the different spin densities, and the way the distribution changes with flow rate is easy to follow. Such reactors are most efficient in the trickle flow regime when the liquid falls in rivulets from one catalyst particle to the next. The rates of reaction also depend upon the shape and size of the catalyst particles and the way they pack in the reactor, and these factors can now be studied directly. Packed reactors through which two-phase liquid systems are being passed can also be examined and the way the phases flow and co-dissolve can be observed. The distribution of flow rates in the interstices of the reactor can be measured using techniques already discussed. An example is given in Fig. 10.12.

The deactivation of catalyst beads can also be followed by imaging. A catalyst bead, whose pores may have become partially blocked by

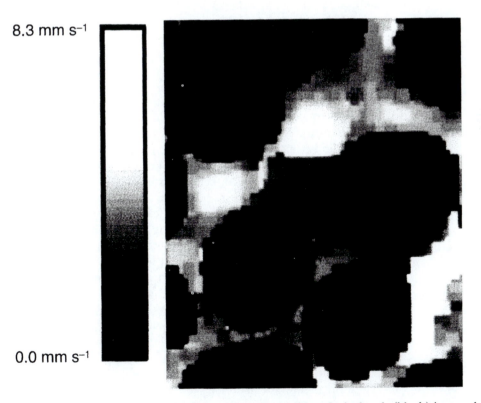

8.3 mm s⁻¹

0.0 mm s⁻¹

Figure 10.12 Flow rates of a water/hydrocarbon mixture through the beads (black) in a packed bed reactor. The rates are calibrated by the shade scale at the left. (From Gladden (1998) *Chem. in Britain*, **34**, No. 3, 35, with permission.)

carbon deposits is soaked in a solvent and then the cross-section examined by NMR microscopy and the points where there is a solvent signal show the areas where the pores are still open. The distribution of the carbon in a partially deactivated pellet is not as expected from current theory and this observation should lead to new developments in the theory of deactivation.

10.7 BIOMEDICAL NMR

In addition to the mobile protons in a living organism, there are molecules containing ^{13}C and ^{31}P nuclei that can also be used for study of how the chemistry of life operates. Early, pre-imaging experiments with whole, anaesthetized, small animals in large-bore magnets were promising, but signals were obtained from all parts of the animal. It quickly became clear that it would be of most use if signals could be localized to a specific known organ. One approach is to create a specially contoured magnetic field that has the correct value for resonance over only a small, controllable volume. Alternatively, a coil may be placed on the body surface. This coil produces a radiofrequency field that penetrates below the surface by approximately its own radius and so can stimulate responses from nuclei inside the body. This is called 'topical NMR'. It is possible, for instance, to follow the ^{31}P signals of nucleotide phosphates and inorganic phosphate in muscle and monitor how these respond to different conditions imposed on the muscle or to various disease conditions. Figure 10.13 demonstrates the effect on the ^{31}P spectrum of a subject's arm of applying a tourniquet. The normal spectrum shows peaks for the three phosphorus atoms in adenosine triphosphate ATP (I, II and III), for phosphocreatine (IV) and for inorganic phosphate (V). Application of the tourniquet leads to oxygen starvation and breakdown of organic phosphate to inorganic phosphate. The effects of exercise on muscle chemistry, the spectra of skin tumours or the brain can all be obtained in this way. It is, for instance, possible to follow changes during treatment of a baby suffering from birth asphyxia using such techniques.

^{13}C spectroscopy offers wider dispersion of resonances in work with non-homogeneous biological samples and ^{13}C spectra can be obtained routinely with surface coils using modern equipment. Proton irradiation is used and is applied using a second surface coil. Power dissipation has to be limited ($8\,W\,kg^{-1}$ is the recommended maximum dissipation) and the WALTZ technique is ideal for this reason. Tracer studies using ^{13}C enriched molecules are very informative. For instance, it is possible to follow the liver metabolism in anaesthetized rats. The rat is perfused with either [1-^{13}C] glucose or [1,2-^{13}C] glucose and the evolution of the spectra observed as a function of time. In fact, difference spectra are used in which the spectrum obtained before the start of perfusion is subtracted from those obtained under perfusion. A steady increase is seen with [1-^{13}C] glucose perfusion over about

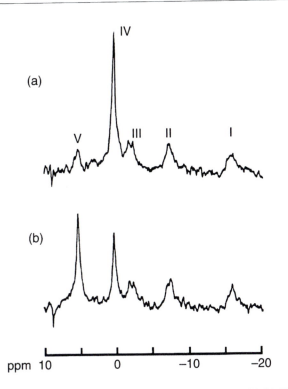

Figure 10.13 The ^{31}P topical NMR spectra obtained at 32.5 MHz using a single-turn coil placed on the surface of a subject's arm, in a contoured magnetic field. (a) The normal spectrum (64 2-s scans). Peak assignments are given in the text. (b) The same subject 50 min after the application of a tourniquet to the upper arm. (Reproduced with permission from Gordon (1981) *Eur. Spectrosc. News*, **38**, 21.)

130 minutes of the concentrations of the metabolites glycogen, α and β forms of glucose, glycerol and after about 87 minutes, resonances appear due to glutamate/glutamine. If [1,2-^{13}C] glucose is used for infusion, we have the situation that metabolites may be formed with the ^{13}C–^{13}C bond intact and so showing ^{13}C–^{13}C coupling. On the other hand, if this bond is ruptured by the metabolic pathway then this coupling will disappear. Two spectra are shown in Fig. 10.14 obtained with the two differently labelled glucose molecules. The α and β forms of glucose are evident (resonances 3,9 and 2,8 respectively) as is a doublet resonance of glycerol which confirms that the pair of ^{13}C atoms from the glucose have been incorporated into the glycerol via the glycolytic intermediate [2,3-^{13}C$_2$] dihydroxyacetone phosphate.

Our final example returns to imaging with the measurement of the diffusion of metabolites *in vivo*. This is a combination of the voxel selection sequence PRESS with the addition of diffusion weighting gradient pairs as shown in Fig. 10.15. PRESS is designed to select one voxel in the sample and produce a one-dimensional spectrum. The RF pulses are all selective, the first being 90° and the second two 180°.

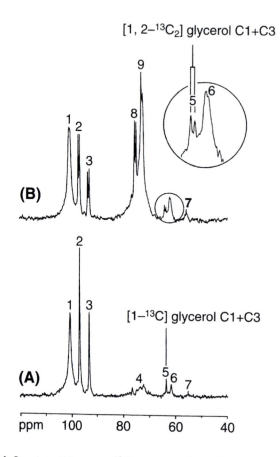

Figure 10.14 *In vivo* difference ¹³C spectra of rat liver after 125 minutes of infusion of either [1-¹³C] glucose (A) or [1,2-¹³C] glucose (B). The doublets indicate C–C bonds which have been metabolized without bond fission. Resonances 1, 4 and 6 arise from glycogen, the others are named in the text. (From Küstermann *et al.* (1996) Bruker Report 143, 33, with permission.)

A double echo is thus formed and the second is collected. Each RF pulse is accompanied by a gradient pulse in one of the three orthogonal axes so that we select three orthogonal slices and the voxel where they intersect is the only part of the sample to give a signal. In addition, each 180° pulse is bracketed by a gradient pair which encodes the diffusion. In order to suppress the water resonance, the sequence shown can be preceded by a CHESS sequence. A set of results are shown in Fig. 10.16 for a 45 µl volume localized in the brain of an anaethetized rat. On the left are the spectra with the CHESS pulses omitted which give essentially only the water resonance. On the right, the water is suppressed and the metabolites are now visible. The spectra are taken with the diffusion gradients *G* increasing up the series. It will be evident that water diffuses the fastest and that

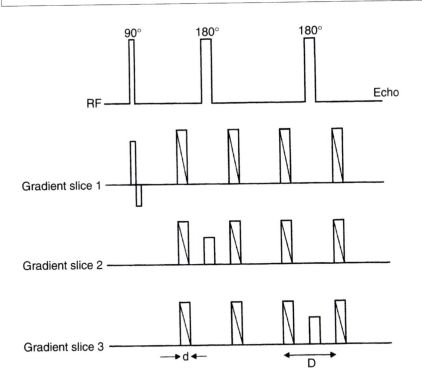

Figure 10.15 The PRESS sequence for obtaining the spectrum of a single voxel, modified to diffusion weight the spectra so obtained. The PRESS pulses are open, the diffusion weighting pulses are marked with a diagonal. D is the interval between pairs of these pulses and d is the pulse duration.

the metabolites diffuse appreciably more slowly. Diffusion in the presence of living cells is a complex process and may be anisotropic due to the structures present. Evidently, the type of experiment described, which is in its infancy, can give access to much detailed information on the behaviour of living systems. The sensitivity is incredible when it is recalled that the molecules investigated move only micrometres during the experiment.

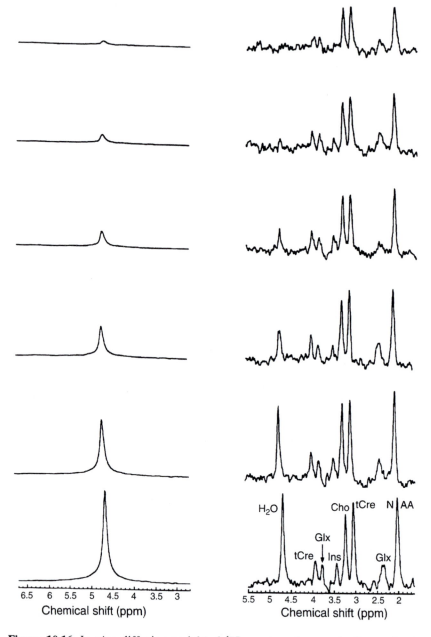

Figure 10.16 *In vivo* diffusion weighted ^{1}H spectra of a voxel of a rat brain showing water (left) and (right) the water suppressed by a CHESS sequence to show the metabolites. The spectral conditions were: TE = 136 ms; repetition rate = 3 s; d = 8 ms; D = 26.2 ms. The metabolites are: tCre, creatine and phosphocreatine; Glx, glutamine/glutamate; Ins, myoinositol; Cho, the *N*-methyl protons of choline; NAA, the methyl protons of *N*-acetyl aspartate. (From K. Nicolay *et al.* (1995) *NMR Biomed.*, **8**, 365, with permission, copyright (1995) John Wiley & Sons Ltd.)

High-resolution solid-state NMR $\boxed{\textbf{11}}$

We have already mentioned in Chapter 4 that, in the solid state, the relaxation time T_1 is long due to the lack of modulation of the dipole–dipole interaction and T_2 is short due to mutual spin flips occurring between pairs of spins. In a static solid, each nucleus produces a rotating magnetic field as it precesses in the applied magnetic field, and this can cause direct exchange of energy between nuclei. The lifetimes of the spin states are thus reduced and so T_2. In addition, each spin has a static field component that influences the Larmor frequencies of its neighbours. An individual nucleus will experience the fields of several neighbours, but their spin directions will vary randomly, so that there will be a range of frequencies that will add to the line broadening due to the rapid rate of relaxation. Finally, particularly for the heavier nuclei, including ^{13}C, there will exist a chemical shift anisotropy, which will also contribute to the broadening, assuming that the sample is a powder or a glass and not a single crystal, because the chemical shift varies with orientation relative to the $\boldsymbol{B}_0$ direction. Thus solid materials, particularly if they contain nuclei with high magnetic moments such as ^1H or ^{19}F, will have broad, structureless resonances, which will not permit the type of investigation that we have shown can be carried out in the liquid phase. This state of affairs has proved a challenge to the NMR community, who have over the last two and a half decades found means to render ineffective the apparent physical restraints to the spectroscopic examination of solids at high resolution.

Before discussing this work in detail, however, it is necessary to mention two useful aspects of the broad lines. In the first place, because the broadening is determined by the dipole–dipole interactions, it is sensitive to the distance separating interacting spins. The spectrum of a solid can thus be used to obtain internuclear distances, which in the case of the proton are difficult to obtain by other means. The molecules studied must be static in the solid state and must be sufficiently simple, preferably an isolated spin pair, that the resonance width can be interpreted in terms of a single, principal distance. If the molecules reorient in the solid in some way, then this modulates the internuclear interaction and its magnitude is reduced, and this is the second property that proves useful. If the linewidth of a solid material is measured as a function of temperature from very low temperatures, it is often found that there are quite rapid changes in linewidth at certain transition temperatures. These mark, if the temperature is being gradually increased, the onset of motion within the solid lattice. Figure 11.1 shows an example for the solid complex adduct of boron trifluoride with trimethylamine, $Me_3N{\rightarrow}BF_3$. The ^1H resonance linewidth below 80K is 85 kHz (the old unit of 1 gauss is equivalent to 4250 Hz in this case). Heating from 68 to 103K reduces the linewidth to about 21 kHz, and this corresponds to the onset of rotation of the methyl groups around the C–N bonds. Further narrowing to 13 kHz occurs on raising the temperature to 150K owing to the onset of rotation of the whole NMe_3 moiety around the B←N bond. Finally, just below 400K, the line narrows to a few hundred hertz as the whole molecule starts to rotate and diffuse isotropically within the still solid crystal. The ^{19}F resonance can also be examined and is found to be broad only below 77K and the BF_3 rotates around the B←N axis at all higher temperatures. Of course, when the sample melts, the linewidth falls to a fraction of a Hertz.

Such studies using broad lines are, however, of relatively limited application, and solid-state NMR formed only a small part of the total

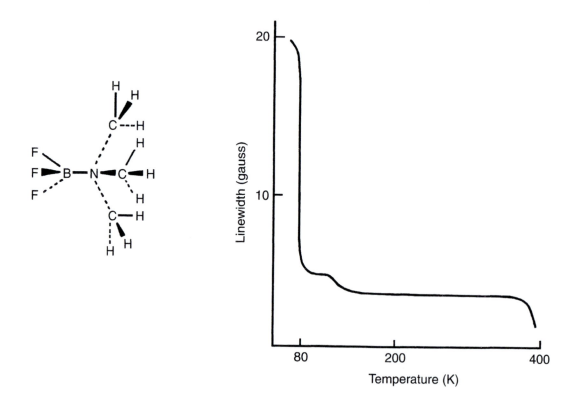

Figure 11.1 The proton resonance linewidth of the methylamine protons in the solid adduct $(CH_3)_3N \rightarrow BF_3$ as a function of temperature. Note that 1 gauss = 10^{-4} tesla = 4250 Hz. (After Dunnell (1969) *Trans. Faraday Soc.*, **65**, 1153, with permission.)

NMR work that was undertaken. The situation has now changed dramatically with the application of the modern techniques to be described below. Both $I = 1/2$ and quadrupolar nuclei are studied, though the treatment of these two classes of nuclei is rather different, and we will have to discuss them separately. One technique is common to both, one which has revolutionized solid-state NMR more than any other, and we will describe this first.

11.1 MAGIC ANGLE SPINNING

The magnetic field produced by a nucleus with magnetic moment μ at a second nucleus a distance r away will, in general, have a component in the z or B_0 direction, which influences the frequency of the second nucleus and also couples the two spins. The z component B_z is given by

$$B_z = \frac{K\mu}{r^3} (3\cos^2\theta - 1)$$

where K is a constant and θ is the angle between the direction of B_0 and the line joining the two nuclei. At one particular angle, shown in Fig. 11.2, B_z is zero. This is the angle for which $3\cos^2\theta - 1$ is zero or $\theta = 54°44'$. It is the angle at which all dipolar interactions disappear. In a real sample, of course, which typically will be a powder, the internuclear vectors take all possible angles and the trick is to make them behave as if all had $\theta = 54°44'$. This is done by mounting the sample in the rotor of a small air-driven turbine whose axis is inclined at an angle of $54°44'$ to the magnetic field direction. Means are provided to adjust the angle so as to obtain the optimum results. The turbine is rotated at high speed, and this gives all the internuclear vectors an average orientation at the rotor angle, which produces dramatic changes in linewidth. This is illustrated in Fig. 11.3. The angle has to be adjusted quite finely to obtain the best results and is accordingly called the 'magic angle', and the technique is called 'magic angle spinning', usually abbreviated to MAS. The chemical shift anisotropy also follows a $3\cos^2\theta - 1$ law, and broadening due to this cause is removed as well by MAS. Quadrupole broadening for nuclei with $I > 1/2$ is also reduced by MAS, but the situation is more complex for the quadrupolar nuclei, as we shall see. The spinning speed is important, since if the static linewidth of the resonance to be studied is F Hz, then the spinning speed must be greater than this if all the broadening interactions are to be nullified. Since the basic physical strengths of materials are limited, there is a limit to the centrifugal forces that they will withstand, and so a limit to the speed of spinning. Currently, the normal limiting speed is around 35 kHz for small ceramic rotors. It is also useful to know the speed at which a rotor is spinning, and these are provided with marks on the rotor cap that can be detected by a suitable frequency counter.

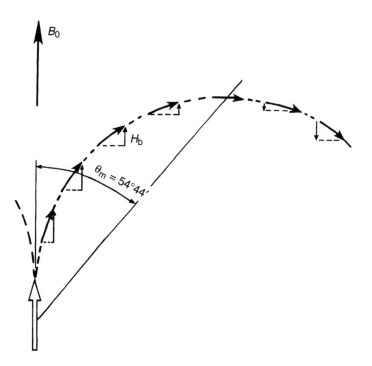

Figure 11.2 A line of the magnetic field originating from a magnetic dipole has zero z component at a point situated on a line originating at the centre of the dipole and at an angle of 54°44′ to the direction of the dipole. (From Bruker CXP Application Notes, with permission.)

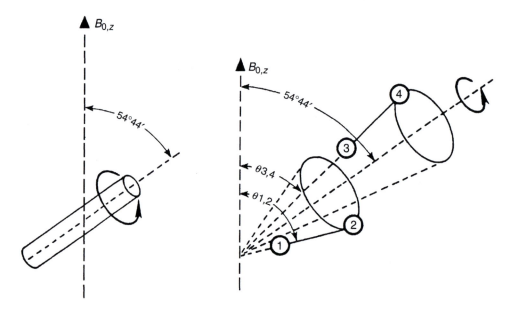

Figure 11.3 Showing how a solid sample is mounted for magic angle spinning and how this gives the internuclear vectors an average orientation at the spinning angle. (From Bruker CXP Application Notes, with permission.)

11.2 SPIN-1/2 NUCLEI WITH LOW MAGNETOGYRIC RATIOS

Common examples of compounds within this class are ^{13}C in organic compounds or ^{31}P in inorganic or organic phosphates. If protons are present, they, of course, cause strong dipolar broadening and their influence has to be removed. This can be achieved by high-power double irradiation, high power being required because the proton resonance will be of very large width. The ^{13}C spectra of adamantane (Fig. 11.4) show the improvements in resolution that can be obtained. The normal solid-state spectrum has a linewidth of some 5000 Hz, and the two types of carbon present in the molecule do not have resolved resonances. MAS alone reduces the linewidth to 200 Hz and enables the two types of ^{13}C to be distinguished (Fig. 11.4(a)). High-power decoupling of the protons alone also reduces the linewidth to 450 Hz in the case illustrated (Fig. 11.4(b)). Combination of the two techniques, however, reduces the linewidth to 2 Hz, which is little worse than in liquid-state samples (Fig. 11.4(c)). Obtaining solid-state spectra

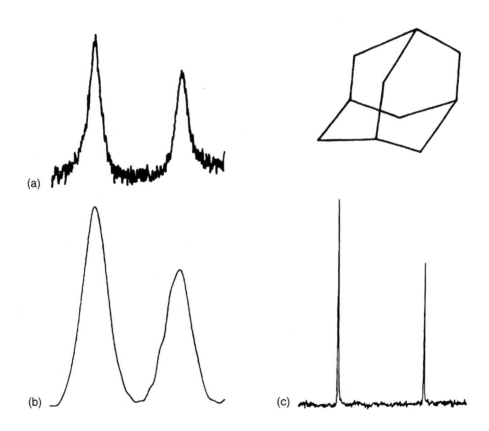

Figure 11.4 The ^{13}C spectra of solid adamantane (formula inset). (a) With MAS. (b) With high-power proton double irradiation. (c) With both MAS and double irradiation. (From Bruker CXP Application Notes, with permission.)

in this way has certain drawbacks. The rather long ^{13}C T_1 values mean that pulse rates have to be sufficiently slow so as not to saturate the resonances, and the natural insensitivity of the nucleus means that long accumulation times are needed. These disadvantages can be circumvented by a modification of the double-resonance technique, which permits the exchange of polarization between the ^{1}H and ^{13}C spins, and which is called cross-polarization (CP); when combined with MAS, the whole is abbreviated to CP-MAS.

In order that energy exchange shall be possible between the two nuclear species, we must introduce components of motion with the same frequency for each. This is done as follows. Referring to Fig. 11.5, we first prepare the protons for the cross-polarization by applying a short 90° pulse, which swings the proton magnetization into the xy plane. We will call this field B_{1H}. We then change the phase of B_{1H} by 90° (and perhaps reduce its intensity also), so that the B_{1H} vector becomes parallel with the magnetization in the xy plane, which therefore remains in the xy plane and stays there locked to B_{1H}. This second pulse is known as a spin locking pulse. The magnetization precesses in the xy plane at the Larmor frequency, and can be thought of as also precessing around B_{1H} at a frequency of $\gamma_H B_{1H}/2\pi$ Hz, behaving as if it was very strongly polarized in the weak B_{1H}. Reference to equation (1.3) will show that such a polarization, correct under normal circumstances for B_0, can only be attained if the temperature is very low for a weak field B_{1H}, and we can consider that the spin locking has cooled the spins, which can now act as an energy sink. Energy transfer is obtained by applying a long pulse at the ^{13}C frequency, B_{1C}, which has an amplitude such that the ^{13}C nuclei precess around it at a frequency of $\gamma_C B_{1C}/2\pi$ Hz which is equal to $\gamma_H B_{1H}/2\pi$ Hz. This is known as the Hartmann Hahn matching condition. The identical frequency components allow energy transfer, which follows an exponential curve, and when this has reached a maximum the B_{1C} is cut off and is followed by a ^{13}C FID of enhanced intensity. The B_{1H} field remains on during this time and provides the decoupling of the protons from the carbon nuclei. The experiment can be repeated when the proton spins have reached equilibrium again and so the effective relaxation time is the shorter one of the proton system. The cross-polarization increases the ^{13}C population difference by the factor γ_H/γ_C and so produces a useful improvement in signal strength. An example of a ^{13}C CP-MAS spectrum is shown in Fig. 11.6 for a more complex molecule, the steroid deoxycholic acid, together with a partial assignment of the resonances. Deoxycholic acid forms inclusion compounds, and, when the guest molecule is ferrocene, $[Fe(\eta^5\text{-}C_5H_5)_2]$, the methyl singlets due to C18, C19 and C21 become doublets due to differentiation of the deoxycholic acid molecules in the solid lattice. The cross-polarization technique is much used with low-γ nuclei where the compound studied contains hydrogen; ^{13}C, ^{15}N, ^{29}Si, ^{31}P and ^{113}Cd are some examples of $I = 1/2$ nuclei studied. Quadrupolar nuclei can also benefit from the CP technique.

If the spinning speed is appreciably lower than the width of the static resonance of the compound studied, then sidebands are produced separated by the spinning speed. Figure 11.7 shows the solid-state ^{31}P spectra of aminomethanephosphonic acid, $H_2NCH_2PO_3H_2$, which has a zwitterionic structure. All the three spectra shown benefited from

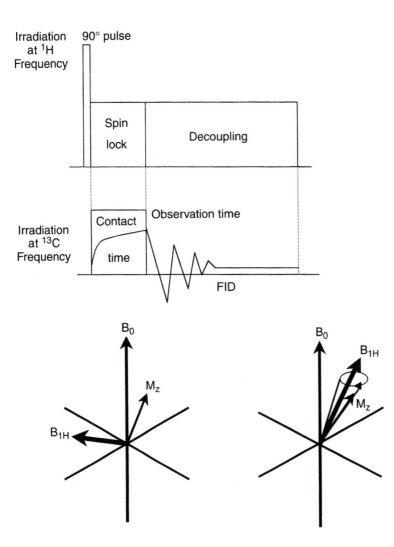

Figure 11.5 The timing of the various events in a cross-polarization experiment. A 90° pulse at the proton frequency is followed by a long spin locking pulse changed in phase from the initial pulse by 90°, which prevents the normal proton relaxation processes. Energy is caused to flow from the assembly of ^{13}C nuclei to the cooled proton nuclei by applying a long pulse at the ^{13}C frequency, which introduces a precession frequency equal to that already established for the protons. When equilibrium is reached, this field is switched off and is followed by a ^{13}C FID of enhanced intensity. Note that the contact time (about 2 ms) is much shorter than the observation time (about 50 ms). (From Oldfield *et al.* (1984) *J. Magn. Reson. Chem.*, **60**, 467.)

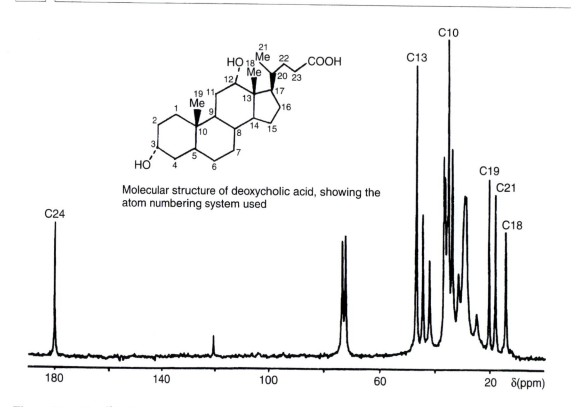

Figure 11.6 The ^{13}C CP-MAS spectrum of deoxycholic acid. The ^{13}C frequency was 50.32 MHz and that for ^{1}H was 200 MHz. Some 350 mg of sample was packed into the MAS rotor and some 800 transients were acquired. The contact time was 1 ms and recycle delay 3.5 s. (From Heys and Dobson (1990) *Magn. Reson. Chem.*, **28**, S37–46, copyright (1990) John Wiley and Sons Ltd, reprinted with permission.)

CP, but the uppermost one (a) was from a static sample and shows the shielding anisotropy of the phosphorus nucleus, which does not have axial symmetry and so has three principal components σ_{11}, σ_{22} and σ_{33} (see Chapter 2). MAS at 813 Hz (b) produces a group of narrow lines separated by the spinning frequency. The number of lines is reduced on increasing the spinning speed to 2950 Hz (c), and it is evident that only one does not move, so that this (arrowed) is the true resonance with the isotropic chemical shift and the remainder are spinning sidebands. It is particularly important to note that at low spinning speeds, the envelope enclosing the sidebands approximates the static lineshape and so retains the form of the shielding or chemical shift anisotropy. The ^{31}P spectrum of *N,N*-dimethylamino-methanediphosphonic acid, $Me_2NHC(PO_3H_2)_2H$, is also shown for a spinning speed of 1740 Hz (d). In this case, there are two resonances that are not sidebands, since the two phosphorus atoms in a single molecule are not related by symmetry due to the crystal structure. In solution, of course, only a singlet ^{31}P resonance is produced by this compound.

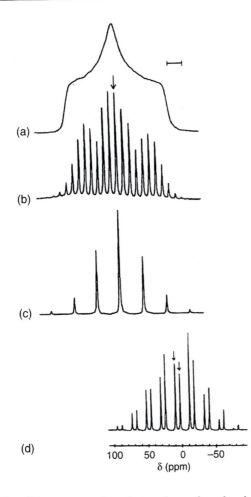

100   50   0   −50
δ (ppm)

Figure 11.7 The ^{31}P spectra of aminomethanephosphonic acid. (a) Static sample showing the shielding powder pattern. The ^{1}H–^{31}P dipole interaction is eliminated by double resonance. (b), (c) With MAS at 813 Hz and 2950 Hz. The arrow is at 18.3 ppm and the bar represents 2 kHz. (d) MAS spectrum of N,N-dimethylaminomethanediphosphonic acid showing the existence of two crystallographically differentiated sites. The chemical shift scale applies only to this spectrum. The arrows show the centre band lines with the isotropic chemical shifts. The other lines are spinning sidebands. (From Harris *et al.* (1989) *Magn. Reson. Chem.*, **27**, 470, with permission, copyright (1989) John Wiley and Sons Ltd, reprinted with permission.)

Such high-resolution spectra allow access to parameters such as chemical shift anisotropy and permit comparison of molecular structure in solution and crystal. In addition, they permit insoluble substances to be studied at reasonably high resolution, and investigations are being carried out into the structures of materials such as plastics or coals. The latter have been notoriously difficult to study without degrading the coal structure, and this is now possible using NMR. An example is shown in Fig. 11.8 for both a plastic and a coal and, while the resolution for the coal is not exceptional, it has to be remembered that it is a most complex mixture of structures and that it is very useful to be able to distinguish aromatic and aliphatic resonances and perhaps to be able to do this quantitatively.

It will be no surprise to find that two-dimensional techniques have also been found useful for the study of solid state systems. We give an example of an EXSY experiment used to study motion in the solid polymer $[(Me_3Sn)_4Ru(CN)_6]_\infty$; in which the tin coordination is

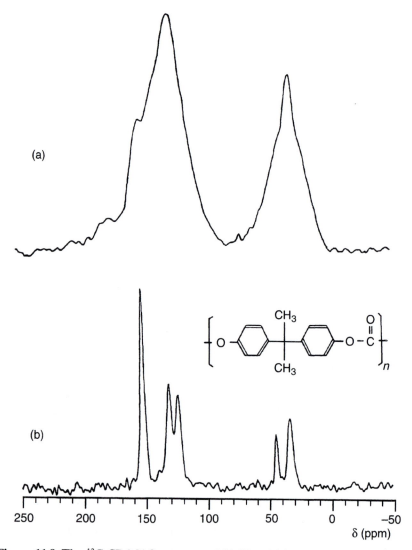

Figure 11.8 The ^{13}C CP-MAS spectra at 15 MHz of (a) a sub-bituminous coal and (b) polycarbonate solids. (From Wind (1991) *Modern NMR Techniques and Their Application in Chemistry,* eds Popov and Hallenga, Marcel Dekker Inc., New York, p. 186, with permission.)

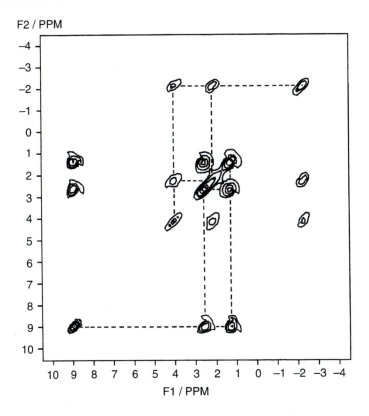

Figure 11.9 Decoupled ^{13}C proton two-dimensional exchange spectrum at 75.4 MHz and with CP-MAS of $[(Me_3Sn)_4Ru(CN)_6]_\infty$. The off-diagonal signals link methyl carbon atoms belonging to the same Me_3Sn group, there being two sets of three as shown by the dashed lines. (Harris (1995) *Frontiers of Analytical Spectroscopy*, eds Andrews and Davies, Royal Society of Chemistry, p. 77.)

trigonal bipyramidal with the methyl groups in equatorial positions. The crystal symmetry is such that there are two distinguishable $SnMe_3$ groups and so six methyl ^{13}C signals. However, the $SnMe_3$ groups can rotate about the axial bonds and line shape calculations could in principle give kinetic data, except that it is not known which three signals belong to a given $SnMe_3$ group. This can be determined by the two-dimensional exchange experiment, the results of which are shown in Fig. 11.9 and where it will be seen that the cross peaks neatly connect two sets of three carbon resonances on the diagonal.

11.3 *I* = 1/2 NUCLEI WITH HIGH MAGNETOGYRIC RATIOS

Here we are thinking specifically of the nuclei ^{1}H or ^{19}F, where the homonuclear interactions are very strong and so difficult to remove by MAS. ^{1}H spectra are the most difficult to deal with, since not only are the static linewidths very large, but the chemical shifts are small, so making big demands on the resolution ability of the system. MAS at the very highest spinning speeds and using the highest magnetic fields to maximize the chemical shift dispersion is capable of reducing a static linewidth of the order of 10 kHz to around 1500 Hz, and this can give resolvable resonances though the actual improvement obtained is very sample dependent. Alternatively, the spins can be swung around by a succession of pulses so that they appear to adopt the magic angle in the rotating frame. A typical sequence, called MREV-8, is shown in Fig. 11.10 and has the effect that the magnetization is shifted quickly between the three orthogonal axes, which, as it will be remembered from section 4.3 and Fig. 4.7, are placed at the magic angle from their three-fold symmetry axis. The spins thus hop around the magic angle axis and their dipole–dipole interaction is much reduced, though the chemical shift anisotropy and heteronuclear

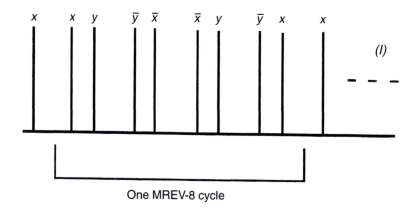

One MREV-8 cycle

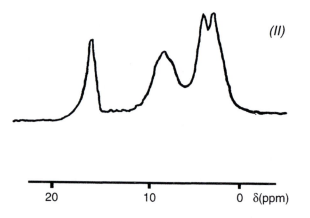

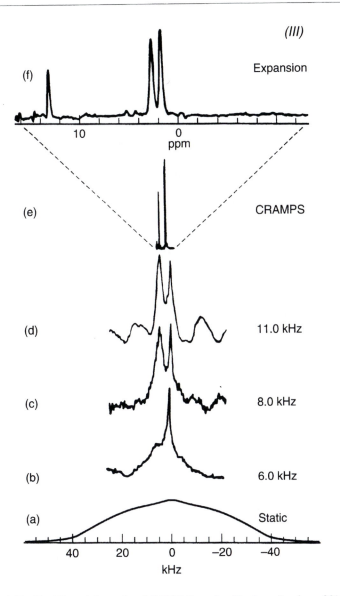

Figure 11.10 (I). The eight-pulse MREV-8 cycle. Each pulse is a 90° pulse, which rotates the magnetization around the x or y axes in one direction or its opposite. The large spacings are double the length of the small spacings. The nuclear signal is sampled between pulses. (II) The ^{1}H CRAMPS spectrum at 200 MHz of aspartic acid HOOCCHNH$_2$CH$_2$COOH. (From Bruker Report 1/1988, with permission.) (III) ^{1}H NMR spectra of solid adipic acid (a) static, showing the broadening due to dipolar interactions, (b, c, d) with MAS at various speeds, none exceeding the linewidth of the static spectrum (40 kHz) and (e) with CRAMPS. (f) is an expansion of (e) and it will be seen that the three types of proton are well resolved. (From Maciel, in Gerstein (1996) *Encyclopedia of Nuclear Magnetic Resonance*, **3**, 1501, eds Grant and Harris, copyright John Wiley and Sons Ltd, with permission.)

interactions are not affected. As might be expected, then, since these will be reduced by MAS, a combination of the two methods proves very successful. This technique is called combined rotation and multiple-pulse spectroscopy (CRAMPS). Resolution of the order of 180 Hz is possible in ^1H spectroscopy. The CRAMPS spectrum of aspartic acid, $HOOCCHNH_2CH_2COOH$, is also shown in Fig. 11.10, where resonances due to the four different types of proton can easily be distinguished.

A second example of a CRAMPS spectrum is shown in Fig. 11.10 for adipic acid ($HOOCCH_2CH_2CH_2CH_2COOH$ with three different types of proton), which also shows the way the spectrum changes from its static shape with the introduction of MAS at various speeds and finally the improvement in resolution gained by introducing the multi-pulse sequence as well as MAS. The pulse sequences are arranged to return the spin system to its original state at the end of each cycle of pulses and the spectrometer output is then sampled, digitized and stored to give the usual FID in memory.

In favourable cases the resolution required can be obtained using only MAS with double irradiation if appropriate. An example is given of the study of the ring inversion of fluorocyclohexane in its solid thiourea inclusion compound. The host thiourea contains tunnels in which the guest molecule resides and many studies have been made of a variety of monosubstituted cyclohexanes in this environment to try to understand the constraints put upon the conformation of the cyclohexane in these tunnels. It has been found that if the substituent is Me, NH_2 or OH then the substituent takes predominately the equatorial position whereas with halide substituents the axial conformer predominates. This work was done using ^{13}C NMR but for the fluoro isomer it is possible to use the much more receptive ^{19}F nucleus. Double irradiation of the protons is required but this presents technical problems due to the proximity of the ^1H and ^{19}F frequencies and the high power used for double irradiation and a specially designed probe is required. The ^{19}F spectra obtained in this way are shown in Fig. 11.11, obtained as a function of temperature to give the typical two site exchange bandshapes. Rate constants were obtained by computer fitting of the bandshapes. The equatorial and axial conformers have chemical shifts of -163.9 and -187.0 ppm respectively. It will be evident in this case that the two conformers have apparently equal populations though careful assessment of the data shows that at lower temperatures, there is a slight preference for the equatorial conformer.

11.4 MAS OF QUADRUPOLAR NUCLEI

Quadrupolar nuclei in the solid state usually have weak dipolar interactions with their surroundings, and this will not concern us here. Where they do exist then a cross-polarization experiment may be

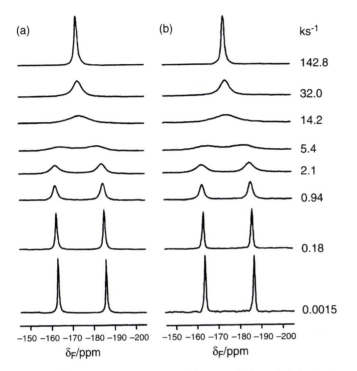

Figure 11.11 ^{19}F spectra of fluorocyclohexane in its solid inclusion compound with thiourea. The temperatures are, from bottom to top: 177, 217, 237, 247, 258, 268, 278 and 300K. (a) is the calculated spectrum and (b) is the actual spectrum. The ring inverts readily at 300K. (From Harris *et al.* (1999) *Magn. Reson. Chem.*, **37**, 15, copyright John Wiley and Sons Ltd, reproduced with permission.)

possible. Quadrupole nuclei are, of course, polarized by the magnetic field but are also subject to any electric field gradient present at their position in the molecular and crystallographic environment, and which can arise from the bonding electrons and, in contrast with the liquid, from more distant electronic distribution. The nature of the interaction depends upon whether the nucleus examined has an integral or half-integral spin. As there are only effectively three important nuclei with integral spin, namely ^{2}H, ^{6}Li and ^{14}N, we will discuss initially only the half-integral spin nuclei. If a single crystal of a solid that contains such a nucleus is placed in a strong magnetic field at some particular orientation of a crystal axis to that field, then we have also determined the way the electric field gradient interacts with the several possible spin states. If the electric field gradient (EFG) is zero, then the energy gap between spin states is the same for any pair: that is, the energy of the transitions possible for a $I = 3/2$ nucleus, $3/2 \leftrightarrow 1/2$, $1/2 \leftrightarrow -1/2$, $-1/2 \leftrightarrow -3/2$ are all equal and a single resonance results. If there is an EFG, then the energy of each spin state is altered. The 1/2 and −1/2 states move in parallel, so that the transition energy is unaltered, but the 3/2 and −3/2 states change in opposite directions and so reduce the energy of one transition and increase the energy of the other. The degeneracy of the three possible transitions is now lifted and we detect three resonances. This is shown schematically in Fig. 11.12. If we change the orientation of the crystal in the magnetic field, we change the orientation of the EFG tensor and so the interaction of the nucleus with the EFG. This alters the energy levels: though the $1/2 \leftrightarrow -1/2$ transition frequency is unchanged, the other two will both be affected, one being increased and one decreased. At certain orientations (say $0°$) they will coincide with the $1/2 \leftrightarrow -1/2$ transition and, as the crystal is rotated, they will move away from this $1/2 \leftrightarrow -1/2$ central line, reach a maximum displacement, return to the central line, cross over, reach maximum again and, at $180°$ rotation for crystals with axial symmetry, again coincide with the central line. This is known as the first-order

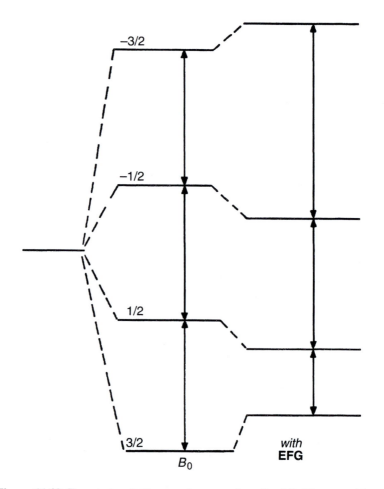

Figure 11.12 Energy level diagram for a nucleus $I = 3/2$. The transition energies are equal if only the magnetic field $\boldsymbol{B}_0$ is taken into account, but the quadrupole interaction causes one to become larger, one to become smaller and one, the central transition, to remain unchanged.

quadrupole effect. An example of how the outer bands move as a function of rotation angle is given in Fig. 11.13, which shows the results for a single crystal of sodium nitrate placed so that its three-fold axis of symmetry could be rotated to describe a plane parallel with B_0. The maximum displacement of the satellites is proportional to the magnitude of the quadrupole coupling constant and to $3\cos^2\theta - 1$, where θ is the angle between the EFG tensor axis (V_{zz}) and the direction of the magnetic field B_0. In a powder sample, where the crystallites are arranged at random, the satellite lines have all possible positions and so smear out into the baseline, leaving only the central line invariant and detectable.

Because of the tensor nature of the EFG, this description is over-simplified and we have to take into account a second-order perturbation of the energy levels and so of the frequencies of all the lines. If we define a quantity ν_Q as a measure of the magnitude of the quadrupole interaction, where

$$\nu_Q = \frac{3e^2qQ}{2Ih(2I-1)}$$

and provided the quantity ν_Q^2/ν_0 is appreciable in magnitude, where ν_0 is the Larmor frequency of the nuclear species examined, then the frequency of the central $1/2 \leftrightarrow -1/2$ transition also depends upon the angle θ but in a more complex way. Thus this second-order shift ν_2 is

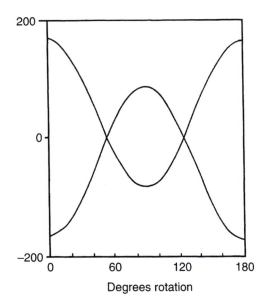

Figure 11.13 The positions of the two satellite lines in the ^{23}Na spectrum of a single crystal of $NaNO_3$ as a function of angle between the B_0 magnetic field direction and the three-fold symmetry axis of the crystal. The origin of the frequency axis is the frequency of the central line. (Derived from Andrew (1958) *Nuclear Magnetic Resonance*, Cambridge University Press, Cambridge.)

$$v_2 = -\left(\frac{v_Q{}^2}{16v_0}\right)\left[I(I+1) - \frac{3}{4}\right](1-\cos^2\theta)(9\cos^2\theta - 1)$$

for an axially symmetric field gradient with $\eta = 0$. These terms have already been defined in Chapter 4. In a powder sample, this second-order angular dependence of the line frequency imparts width, which, it has to be emphasized, is a shift effect, not a relaxation effect as in the liquid phase. The line has breadth $v_{1/2}$ and shift from the isotropic resonance frequency δ_2 of approximately

$$v_{1/2} = \frac{25v_Q{}^2}{18v_0} \quad \text{and} \quad -\delta_2 = \frac{v_Q{}^2}{20v_0}$$

The line shape of a powder sample is thus seen to be a valuable source of information about the nuclear environment, though the centre of the line is not at the chemical shift position and the true chemical shift has to be calculated if a precise value is required. The form of the spectrum is also a function of the spectrometer magnetic field strength, and the second-order effects become less evident as this is increased. Thus high magnetic fields are essential for the high resolution study of quadrupolar nuclei in the solid state, though if the value of the quadrupolar coupling constant is required then this is better obtained at lower magnetic fields.

The question we must next ask is, of course, will magic angle spinning reduce this linewidth? Evidently, if there is any contribution from chemical shift anisotropy or dipolar coupling then this will be reduced or eliminated. Otherwise MAS is not so magic in the case of the quadrupolar nuclei. The second-order term, which does not vary as $3\cos^2\theta - 1$, is not eliminated by MAS, though it is reduced by a factor of about four times. Thus if $\eta = 0$, the width and shift of the resonance under MAS are approximately

$$v_{1/2} = \frac{v_Q{}^2}{3v_0} \quad \text{and} \quad -\delta_2 = \frac{v_Q{}^2}{4v_0}$$

One has to use the term 'approximately' in describing these quantities because the line shape is complex and quite unlike the Lorentzian lines obtained with liquid samples. A theoretical line shape is sketched in Fig. 11.14 for a nucleus of $I = 5/2$ in a powder sample undergoing MAS with $\eta = 0$. Perhaps the most important point to note from this sketch is the fact that the true isotropic chemical shift of the resonance is not at the centre of the resonance but is to high frequency always. In practice, the line shape is rounded off from the rather angular shape shown, and an estimate of the true shift can be made as being the point where the high frequency side of the resonance has lost almost all its intensity. Clearly, this is approximate, and any accurate estimate requires the line shape to be calculated for a particular spectrum. It is, however, a better approximation than that often encountered currently of giving the shift as the centre of the resonance.

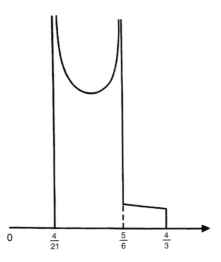

The frequency axis is labelled 0, $\dfrac{4}{21}$, $\dfrac{5}{6}$, $\dfrac{4}{3}$.

Figure 11.14 A sketch of the line shape of a $I = 5/2$ nucleus in a powder sample undergoing MAS. The frequency scale is calibrated in units of $v_Q^2/2v_0$ and the origin of the shift scale is the Larmor frequency in the absence of any second-order quadrupole interaction. (From Akitt (1981) *Prog. NMR Spectrosc.*, **21**, 1, copyright (1981) Elsevier Science with permission.)

This adequately describes the position of the peak on the spectrum but is *not* its chemical shift.

The line shape of a quadrupolar nucleus in a spinning sample is strongly dependent on the angle of the turbine set relative to B_0. In fact, the best angle for reducing the second-order linewidth is not the magic angle, and it is better to choose some other angle. The actual value to use is a compromise between reducing the second-order effects and those effects, if present, that depend upon $3\cos^2\theta - 1$ terms. It is thus necessary to be able to vary the spinning angle, and this has been given the name VAS – variable angle spinning.

Some of the aspects of solid-state NMR that we have just discussed are summarized in Fig. 11.15, which shows the ^{27}Al spectra of the tridecameric aluminium cation already discussed in Chapter 4. This has the formula $[AlO_4Al_{12}(OH)_{24}(H_2O)_{12}]^{7+}$, with one Al in a site of high tetrahedral symmetry and so low quadrupole coupling constant and the remaining 12 in sites of distorted octahedral symmetry and so with high quadrupole coupling constants. The solution-state spectrum contains one narrow line (AlO_4) and a very broad, usually undetectable line 58 ppm to high frequency (AlO_6). The solid-state spectrum obtained at low field is similar, though it must be appreciated that the AlO_6 resonance is broadened by quite different mechanisms in the two phases. As the field and v_0 are increased, the second-order broadening is reduced and the octahedral resonance becomes visible ((c), (b) and (a)). If the spinning angle is changed to 75°, then the AlO_6 linewidth is further reduced (d) and two spinning sidebands are

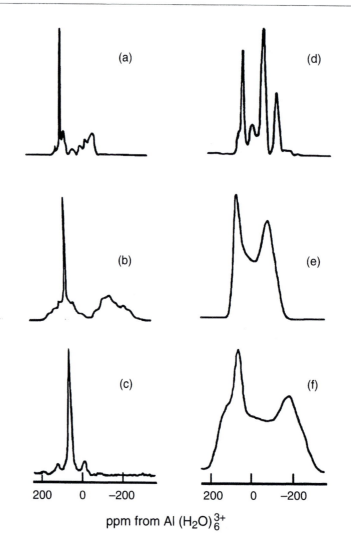

(a)

(d)

(b)

(e)

(c)

(f)

200   0   −200   200   0   −200

ppm from Al (H$_2$O)$_6^{3+}$

Figure 11.15 The ^{27}Al solid-state spectra of the tridecameric aluminium cation at different magnetic fields, with and without sample spinning. Spectrometer frequencies were (c), (f) 39 MHz, (b), (e) 93.7 MHz and (a), (d) 129.7 MHz. Samples (e) and (f) were static, samples (a), (b) and (c) were spinning at the magic angle, and sample (d) was spinning at an angle of 75° to the magnetic field direction. (After Oldfield *et al.* (1984) *J. Magn. Reson.*, **60**, 467, with permission.)

also visible. It is remarkable to note that, in this spectrum, we have better resolution of the octahedral aluminium resonance than we do in solution. The spectra of static samples are also shown and make very evident the improvement in resolution obtained with spinning. The static spectra can nevertheless be used to calculate the quadrupole coupling constants, which are to be used to interpret the spectra of the spinning samples.

A further possibility for improving the resolution of resonances exists in the solid state which is absent for liquid samples. Since the second-order quadrupole effect produces frequency shifts, then sites with similar chemical shifts but different quadrupole coupling constants may well have resonances with centroids that are well separated. A striking example of this is shown in Fig. 11.16, which shows the ^{27}Al MAS spectrum of CaO·3Al$_2$O$_3$·3H$_2$O obtained at 130 MHz. The crystal contains two crystallographic types of four-coordinate aluminium but which have a chemical shift of only 1 ppm, or 130 Hz in the spectrometer used. The quadrupole coupling constants at the two sites are, however, very different, so the line shapes are different and their centroids are separated by some 10 ppm, sufficient to observe the resonances separately and to compute their line shapes. Note also how the isotropic chemical shifts, shown by two marks on the chemical shifts axis near 80 ppm, coincide with essentially zero signal intensity.

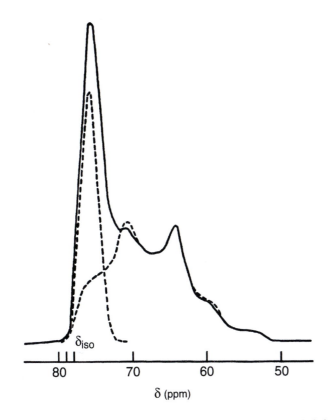

Figure 11.16 The ^{27}Al MAS spectra at 130 MHz of CaO·3Al$_2$O$_3$·3H$_2$O showing the two superimposed signals and their computed line shapes. Note particularly the two close values of the true isotropic chemical shifts, which are marked on the shift axis and correspond with hardly any signal intensity. Note also how two types of chemically very similar aluminium are well differentiated by different quadrupole couple constants. (Reprinted with permission from Muller *et al.* (1986) *J. Chem. Soc., Dalton Trans.*, 1277.)

11.5 SOME APPLICATIONS

Solid-state NMR can usefully be applied to the determination of the state of the cations in the alkalide salts. These substances typically have formulae $MM'L_n$, where M and M' are an alkali metal or metals and L is a strong complexing ligand, either a cryptand or a crown ether capable of enclosing or partially enclosing the alkali-metal cation. The second alkali metal is present as the anion, M^-. The ^{23}Na MAS spectrum of the alkalide Na[C222.Na] is shown in Fig. 11.17. C222 is the cryptand $N(C_2H_4OC_2H_4OC_2H_4)_3N$. Two resonances are observed, one with an isotropic shift near that of the standard at 0.0 ppm, $Na^+(aq)$, and one with a shift significantly to low frequency, as would be expected for the greater electron density on Na^-. If the alkalide contains two different alkali-metal atoms, then the solid-state chemical shifts of each will indicate which is the anion and which the cation.

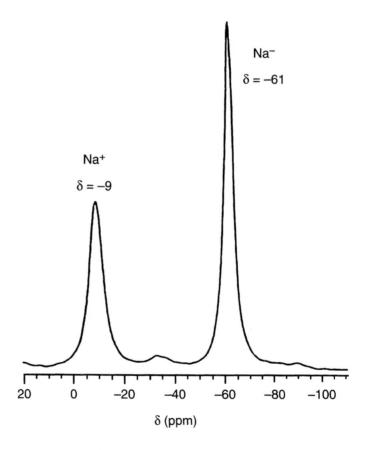

Figure 11.17 The ^{23}Na MAS NMR spectra of the homonuclear alkalide Na[C222.Na]. (From Dye *et al.* (1991) *Modern NMR Techniques and Their Application in Chemistry*, eds Popov and Hallenga, Marcel Dekker Inc., New York, p. 291, with permission.)

The examples shown up to present have all given data comparable with crystallographic work, and the work could in many cases have been achieved equally well by this older technique, though NMR is perhaps less time-consuming and can be used to follow the effect of small changes in sample conditions with relative ease. There is, however, a whole class of compounds where crystallography is of little help, such as disordered solids, glasses and amorphous substances, heterogeneous substances and solids with minor but important components that are not picked up by diffraction experiments. NMR is ideal for the study of such materials and we give several examples below.

11.5.1 The setting of cement

Monocalcium aluminate, $CaO \cdot Al_2O_3$, is the main constituent of high-alumina cement and the ability to investigate its hydration, i.e. how it sets, is of obvious general interest and technological importance. The dry starting material contains four-coordinate aluminium, which becomes six-coordinated upon hydration. There is a chemical shift of some 70 ppm between the two types of aluminium, and so the hydration can be followed through all its stages. A typical ^{27}Al MAS spectrum is shown in Fig. 11.18, together with plots showing how the proportion of octahedral aluminium changes with time and also the amount of heat of reaction evolved with time. Setting is more rapid at high temperatures and the conversion to the octahedral form is more complete. The rate of reaction also varies quite widely at the two higher temperatures and a series of steps are observed. The initial reaction is believed to produce a mixture of phases, which covers the unreacted material and causes the reaction to slow down. This covering then suffers transformation to new phases, which expand and loosen the coating and so permit reaction to proceed again.

11.5.2 Zeolites

These substances are formed of networks of aluminosilicates that contain pores of certain fixed sizes and are active as catalysts. They are produced by crystallization from a gel formed upon mixing, say, an aluminate salt with a soluble silicate, and the structure of the zeolite formed depends upon the nature of the components used to form the gel. The ratio of silicon to aluminium present in the zeolite can be varied within certain limits and the silicon is always in excess. The solids thus contain Si–O–Si and Si–O–Al linkages but not Al–O–Al linkages, which appear to be forbidden. These substances contain two magnetically active nuclei, ^{29}Si with $I = 1/2$ and the quadrupolar ^{27}Al with $I = 5/2$, and both are used extensively in their study. The ^{29}Si spectra may contain up to five resonances, which correspond to tetrahedral SiO_4 units with zero, one, two, three or four attached aluminium atoms. A spectrum of zeolite Na–Y with Si/Al = 2.61 is shown in Fig. 11.19, where the resolution of the different environments is seen to be

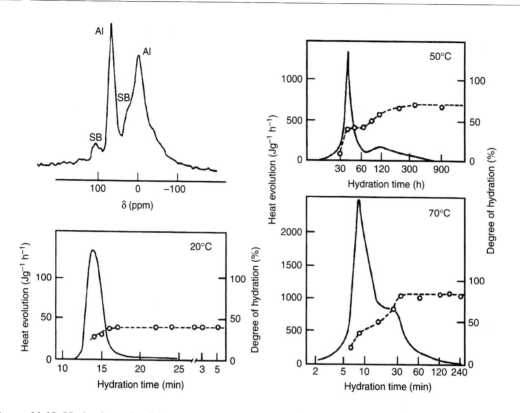

Figure 11.18 Hydration of calcium aluminate. A typical ^{27}Al MAS spectrum is shown at the top left of the figure. The three other plots show the progress of hydration with time at three different temperatures. The full curves show the heat evolution and the broken curves show the percentage of six-coordinate aluminium formed. SB on the spectrum signifies spinning sidebands. (Reprinted with permission from Rettel *et al.* (1985) *Br. Ceram. Trans. J.* **84,** 25.)

excellent. With care, the intensities are quantitative and the pattern allows the Si/Al ratio to be calculated. In natural zeolites, this ratio is always less than about 5, but materials with much lower aluminium contents can be synthesized. The series of highly siliceous zeolites called ZSM-5 with Si/Al typically 31 are a well-known range of catalysts with extra stability conferred by the high silicon content. The ^{29}Si spectra of such substances are simple with essentially a single line due to Si(OSi)$_4$ units. A second material isostructural with ZSM-5 but with only a trace of aluminium and called silicalite is also known. If the aluminium content is particularly low, then the single-line ^{29}Si spectrum is found to be resolved into a group of lines, which arise from the various crystallographic sites in the as-yet uncertain structure of this material. This remarkably well-resolved spectrum is also shown in Fig. 11.19. Because the aluminium content of silicalite is so low, it was argued in order to be able to patent its use, that it was not a zeolite and that any aluminium was present as alumina impurity. ^{27}Al is a good, receptive nucleus and can be detected at quite low levels in solids, which are concentrated states of matter. The ^{27}Al MAS spectrum of the same sample of silicalite is shown in Fig. 11.19, and it is evident that the ^{27}Al is detectable, even though an accumulation time of over two days was required to collect the 176 214 FIDs needed. The chemical shift is diagnostic for tetrahedrally coordinated aluminium, so that the aluminium is to be found within the silicalite framework and, further, there are at least two different aluminium environments. Note, again, that it is the peak centroids that are indicated on the figure. The same type of structure has since been observed in ZSM-5 that has been thoroughly de-aluminated to reach Si/Al = 800.

An alternative approach to these catalysts is to take, for instance, zeolite Y and subject it to what is known as decationation and ultrastabilization. The ammonium form of the zeolite is subjected to heat treatment under vacuum conditions, when it loses ammonia and water. The resulting crystalline material has much greater stability than the starting material and is a good catalyst used for hydrocracking in the petroleum industry. It has a much reduced ion-exchange capacity and this indicates that aluminium has been lost from the framework, the vacancies created being reoccupied by silicon. The aluminium remains but can subsequently be leached out of the solid catalyst.

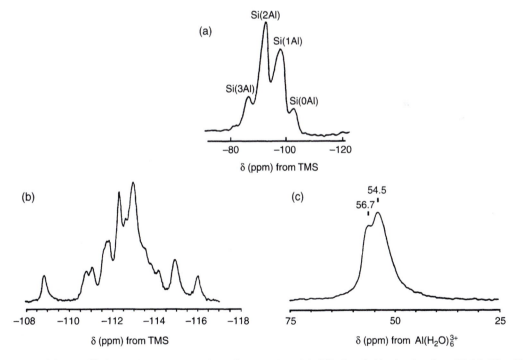

Figure 11.19 (a) The ^{29}Si MAS spectrum of zeolite Na–Y with Si/Al = 2.61 obtained at 79.8 MHz. Five Si environments are indicated. (b) The ^{29}Si MAS spectrum of silicalite with Si/Al > 1000 obtained at 99.32 MHz. The resonances are all from SiO$_4$ with no directly linked Al. (c) The ^{27}Al MAS spectrum of the same sample taken at 104.2 MHz and the result of accumulating 176 214 FIDs. (From Klinowski and Thomas (1985) *Adv. Catal.*, **33**, 199, and Fyfe *et al.* (1982) *J. Phys. Chem.*, **86**, 1247, and Klinowski *et al.* (1984) *Prog. NMR Spectrosc.*, **16**, 237, copyright (1984) Elsevier Science, reprinted with permission.)

This ultrastabilization process has been studied by both ^{29}Si and ^{27}Al spectroscopy, as shown in Fig. 11.20. The starting material (a) had Si/Al = 2.61, four lines in the ^{29}Si spectrum and all tetrahedral Al. After calcining in air at 400°C for one hour (b), there are evidently fewer aluminium atoms linked to the silicon and some octahedral aluminium has appeared. Si/Al was calculated to be 3.37, a calculation that does not include the octahedrally coordinated metal, since it is based on the ^{29}Si intensities. More drastic treatment, heating in steam at 700°C, produces even greater spectral changes and an Si/Al ratio of 6.89 (c). Repetition of this procedure followed by leaching with nitric acid (d) removes most of the aluminium to give Si/Al = 50 and a single ^{29}Si line, which indicates good crystallinity as is required if the Al vacancies are filled. The octahedral aluminium resonance becomes very narrow on leaching and represents remaining Al in the form of $[Al(H_2O)_6]^{3+}$ free to rotate in lattice cavities. These changes can also be achieved by treatment with $SiCl_4$ vapour or the aluminium can be put back into the structure using $AlCl_3$ vapour, both processes having been monitored by ^{27}Al MAS.

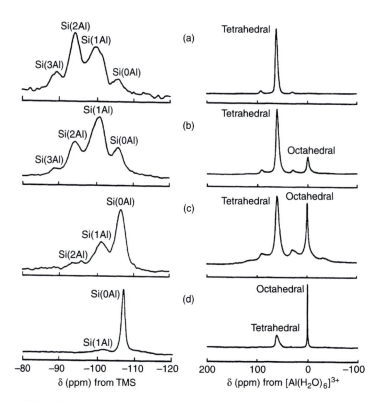

Figure 11.20 High-resolution MAS spectra monitoring the ultrastabilization of zeolite-Y as described in the text. The left-hand spectra are ^{29}Si at 79.8 MHz and those on the right are ^{27}Al at 104.2 MHz. (From Thomas and Klinowski (1985) *Adv. Catal.*, **33**, 199 and from Thomas and Klinowski (1982) *Nature*, **296**, 533–6, copyright (1982) Macmillan Magazines Ltd, reprinted with permission.)

11.6 DEUTERIUM, AN INTEGRAL-SPIN NUCLEUS

The patterns obtained in the solid state with the nucleus ^2H, for which $I = 1$, are somewhat different from those described above. A nucleus with $I = 1$ has three energy levels and two degenerate transitions in the absence of any quadrupole coupling. The interaction of the nucleus with the magnetic field and the electric field gradient causes the three energy levels to be modified, so that there are two transition frequencies disposed symmetrically about the isotropic chemical shift value and with a frequency difference that is proportional to the quadrupole coupling constant and to the orientation of the bond to deuterium (and so the electric field gradient) relative to the magnetic field. In a powder sample all orientations exist and the resulting ^2H spectrum, shown in Fig. 11.21, has a particular shape and is known as a Pake powder pattern. The separation of the sharp edges of this spectrum is three-quarters of the value of the quadrupole coupling constant and usually lies in the range 120–150 kHz. If the moiety in which the deuterium lies is capable of rotation in the solid, then the width of the Pake pattern will be reduced and the extent of the reduction and the shape of the pattern will depend upon the details of the motion. The motion has to be fast relative to the value of the quadrupole coupling constant. These comments will be illustrated by reference to the ^2H spectra of deuteriobenzene absorbed on graphite to form a multilayer some ten molecules thick. Spectra were obtained at several temperatures and are shown in Fig. 11.22. At 298K a singlet narrow signal is observed, and shows that the absorbed benzene is reorienting as if it were in the liquid phase. Indeed, this is found to be the case even if the amount of benzene absorbed is reduced until it forms a monolayer. At 170K the spectrum is of mixed form, with a Pake pattern and a minor singlet. This latter is due to the absorbed monolayer, which is still undergoing fast two-dimensional motion, while the Pake spectrum arises from benzene crystallites that form further out from the surface. The splitting between the sharp edges of the Pake pattern is 70 kHz, and this is typical of the value found for benzene undergoing fast rotation around its hexad axis. The spectra obtained at 90K were run using two different sets of conditions. In one, the

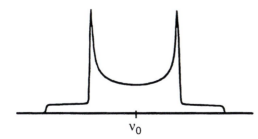

Figure 11.21 The shape of the ^2H resonance in a solid powder sample, or Pake powder pattern.

pulse repetition rate was slow, so that all the deuterons were detected. In the other, the pulse repetition rate was much faster, so that the component with the longer relaxation time T_1 was saturated and effectively was removed from the spectrum. In the first case, two components are observed, both Pake patterns, and with splitting of 70 and 140 kHz. In the second case, the spectrum of the less mobile phase has disappeared and the 70 kHz pattern remains. Note, though, that this is not the same as that seen at 170K, since there is a distinct asymmetry in the base. This is due to the now solid absorbed monolayer rotating only around its hexad axis, and it follows that the broader pattern arises from the now static benzene crystallites.

This type of spectroscopy is also much used in the study of liquid-crystal phases, where the Pake-type patterns obtained from these partially ordered materials can give much information about the degree of order and the rates of motion, and how these change with the experimental conditions.

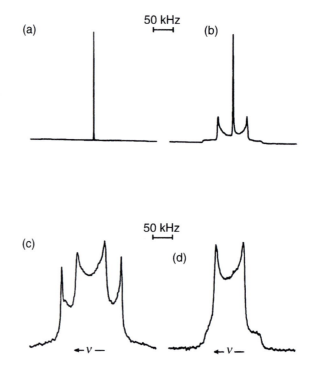

Figure 11.22 The ^2H NMR spectra of deuteriobenzene (benzene-d_6) absorbed on graphite to a thickness of 10 molecular layers: (a) temperature 298K; (b) at 170K; (c) at 90K and time between read pulses of 10 s; (d) at 90K but with the time between pulses 0.2 s so that the broad component is saturated. (From Boddenberg and Grosse (1987) *Z. Naturforsch*, **42a**, 272–4.)

11.7 QUESTIONS

11.1. Figure 11.23 shows the ^{13}C NMR spectrum at 9.4 T of a solid copper cyanide sample enriched in both ^{13}C and ^{15}N. The sample is stationary. The shape of the spectrum shows principally the chemical shift anisotropy though there are also contributions from coupling (direct and indirect) between the nuclei present. Calculate the approximate principal chemical shifts and the isotropic chemical shift of the ^{13}C given that TMS is at 0 ppm. What information does this spectrum give about the structure of $(CuCN)_n$?

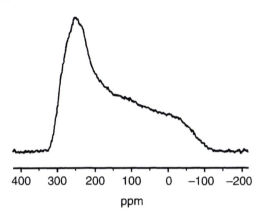

Figure 11.23 The ^{13}C NMR spectrum at 9.4 T of solid copper cyanide enriched in ^{13}C and ^{15}N. The sample was non-spinning. (After R.E. Waylishen *et al.* (1999) *J. Am. Chem. Soc.*, **121**, 1528, with permission, copyright (1999) American Chemical Society.)

11.2. Figure 11.24 shows the ^{31}P CP NMR spectrum of a stationary powder sample of $(Ph_2P)_2CH_2$. As in the previous example, calculate the principal chemical shifts and the isotropic chemical shift of ^{31}P.

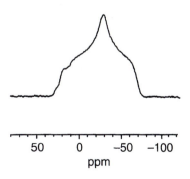

Figure 11.24 The ^{31}P CP NMR spectrum of a non-spinning powder sample of $(Ph_2P)_2CH_2$. (After R.E. Waylishen *et al.* (1999) *Inorg. Chem.*, **38**, 639, with permission, copyright (1999) American Chemical Society.)

Bibliography

NMR is a subject of sufficient importance to have a considerable literature devoted entirely to various aspects of both its technology and its use. The following list represents only a fraction of the total available but is sufficient to give an entry into the field.

COMPREHENSIVE WORKS FOR GENERAL REFERENCE

Abragam, A. (1961) *The Principles of Nuclear Magnetism*, Oxford University Press, Oxford, reprinted 1983, ISBN 019852014X.

Grant, D.M., Harris, R.K. (ed.) (1996) *Encyclopedia of Nuclear Magnetic Resonance*, Wiley, ISBN 0471938718.

Jackman, L.M. and Sternhell S. (1969) *Applications of Nuclear Magnetic Resonance Spectroscopy in Organic Chemistry*, Pergamon Press, Oxford, ISBN 0080125425.

Pople, J.A., Schneider, W.G. and Bernstein, H.J. (1959) *High Resolution Nuclear Magnetic Resonance*, McGraw-Hill, New York, ISBN 70505160.

INTRODUCTORY TEXTS, SOME WITH A VARIETY OF PROBLEMS

Fisher, J., Loftus, P. and Abraham, R. J. (1988) *Introduction to NMR Spectroscopy*, Wiley, Chichester, ISBN 0471918938.

Günther, H. (1995) *NMR Spectroscopy : Basic Principles, Concepts, and Applications in Chemistry*, 2nd edn, Wiley, New York, ISBN 047195201X.

Harris, R.K. (1986) *Nuclear Magnetic Resonance Spectroscopy*, Addison Wesley Longman Higher Education, ISBN 0582446538.

Sanders, J.K.M. and Hunter, B.K. (1993) *Modern NMR Spectroscopy*, 2nd edn, Oxford University Press, Oxford, ISBN 0198555679.

Sanders, J.K.M., Constable, E.C. and Hunter, B.K. (1995) *Modern NMR Spectroscopy : A Workbook of Chemical Problems*, 2nd edn, Oxford University Press, Oxford, ISBN 0198558120.

THE TECHNIQUES OF NMR

Braun, S., Kalinowski, H.-O., and Berger, S. (1996) *150 and More Basic NMR Experiments: A Practical Course*, 2nd edn, Wiley, New York, ISBN 3527295127.

Canet, D. (1996) *Nuclear Magnetic Resonance: Concepts and Methods*, Wiley, New York, ISBN 0471961450.

Derome, A.E. (1987) *Modern NMR Techniques for Chemistry Research*, Pergamon, Oxford, ISBN 0080325149.

Fukushima, E. and Roeder, S.B.W. (1998) *Experimental Pulse NMR*, 10th edn, Addison Wesley, Reading, MA, ISBN 0201627264.

Martin, M.L., Delpeuch J.J. and Martin, G.J. (1980) *Practical NMR Spectroscopy*, Heyden, London, ISBN 0855014628.

Müllen, K. and Pregosin, P.S. (1976) *FT NMR Techniques – A Practical Approach*, Academic Press, New York, ISBN 0125104502.

Shaw, D. (1984) *Fourier Transform NMR Spectroscopy*, 2nd edn, Elsevier, Amsterdam.

Neuhaus, D. and Williamson, M. (1989) *The Nuclear Overhauser Effect*, VCH, Weinheim, ISBN 3527266399.

SHIMMING

Conover, W.W. (1983) *Top. Carbon-13 NMR Spectrosc.*, **4**, 37–51.

SAM 1.0, Shimming Simulation Software Package for IBM-PC Compatible Computers, ACORN NMR, 46560 Fremont Blvd., Fremont, CA 94538–6482.

DYNAMIC NMR

Jackman, L.M. and Cotton, F.A. (eds) (1975) *Dynamic NMR Spectroscopy*, Academic Press, New York.

Kaplan, J.I. and Fraenkel, G. (1980) *NMR of Chemically Exchanging Systems*, Academic Press, New York.

Sandström, J. (1982) *Dynamic NMR Spectroscopy*, Academic Press, London, ISBN 0126186200.

Ōki, M. (1985) *Applications of Dynamic NMR Spectroscopy to Organic Chemistry*, VCH, Deerfield Beach, FL, ISBN 0895731207.

TEMPERATURE CALIBRATION

Van Geet, A.L. (1968) *Anal. Chem.*, **40**, 2227; *ibid*, 1970, **42**, 679.

PULSED FIELD GRADIENTS

Berger, S. (1997) *Prog. NMR Spectrosc.*, **30**, 137–56.
Norwood, T.J. (1994) *Chem. Soc. Rev.*, **23**, 59.

ORGANIC SPECTRAL INTERPRETATION

Atta-ur-Rahman, Choudhary, M.I., Ernst, R.R. and Jackman, L.M. (1995) *Solving Problems With Nmr Spectroscopy*, Academic Press, San Diego, ISBN 0120663201.
Breitmaier, E. (1993) *Structure Elucidation by NMR in Organic Chemistry: A Practical Guide*, Wiley, Chichester, ISBN 0471937452.
Duddek, H. (1998) *Structure Elucidation by Modern NMR; A Workbook*, Springer-Verlag, Darmstadt, ISBN 3798509301.
Lambert, J.A., Shurvell, H.F., Lightner, D.A., and Cooks, R.G. (1998) *Organic Structural Spectroscopy*, Prentice Hall, ISBN 0132586908.
Pretsch, E. and Clerc, J.-T. (1997) *Spectra Interpretation of Organic Compounds*, Wiley, ISBN 3527288260.
Williams, D.H. and Fleming, I. (1995) *Spectroscopic Methods in Organic Chemistry*, 5th edn, McGraw-Hill, London, ISBN 0077091477.

RELAXATION

Wehrli, F.W. (1976) *Topics in ^{13}C Nuclear Magnetic Resonance Spectroscopy*, (ed. G.C. Levy), Vol. 2, Chap. 6, Wiley-Interscience, New York.

TWO-DIMENSIONAL NMR SPECTROSCOPY

Brey, W.S. (ed.) (1988) *Pulse Methods in 1D and 2D Liquid-Phase NMR*, Academic Press, San Diego, ISBN 0121331555.
Croasmun, W.R. and Carlson, R.M.K. (eds) (1994) *Two-Dimensional NMR Spectroscopy*, VCH, Weinheim, ISBN 1560816643.
Ernst, R.R., Bodenhausen, G. and Wokaun, A. (1987) *Principles of Nuclear Magnetic Resonance in One and Two Dimensions*, Clarendon Press, Oxford, ISBN 0198556470.
Freeman, R. (1987) *A Handbook of NMR*, Longman, Harlow.
Freeman, R. (1996) *Spin Choreography: Basic Steps in High Resolution NMR*, Spektrum Academic Publishers, ISBN 1901217043.
Friebolin H. (1998) *Basic One- and Two-Dimensional NMR Spectroscopy*, 3rd edn, Wiley, New York, ISBN 3527295135.
Popov, A.I. and Hallenga, K. (eds) (1991) *Modern NMR Techniques and Their Application in Chemistry*, Marcel Dekker, New York.

MULTINUCLEAR NMR SPECTROSCOPY

Berger, S., Kalinowski, H.-O. and Braun, S. (1996) *NMR Spectroscopy of the Non-Metallic Elements*, Wiley, New York, ISBN 0471967637.

Harris, R.K. and Mann, B.E., *NMR and The Periodic Table* (1978) Academic Press, London, ISBN 0123276500.

Mason, J. (ed.) (1987) *Multinuclear NMR*, Plenum Press, New York, ISBN 0306421534.

Pregosin, P.S. and Kunz, R.W. (1979) ^{31}P and ^{13}C *NMR of Transition Metal Phosphine Complexes*, Springer-Verlag, Berlin, ISBN 3540091637.

Pregosin, P.S. (ed.) (1991) *Transition Metal Nuclear Magnetic Resonance (Studies in Inorganic Chemistry, No. 13)*, Elsevier, Amsterdam, ISBN 044488176X.

COLLECTIONS OF NMR DATA

Mann, B.E. and Taylor, B.F. (1981) ^{13}C *NMR Data for Organometallic Compounds*, Academic Press, London, ISBN 0124691501.

Pouchert, C.J. and Behnke, J. (1993) *The Aldrich Library of ^{13}C and ^1H FT-NMR Spectra*, Aldrich Chemical, Milwaukee, WI, ISBN 0941633349.

Tebby, G. (1991) *CRC Handbook of Phosphorus-31 Nuclear Magnetic Resonance Data*, CRC Press, ISBN 0849335310.

SOLID STATE NMR SPECTROSCOPY

Stejskal, E. O. and Memory, J.D. (1994) *High Resolution NMR in the Solid State: Fundamentals of CP/MAS*, Oxford University Press, USA, ISBN 0195073800.

Klinowski, J. and Kolodziejski, M. (1999) *Solid-State NMR Techniques*, Imperial College Press, ISBN 1860940897.

IMAGING

Bushong, S.C. (1995) *Magnetic Resonance Imaging: Physical and Biological Principles,* C.V. Mosby, St Louis, ISBN 0815113420.

Callaghan, P.T. (1993) *Principles of Nuclear Magnetic Resonance Microscopy*, Clarendon Press, Oxford, ISBN 0198539975.

Kimmich, R. (1997) *NMR: Tomography, Diffusometry, Relaxometry*, Springer-Verlag, Berlin, ISBN 3540618228.

Mansfield, P. and Morris, P.G. (1982) *NMR Imaging in Biomedicine*, Academic Press, New York, ISBN 0120255626.

Mansfield, P. and Hahn E.L. (1991) *NMR Imaging*, Cambridge University Press, ISBN 0521404606.

Rajan, S.S. (1997) *Magnetic Resonance Imaging: A Conceptual Overview*, Springer-Verlag, Berlin, ISBN 0387949119.

The student may also care to read the following few original short papers which summarize the early and unexpected results that heralded the development of NMR as a subject useful to chemists:

Arnold, J.T., Dharmatti, S.S., and M.E. Packard, (1951) First observation of chemical shifts in a single chemical compound, *J. Chem. Phys.*, **19**, 507.
Dickinson, W.C. (1950) Observed chemical shifts in fluorine compounds and noted the effect of exchange, *Phys. Rev.*, **77**, 736.
Gutowsky, H.S. and D.W. McCall, (1951) An early observation of spin–spin coupling, *Phys. Rev.*, **82**, 748.
Gutowsky, H.S., D.W. McCall and Slichter, C.P. (1951) A theory of spin–spin coupling, *Phys. Rev.*, **84**, 589.
Proctor, W.G. and Yu, F.C. (1950) ^{14}N chemical shift between $[NH_4]^+$ and $[NO_3]$, *Phys. Rev.*, **77**, 717.

It should be remembered in reading the two Gutowsky papers that the hertz separation of the ^{31}P doublet and the ^{19}F doublet are the same. The gauss separation can be calculated from $\Delta B_0 = (J/v_0)B_0$ and is greater for ^{31}P since v_0 is smaller for the fixed field used.

Answers to Questions

Outline answers are given here. More detailed answers can be found at http://www.thorneseducation.com

CHAPTER 1

1.1   250, 1000, −1000, −250 Hz.

1.2   5000 Hz, 20 ms.

1.3   Null point is 25 000 Hz from the carrier. Shorten the pulse length. Move the carrier frequency closer to the signal.

1.4 $M_z = M_{xy} = 0.707M$. $M_z = 0$, $M_{xy} = M$. $M_z = -M$, $M_{xy} = 0$.

1.5 ^{32}S, ^{40}Ar, ^{64}Zn.

1.6   4000.

1.7   54.74°.

1.8   48.2.

1.9   2130 Hz, 3.2×10^{-10}.

1.10 2 Hz.

CHAPTER 2

2.1 $CHCl_2$.

2.2   δ 7.30, 2190 Hz, δ 7.30.

2.3   Worse.

2.4   −15950 ppm.

2.5   Ring current.

2.6 $^{35,37}Cl$ isotopomers.

2.7   The sign of the susceptibility correction is dependent on the alignment of the magnetic field.

2.8   126.241 609 MHz, δ 981.

2.9   δ 4519.7, δ −204.0, δ 0, δ −4701.5.

2.10 $\dfrac{\delta_{obs}^{\perp} - \delta_{obs}^{\parallel}}{2\pi} = \chi_{ref} - \chi_{sample}.$

CHAPTER 3

3.1 δ 3.43, δ 1.67, 8 Hz.
3.2   2.97 : 1. 1 : 4 : 6 : 4 : 1 quintet.
3.3   14.3 Hz.
3.4   1 : 6 : 15 : 20 : 15 : 6 : 1 septet.
3.5   8.6 Hz, δ 6.479, δ 6.394,   1.86 : 0.14,   3244.2 Hz,   3235.6 Hz, 3200.9 Hz, and 3192.3 Hz, δ 6.437, 1.20 : 0.80.
3.6   Unresolved coupling to the *ortho*-phenyl protons.
3.7 A_2X_3 or A_2B_3, A_2X_2, $[AX]_2$ or $[AB]_2$, ABC or ABX or AMX, AX_2 or AB_2, ABC or ABX or AMX, $[AX_9]_2$, $[AX]_3$.
3.8   Doublet of quartets of triplets.
3.9 $[IrH_5(PEt_2Ph)_2]$, *mer-trans*-$[IrH_3(PEt_2Ph)_2(AsMe_2Ph)]$.
3.10   2, 3, 4.
3.11   –3.3 ppm.
3.12   125 Hz, 19 Hz, –0.17 ppm.
3.13   *cis-mer*-$[RuH_2(PPh_3)_3(CO)]$.
3.14   Diastereotopic protons. Triplet.
3.15   Monomeric, $[Li(Me_2NCH_2CH_2NMe_2)Bu^n]$.
3.16   Each signal is two doublets of quartets. There are two doublets as the ^{107}Ag and ^{109}Ag have different coupling constants.
3.17 $\overline{CHMe_2}$ and H^2.
Ex. 1 $\overline{CH{=}CHCH_2SO_2}$.
Ex. 2 CF_3CH_2OH.
Ex. 3 $CF_2HCF_2CH_2OH$.

CHAPTER 4

4.1   22 s.
4.2   1 : 1.15. Note ^1H and ^{13}C relax ^1H, whereas only ^1H relaxes ^{13}C.
4.3   Since the z coordinate is zero, then no terms can cancel and the EFG is proportional to $-3qr^{-3}$. For zero EFG, each term of the sum has to be zero, and if z is the coordinate of the displaced charge, then $3z^2 - r^2 = 0$ and $z = \pm r/\sqrt{3}$.
4.4   46.9 s, 11.7 s.
4.5   1 : 0.103.
4.6 3.5×10^{-13} s.
4.7 ^{14}N coupling. Cooling increases the rate of ^{14}N relaxation and decouples it.
4.8   Preferential rotation about the long axis. The *sp* carbon relax by CSA, which depends on B_0^2.
4.9   Scalar relaxation by ^{79}Br.
4.10 ^{103}Rh relaxes by CSA.
4.11 $R_2 = 282.74$ s^{-1}, $T_2 = 0.0035$ s, $R_{2SR} = 280.51$ s^{-1}, $T_{2SR} = 0.0036$ s.

CHAPTER 5

5.1 Dwell time = 56 μs, 32K, 0.910 s, not fully relaxed, maximum pulse width = 56 μs.

5.2 M_z, $0.368M_z$.

5.3 Small error in frequency. Poor shimming.

5.4 $4\sqrt{2}$.

5.5 68.4°.

CHAPTER 6

6.1 In homonuclear decoupling, time sharing prevents interference of the decoupler frequency with the receiver.

6.2 z direction.

6.3 0.000 117 T.

CHAPTER 7

7.1 Intra-molecular exchange.

7.2 10 Hz, 1 : 100.

7.3 $\Delta H^\ddagger = 51.7$ kJ mol^{-1}, $\Delta S^\ddagger = -5.2$ J K^{-1} mol^{-1}.

7.4 1,2-shift, 31.42 s, 45.5 kJ mol^{-1}.

7.5 C^a at δ 21.01, C^b at δ 128.37, C^c at δ 128.54, C^d at δ 31.01.

7.6 2.6 s^{-1}, 53.0 kJ mol^{-1}.

7.7 Coalescence temperature depends on the frequency separation of the signals which is field dependent. 1.5 Hz, 1.5 Hz, 1.14 Hz, 4.55 Hz.

CHAPTER 8

Ex. 5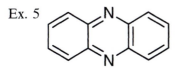

Ex. 6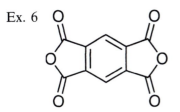

CHAPTER 9

9.1   OSnBu$^t_2$O(μ-F), δ –232, $^1J(^{119}$Sn^{19}F) = 1800 Hz; (μ-F)$_2$SnBu$^t_2$, δ –282, $^1J(^{119}$Sn^{19}F) = 3300 Hz. δ –163 and –165: *cis*- and *trans*-isomers of [Bu^t(F)Si(OSnBu$^t_2$O)$_2$Si(F)Bu^t].

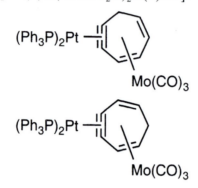

9.2 $\delta(^{195}$Pt) = –4481; $\delta(^{31}$P) = 22.1, 23.8; $^1J(^{195}$Pt,^{31}P) = 3325, 3415 Hz; $\delta(^{195}$Pt) = –4449; $\delta(^{31}$P) = 22.9; $^1J(^{195}$Pt,^{31}P) = 3325 Hz; $\delta(^{195}$Pt) (21.4) = –1607, –1575.

9.3   Five coordinate P group is chiral, hence C$_5$H$_4$R CH groups are diastereotopic.
        $\delta(^1$H) = 4.73(2). 5.17(1), 5.23(1).
        $\delta(^{13}$C) = 92.3 (J = 255 Hz); 85.8, 86.4, 94.6, 95.9 (J = 15 for all).

9.4   P^2 and P^3 form an AB pattern centred on 30 700 Hz with J_{AB} = 490 Hz, J_{AM} = 55 Hz, J_{AX} = 20 Hz, J_{BM} = 150 Hz.

9.5. $^3J(^{195}$Pt^{1}H) = 7.5, 12 Hz. Viewed from platinum, the methyls are inequivalent.

9.6.

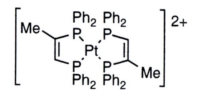

       [AX]$_2$ J_{AA}' large. Two virtual triplets.

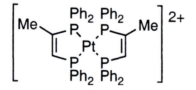

       [AX]$_2$ J_{AX} large. Two N doublets with fine structure.

9.7 ^{203}Tl, ^{205}Tl csa relaxation.

9.8. δ 92.6: $^1J(^{195}$Pt^{31}P) = 2157 Hz; $^3J(^{195}$Pt^{31}P) = 286 Hz; $^2J(^{199}$Hg^{31}P) = 1632 Hz; $^4J(^{31}$P^{31}P) = 100 Hz.
        δ 62.5: $^1J(^{195}$Pt^{31}P) = 3090 Hz; $^2J(^{199}$Hg^{31}P) = 218 Hz.

9.9. H^2 δ 9.39; H^5 δ 8.38; $H^{5'}$ δ 8.30; $H^{6''}$ δ 10.1; $H^{6'''}$ δ 6.5. Use COSY.

9.10. Three $-1 : -1 : 1 : 1$ quartets centred at δ −414 (1), −420 (2), −422 (1).

9.11. $^1J(^{103}Rh^1H) = 11$ Hz produces $1 : -1$ doublets. $^1J(^{103}Rh^{31}P)$ (PPh$_3$) = 109 Hz (triplet). $^1J(^{103}Rh^{31}P)$ {P(OMe)$_3$} = 120 Hz (doublet).

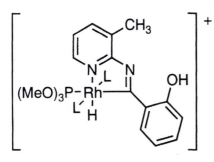

9.12. Me are diastereotopic. $\Delta G^{\ddagger} = 63$ kJ mol^{-1}.

9.13. NMe(CH$_2$)$_2$ exchange. $\Delta G^{\ddagger} = 66.5$ kJ mol^{-1}.

9.14. The −80°C AMM′XX′ pattern collapses to AM$_2$XX′ at 40°C. The N doublet does not change and remains sharp.

9.15. Restricted rotation of the phenyl. At 25°C, *ortho* H form a broad hump, *meta* and *para* are sharp. At −60°C there are two *ortho* and two *para* signals.

9.16. BBN δ 11, B^1 δ 5, B^{2,4} δ −29, B^3 δ 3, B^{5,10} δ −2, B^{6,9} δ 17, B^{7,8} δ −9.

9.17. H^a δ 7.55, H^b δ 6.7, H^c δ 6.1, H^d δ 6.8, H^e δ 4.6, H^o δ 7.15, H^m δ 6.9, H^p δ 6.05. Restricted agostic phenyl rotation.

9.18. H^1 δ −23.3, H^2 δ −21.6, H^3 δ −24.15, H^4 δ −20.3.

9.19. H^1 δ −9.9, H^2 δ −10.4, H^3 δ −7.3. $\Delta G^{\ddagger} = 44$ and 56 kJ mol^{-1}.

9.20. PH. Doublet of triplets of doublets. $^1J(^{31}P^1H) = 310$ Hz; $^3J(^{31}P^1H) = 5$ Hz; $^3J(^1HRuP^1H) = 4$ Hz.

Hydride at δ −8.5. Doublet of triplets of doublets of doublets. $^2J(^1HRu^1H) = 3$ Hz; $^2J(^{31}P^1H)$ (PPr$^i_3$) = 10 Hz; $^2J(^{31}P^1H)$ (PHPh$_2$) = 16 Hz.

Hydride at δ −9.6. Doublet of triplets of doublets. $^2J(^1HRu^1H) = 3$ Hz; $^2J(^{31}P^1H)$ (PPr$^i_3$) = 10 Hz; $^2J(^{31}P^1H)$ (PHPh$_2$) = 43 Hz.

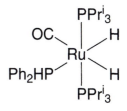

CHAPTER 11

11.1. δ$_{||}$ = −84, δ$_{\perp}$ = +267, δ$_{iso}$ = + 151. Axially symmetric, so linear.

11.2   δ$_{11}$ = 26, δ$_{22}$ = −28, δ$_{33}$ = −70, δ$_{iso}$ = −24.4.

Index